AF836302

A Promenade in Mathematical Proofs

With Comprehensive Review of Proof Techniques

Abdelouahed Hamdi
Noureen A. Khan

Bassim Hamadeh, CEO and Publisher
John Remington, Managing Executive Editor
Gem Rabanera, Senior Project Editor
Susana Christie, Senior Developmental Editor
Alia Bales, Production Manager
Jess Estrella, Senior Graphic Designer
JoHannah McDonald, Licensing Coordinator
Natalie Piccotti, Director of Marketing
Kassie Graves, Senior Vice President, Editorial

Copyright © 2024 by Cognella, Inc. All rights reserved. No part of this publication may be reprinted, reproduced, transmitted, or utilized in any form or by any electronic, mechanical, or other means, now known or hereafter invented, including photocopying, microfilming, and recording, or in any information retrieval system without the written permission of Cognella, Inc. For inquiries regarding permissions, translations, foreign rights, audio rights, and any other forms of reproduction, please contact the Cognella Licensing Department at rights@cognella.com.

Trademark Notice: Product or corporate names may be trademarks or registered trademarks and are used only for identification and explanation without intent to infringe.

Cover image copyright © 2019 iStockphoto LP/25ehaag6.

Printed in the United States of America.

cognella® | ACADEMIC PUBLISHING
320 South Cedros Ave., Ste. 400, Solana Beach, CA 92075

Contents

Preface

If we asked a first-year mathematics student and a graduate researcher to define mathematics, they would certainly have vastly different responses. A novice in advanced mathematics may view the subject as a study of numbers, functions, and graphs. The question is certainly too vague to be even considered by a researcher at the end of their doctoral studies, and many assumptions must be made before a hypothesis can even be proposed! Mathematics students have to take the leap from seeing math as just a set of numbers and functions that can be solved with a calculator to seeing it as the way we quantify our understanding of the universe.

In this textbook the assertion is made that proof writing is as essential as an understanding of mathematics because writing mathematical proofs is just as important as knowing the subject. The text is intended for students learning advanced mathematics who have already completed arithmetic algebra, linear algebra, and calculus courses. In the opening chapter, we introduce the reader to the concept of mathematical writing. A topic that many may find challenging, as it is an unconventional approach to mathematics for many novice students. The goal of Chapter one is to introduce the reader to delivering their ideas about problem solving into mathematical proof writing. After one becomes literate in mathematics writing, they need the proper grammar, which the applications of logic to math, which the reader will learn about in Chapter 2. Mathematical logic is essential for understanding the ideas discussed in the following chapters.

The reader will then dive into the theory of sets in Chapter 3. A few different kinds of sets are discussed, including power sets and indexed sets, as well as the concepts of Sets, subsets, and the complement of sets. We also discuss set operations and the Cartesian product of sets. In Chapter 4, we use the logic learned in Chapter 2 to introduce a variety of proof techniques, such as direct, contrapositive, and contradiction proof methods. The reader will also learn about proofs by cases, counterexamples, equivalences, and a wide range of examples for each type of proof.

Chapter 5 is mostly example-driven, serving as an opportunity for the reader to apply the ideas of induction learned up to this point to a variety of topics, such as calculus, set theory, number theory, and linear algebra. This chapter includes examples from nearly every upper-division mathematics course taught in College. In Chapter 6, we introduce mathematical induction; a very powerful technique used in fields such as computer science and computer logic. In addition to inductive

reasoning, we will discuss strong induction, since it builds on the logic concepts discussed in Chapter 2 and provide more rigor proof techniques.
In Chapter 7, we discuss sequences, their variations, and the definitions of convergence and divergence. Divergence and convergence are prevalent outside of sequences, as they are used in advanced mathematics extensively and are essential in proof techniques readers will encounter in higher-level courses.
We will cover relations and functions in Chapters 8 and 9, respectively. In chapter 8, we discuss the concepts of domain and range, the three properties of relations (reflexivity, symmetry, and transitivity), equivalence classes and partitioning, and relationships between sets such as congruence modulo n. The concepts of domain and range discussed in Chapter 8 are expanded in Chapter 9 by formally defining a function and introducing different properties of functions, such as surjectivity, injectivity, and bijectivity. This is expended upon with the examples of even and odd functions. Composition of functions and their inverses follow, as do equivalence, finite sets, cardinality, and countability.

The practice problems at the end of each chapter are crafted carefully to help you grasp the chapter material comprehensively. As a bonus, every chapter concludes with elegant solutions to selected practice problems to aid in a deeper understanding of the chapter and extended topics. The practice problems are highly recommended so the reader can apply their knowledge of each chapter's content. As we know from elementary courses, examples are often the best teachers, and practice does indeed make perfect. The textbook uses the concept of practice-based learning, a student-centered approach in which students gain knowledge of mathematics by doing a large variety of examples and practicing problems. The goal is to provide as much hands-on work as possible while exploring the topics in this book. The text includes a significant number of solved practice problems, which we believe will help students be successful in the course.

Abdelouahed Hamdi & Noureen A. Khan

Acknowledgments

I would like to extend my heartfelt appreciation to my family for their unwavering support and patience throughout the extensive hours, days, and months I dedicated to this project. I am immensely grateful to my colleagues at Qatar University's Department of Mathematics, Physics, and Statistics, with special thanks to Ms. Samar Jaafar and Dr. Akram Taati, whose meticulous proofreading significantly enhanced the accuracy of this book. Additionally, I'd like to convey my thanks to my Math 220 students, whose enthusiasm served as a motivating force in completing this work. My passion for discovering new, efficient, and ingenious solutions remains insatiable when it comes to learning.

Abdelouahed Hamdi [1]

While writing a book has proven to be challenging, the experience has been gratifying in more ways than I could have expected. This book could not have been completed without the help of my coauthor. When I visited Qatar University as a Fulbright Scholar, Dr. Hamdi was the first colleague and friend I collaborated with. I am very grateful to the Fulbright U-S- Scholar Program for providing me with the opportunity to facilitate this collaboration; it would not have been possible without them. This project required a lot of family time, and I'd like to thank my husband, Aftab, for being patient with me. I extend my sincere gratitude to my son, Abdullah, who helped in checking my proofs for errors. Because of the COVID-19 outbreak, he was able to proofread the manuscript while studying from home, which was a real blessing to me.

Noureen A. Khan [2]

[1] Dr. Abdelouahed Hamdi, professor of mathematics. Mathematics Program, Department of Mathematics, Statistics and Physics. College of Art and Science. Qatar University. P.O. Box: 2713, Doha – Qatar.

[2] Noureen A. Khan, professor of mathematics Department of Mathematics and Information Sc., University of North Texas at Dallas, 7300 University Hills Blvd. Dallas, TX 75241, USA.

1. Preliminaries

Chapter Key Terms

(a) *mathematical writing*
(b) *identities*
(c) *absolute value*
(d) *quadratic Equations*
(e) *discriminant*
(f) *Vieta's Rule*
(g) *parabolas*

The purpose of this chapter is to introduce the basic information needed to write mathematical proofs. Students often find math writing challenging, and they struggle to become proficient at describing their ideas of writing the proof. We will provide skill-building examples on how to present mathematical ideas.

1.1 Mathematical Writing

Students often find mathematics writing challenging, and they struggle to become proficient in describing their ideas for writing the proof. Our primary goal is to provide authentic and logical methods to prove a given mathematical assertion. In this chapter, we provide skill-building examples for presenting mathematical ideas. The mathematical writing is a technique or procedure that allows us to understand mathematical structures, prove mathematical statements, and even discover new mathematical results. In general, mathematical proof writing must

- comply with existing conventions and notations used in proofs;
- validate reasoning of utilizing the tools or formulae being used; and
- contain statements of well-defined concepts, notions, and expressions.

Before we begin, here is a list of commonly used sets that appear throughout this book:

- $\mathbb{N} = \{0,1,2,3,...,100,101,...\}$: the set of natural numbers;
- $\mathbb{Z} = \{...,-100,...,-3,-2,-1,0,1,2,3,...,100,...\}$: the set of integers;
- $\mathbb{Q} = \left\{\frac{p}{q},\ \ p,q \in \mathbb{Z},\ q \neq 0.\right\}$: the set of rational numbers and
- $\mathbb{R} = (-\infty,\ \infty)$: the set of real numbers.

$$\mathbb{N} \subset \mathbb{Z} \subset \mathbb{Q} \subset \mathbb{R}$$

- Numbers that are not rational are called irrational numbers. For instance $\sqrt{2}$, $\sqrt{3}$, e, and π.

For example,

- ∘ to introduce an element x describing a given set A

$$\text{let } x \in A \quad \text{or we can write} \quad \text{for all } \ x \in A;$$

- ∘ when finding the roots of an equation,

$$2x^2 - 3x = 0 \iff 2x - 3 = 0 \iff x = 3/2.$$

It will be considered as a mistake to divide both sides of the equation by x. Instead, when we factor x out first, we obtain the correct roots:

$$2x^2 - 3x = 0 \iff x(2x-3) = 0 \iff x \in \{0,\ 3/2\}.$$

- When working with algebraic expressions and functions, the domain is another place where mistakes can easily be made. For example, when solving for x in the following equation:

$$\sqrt{x-2} = 1 - x \Longrightarrow x - 2 = (1-x)^2 \quad \text{where} \quad x - 2 \geq 0 \text{ and } \ 1 - x \geq 0$$

That is, $x \geq 2$ and $x \leq 1$, which is impossible. So the solution set is $\emptyset$.

- It is not appropriate to write or say that:

$$\text{The function } f(x) \text{ to } \ x\sin(x) + \ln(x^2+1)$$

Instead, one should write

$$\text{The function } f \ \text{ or the function } f : x \longmapsto x\sin(x) + \ln(x^2+1).$$

We note the following.

1. The symbol $\longmapsto$ links two numbers.

2. The symbol $\rightarrow$ links two sets.

In general, we may write

$$\begin{aligned} f : \mathbb{R} &\longrightarrow \mathbb{R} \\ x &\longmapsto f(x). \end{aligned}$$

Many Greek letters have been adopted into common use for constants in equations for a number of reasons. Greek letters date back to the earliest philosophers, including Aristotle, Diophantus, and many others, who used Greek letters as symbols to represent various variables. The use of Greek letters has been maintained through the ages, even if later civilizations used their own scripts, as people tended to use what was already established.The continuation of the use of Greek alphabet symbols as constants and sometimes variables in equations is really a matter of convenience. We conclude this discussion with a table of Greek characters which are used frequently in mathematics.

α	β	γ	δ	ε	ζ	η	θ	ι	κ	λ
Alpha	Beta	Gamma	Delta	Epsilon	Zeta	Eta	Theta	Iota	Kappa	Lambda

μ	ν	ξ	π	ρ	σ	τ	υ	Φ, φ	ψ	χ	ω
Mu	Nu	Xi	Pi	Rho	Sigma	Tau	Upsilon	Phi	Psi	Khi (KI)	Omega

Δ	Γ	Π	Σ	Ω	Λ	Φ
DELTA	GAMMA	PI	SIGMA	OMEGA	LAMBDA	PHI

1.2 Identities

In this section, we will recall many useful famous identities.

1.2.1 Main Identities

1. $(x+y)^2 = x^2 + 2xy + y^2$
2. $(x-y)^2 = x^2 - 2xy + y^2$
3. $x^2 - y^2 = (x-y)(x+y)$
4. $(x+y)^3 = x^3 + 3x^2y + 3xy^2 + y^3 = x^3 + y^3 + 3xy(x+y)$
5. $(x-y)^3 = x^3 - 3x^2y + 3xy^2 - y^3 = x^3 - y^3 - 3xy(x-y)$
6. $x^3 + y^3 = (x+y)(x^2 - xy + y^2), \quad x^3 - y^3 = (x-y)(x^2 + xy + y^2)$
7. $x^n - y^n = (x-y)[\, x^{n-1} + x^{n-2}y + x^{n-3}y^2 + \ldots + xy^{n-2} + y^{n-1} \,], \quad n \in \mathbb{N}$
8. $x^n + y^n = (x+y)[\, x^{n-1} - x^{n-2}y + x^{n-3}y^2 - \ldots - xy^{n-2} + y^{n-1} \,]$, n odd

1.2.2 More Identities

1. $(x+y+z)^2 = x^2 + y^2 + z^2 + 2(xy + yz + zx)$
2. $x^3 + y^3 + z^3 - 3xyz = (x+y+z)\left[x^2 + y^2 + z^2 + xy + yz + zx\right]$
3. $x^3 + y^3 + z^3 - 3xyz = \dfrac{x+y+z}{2}\left[(x-y)^2 + (y-z)^2 + (z-x)^2\right]$
4. If $x+y+z=0 \implies x^3 + y^3 + z^3 = 3xyz$
5. If $x+y+z \geq 0 \implies x^3 + y^3 + z^3 \geq 3xyz$

1.3 Absolute Values

The concept of the **absolute value** of a given real number can be introduced in two ways: as a distance between two numbers or as a given definition. Here, we present both ways:

1. The absolute value of a given real number x is equal to the distance between x and 0 and is denoted $|x| = dist(x,0)$. The distance between two numbers is measured on the real numbers line. For instance, $|-3| = dist(-3,0) = dist(3,0) = 3 = |3|$. Thus$|x-y| = dist(x,y) = dist(y,x) = |y-x|$.
2. The absolute value of $x \in \mathbb{R}$ can also be defined as follows:

$$|x| = \begin{cases} x & \text{if } x > 0 \\ 0 & \text{if } x = 0 \\ -x & \text{if } x < 0 \end{cases}$$

For a complete understanding of absolute value, it is important to understand the following properties, see [**3**] for more details.

Let x, y, and $z \neq 0$ be some real numbers, and let n be an integer.

*P*1 $\boxed{|x| \geq 0}$

*P*2 $\boxed{|x| = 0 \Longleftrightarrow x = 0}$

*P*3 $\boxed{|x \times y| = |x| \times |y|}$

*P*4 $\boxed{\left|\frac{x}{z}\right| = \frac{|x|}{|z|}}$

*P*5 $\boxed{|x^n| = |x|^n}$

*P*6 $\boxed{|x+y| \leq |x| + |y| \quad (\textit{Triangle Inequality})}$

*P*7 $\boxed{|x| \leq a,\ a \geq 0 \Longleftrightarrow -a \leq x \leq a \Longleftrightarrow x \in [-a, a]}$

*P*8 $\boxed{|x| \geq a \Longleftrightarrow [x \leq -a \text{ or } x \geq +a] \Longleftrightarrow x \in (-\infty, -a] \cup [a, \infty)}$

*P*9 $\boxed{a < |x| < b \Longleftrightarrow x \in (-b, -a) \cup (a, b)}$

*P*10 $\boxed{|x| = a \Longleftrightarrow x = \pm a}$

*P*11 $\boxed{\sqrt{x^2} = |x|, \quad \sqrt[2n]{x^{2n}} = |x|}$

(R) A common mistake made by beginners is related to *P*11 when dealing with radicals and especially with square roots:

$$\sqrt{x^2} = x, \qquad \sqrt[2n]{x^{2n}} = x.$$

The preceding affirmations are correct if we consider that x is a non-negative number. But if x is negative, then the results are wrong. For instance, $x = -3$, then $\sqrt{(-3)^2} = -3$ is incorrect; it contradicts the fact that the square root is always a non-negative value. Thus the correct answer should involve the absolute value of the outcome as follows:

$$\sqrt{x^2} = |x|, \text{ and } \sqrt[2n]{x^{2n}} = |x|.$$

For example, $\sqrt[6]{(-5)^6} = |-5| = 5$.

(R) Another way to make this mistake is in the following scenario

$$\sqrt{x^2} = (\sqrt{x})^2 \quad \text{obviously it incorrect because } x \text{ can be negative.}$$

This confusion comes from the following correct equality:

$$(\sqrt{x})^2 = \sqrt{x^2} = x \quad \text{if and only if } x \text{ is a non-negative number.}$$

■ **Example 1.1** Solve for x:

$$|x-3|+|x+1|=x+2.$$

The main idea of solving such an equation is to express it without the bars of absolute value. We proceed first as follows:

Step 1. Find the zeros of each absolute value expression:

$$|x-3|=0 \Longleftrightarrow x-3=0 \Longleftrightarrow x=3, \quad |x+1|=0 \Longleftrightarrow x=-1.$$

Step 2. Build a sign table :

x	$(-\infty,\ -1)$	$[-1,\ 3]$	$(3,\ \infty)$
$\lvert x+1\rvert$	$-(x+1)$	$x+1$	$x+1$
$\lvert x-3\rvert$	$-(x-3)$	$-(x-3)$	$x-3$
$\lvert x-3\rvert+\lvert x+1\rvert$	$2-2x$	4	$2x-2$

(a) For $x<-1$ the equation becomes: $2-2x=x+2$, which implies that $x=0>-1$, and then $x=0$ is rejected.

(b) For $-1\leq x\leq 3$ the equation becomes: $4=x+2$ which implies that $x=2\in[-1,3]$.

(c) For $x>3$ the equation becomes: $2x-2=x+2$ which implies that $x=4>3$.

Thus

$$\boxed{S=\{2,\ 4\}}$$

■

Proposition 1.3.1 In property $P6$, the equality holds if $xy>0$.

$$\boxed{xy>0 \Longrightarrow |x+y|=|x|+|y|}$$

Proof. For example $|(-2)+(3)|\neq|-2|+|3|$, and here $(-2)(3)<0$. Now, to prove the equality, we need to discuss the possible two cases:

1. $x>0,\ y>0 \Longrightarrow x+y>0 \Longrightarrow |x+y|=x+y=|x|+|y|$.

2. $x<0,\ y<0 \Longrightarrow x+y<0 \Longrightarrow |x+y|=-(x+y)=(-x)+(-y)=|x|+|y|$.

■

(R) By replacing y with $(-y)$ in the triangle inequality ($P6$), we obtain

$$\boxed{|x-y|\leq|x|+|y|}$$

Proposition 1.3.2 For all $x,\ y\in\mathbb{R}$, we have

$$\boxed{|\pm x\pm y|\leq|x|+|y|,\quad \forall x,y\in\mathbb{R}}$$

Proposition 1.3.3 Let $x_i\in\mathbb{R},\ i\in\{1,2,\ldots,n\}$. Then we have

$$\boxed{|\pm x_1\pm x_2\pm\ldots\pm x_n|\leq|x_1|+|x_2|+\ldots+|x_n|}$$

The inequality in Proposition 1.3.3 is useful in many situations, as shown in the following examples.

■ **Example 1.2** Solve for

$$x: \ |x-2|+|x|+|x+2| = 4+x.$$

We can solve the equation using sign table, but here, we present an original idea based on the use of (1.3.3) in such a way that cancels as much of the unknown as possible.

$$4+x = |x-2|+|x+2|+|x| \geq |+(x-2)-(x+2)|+|x| = 4+|x| \quad \Longrightarrow x \geq |x| \Longrightarrow x \geq 0,$$

and then

$$4+x = |x-2|+|x+2|+|x| = |x-2|+x+2+x \Longleftrightarrow 2-x = |x-2| \Longleftrightarrow x-2 \leq 0 \Longrightarrow x \leq 2,$$

and finally

$$\boxed{S = [0,\ 2]}$$

■

■ **Example 1.3** Solve for x

$$|x-2|+|x-6| = 3.$$

Using (1.3.3), we get

$$|x-2|+|x-6| \geq |-(x-2)+(x-6)| = |-x+2+x-6| = |-4| = 4 > 3.$$

Then,

$$\boxed{S = \emptyset}$$

■

1.4 Quadratic Functions and Equations

Often, students struggle to understand the properties of a quadratic function. Before we develop proof techniques to prove the results that involve quadratic functions, it is useful to master some results related to the `quadratic functions`.

1.4.1 Completing the Square

We begin by showing a classical way of completing the square of a given quadratic algebraic expression ax^2+bx+c, $a \neq 0$ by using the quadratic identities.

$$ax^2+bx+c = a\left(x+\frac{b}{2a}\right)^2 - \frac{b^2-4ac}{4a}, \quad \forall a \neq 0,\ b,\ c \in \mathbb{R}.$$

(R) Another easy way to complete the square by keeping in mind the following:

$$ax^2+bx+c = a(x-\alpha)^2+\beta, \quad \alpha = \frac{-b}{2a}, \quad \beta = \frac{4ac-b^2}{4a}. \tag{1.1}$$

■ **Example 1.4** Complete the squares in the following quadratic expressions:

$$A = -3x^2+7x-2, \quad B = \frac{2x^2}{5} - 2x+1, \quad \text{and} \quad C = x^2+y^2+z^2-xy-yz-zx.$$

We will apply the formula (1.1).

1. Here $a=-3$, $b=7$, and $c=-2$:

$$\alpha=\frac{-7}{-6}=\frac{7}{6},\ \beta=\frac{4(-3)(-2)-49}{4(-3)}=\frac{-25}{-12}=\frac{25}{12}\Longrightarrow \boxed{A=\frac{25}{12}-3\left(x-\frac{7}{6}\right)^2}$$

2. Here $a=2/5$, $b=-2$, $c=1$,

$$\alpha=\frac{2}{4/5}=\frac{5}{2},\ \beta=\frac{\frac{8}{5}-4}{8/5}=\frac{-12}{8}=\frac{-3}{2}\Longrightarrow \boxed{B=\frac{2}{5}\left(x-\frac{5}{2}\right)^2-\frac{3}{2}}$$

3. The last expression may be viewed as a quadratic expression in the variable x or symmetrically in y or z. Let us begin by considering the variable y:

$$C=y^2-(x+z)y+x^2+z^2-zx \hookrightarrow a=1,\ b=-(x+z),\ c=x^2+z^2-zx.$$

Thus

$$\alpha=\frac{x+z}{2},\ \beta=\frac{4(x^2+z^2-zx)-(x+z)^2}{4},$$

i.e.,

$$\beta=\frac{1}{4}\left(3x^2+3z^2-6zx\right)=\frac{3}{4}\left(x^2+z^2-2zx\right)=\frac{3}{4}(x-z)^2.$$

Thus

$$\boxed{B=\left(y-\frac{x+z}{2}\right)^2+\frac{3}{4}(x-z)^2}.$$

And similarly

$$\boxed{B=\left(x-\frac{y+z}{2}\right)^2+\frac{3}{4}(y-z)^2=\left(z-\frac{x+y}{2}\right)^2+\frac{3}{4}(x-y)^2}$$

■

(R) Given a quadratic expression $y=ax^2+bx$, we provide a new perspective to complete the square approach. One can multiply the terms by $4a$ and then simply add the square of the coefficient of x and get a complete square. We illustrate the technique in the following example:

$$\begin{aligned}
y=ax^2+bx &\iff 4ay=4a^2x^2+4abx,\\
&\iff 4ay+\mathbf{b^2}=4a^2x^2+4abx+\mathbf{b^2},\\
&\iff 4ay+\mathbf{b^2}=(2ax+b)^2.
\end{aligned}$$

■ **Example 1.5** Solve

$$4t^2-t-3=0.$$

Let us multiply both sides by $4a=4(4)=16$:

$$\begin{aligned}
3=4t^2-t &\iff 16(3)=(16)4t^2-16t,\\
&\iff 48=64t^2-16t,\\
&\iff 48+(-1)^2=64t^2-16t+(-1)^2,\\
&\iff 49=(8t-1)^2,
\end{aligned}$$

i.e.,

$$8t-1=\pm 7 \Longrightarrow t=\frac{1\pm 7}{8} \Longrightarrow \boxed{S=\{1,\ -3/4\}.}$$

■

■ **Example 1.6** Solve $\dfrac{2x^2}{5}-2x+1=0$.

Let us multiply both sides by $4a=4(2/5)=8/5$:

$$\begin{aligned}-1=\frac{2x^2}{5}-2x \quad &\Longleftrightarrow \quad -\frac{8}{5}=\frac{16}{25}x^2-\frac{16}{5}x,\\ &\Longleftrightarrow \quad 2^2-\frac{8}{5}=\frac{16}{25}x^2-\frac{16}{5}x+2^2,\\ &\Longleftrightarrow \quad \frac{12}{5}=\left(\frac{4x}{5}-2\right)^2,\end{aligned}$$

i.e.,

$$\begin{aligned}\frac{4x-10}{5} &= \pm\sqrt{\frac{12}{5}},\\ 4x-10 &= \pm 5\sqrt{\frac{12}{5}}\\ 4x &= 10\pm\sqrt{60}\\ x &= \frac{10\pm 2\sqrt{15}}{4}\end{aligned}$$

Finally,

$$\boxed{S=\left\{\frac{5\pm\sqrt{15}}{2}\right\}}$$

■

(R) The problem can be solved directly by identifying $a=4$, $b=-1$, and $y=3$ in $4t^2-t-3$. Thus

$$4ay+b^2=(2ax+b)^2 \Longleftrightarrow 48+1=(8t-1)^2 \Longleftrightarrow 49=(8t-1)^2.$$

1.4.2 Discriminant Δ

Now, we consider the more general case of a quadratic equation: $ax^2+bx+c=0,\ \ a\neq 0$.
We will show later that the preceding three cases are covered by the resolution method known as the `quadratic formula` or `the discriminant method`.

If we complete the square as it was done in Example1.1, we get the following:

$$ax^2+bx+c=a\left[\left(x+\frac{b}{2a}\right)^2-\frac{\Delta}{(2a)^2}\right]$$

where $\boxed{\Delta=b^2-4ac}$ is known as the **discriminant** associated with $ax^2+bx+c=0$.

Now we present the cases when the discriminant Δ (Delta) can be used to solve the equation $ax^2+bx+c=0$.

1. If $\Delta=0$, then $Q(x)=ax^2+bx+c$ can be expressed as follows:

$$\boxed{ax^2+bx+c=a\left(x+\frac{b}{2a}\right)^2.}$$

It has two equal roots (or a double root) $x^*=-\dfrac{b}{2a}$.

2. If $\Delta>0$, then $Q(x)=ax^2+bx+c$ can be expressed as follows:

$$\begin{aligned} ax^2+bx+c &= a\left[\left(x+\frac{b}{2a}\right)^2-\left(\frac{\sqrt{\Delta}}{2a}\right)^2\right], \\ &= a\left[x+\frac{b}{2a}-\frac{\sqrt{\Delta}}{2a}\right]\left[x+\frac{b}{2a}+\frac{\sqrt{\Delta}}{2a}\right]. \end{aligned}$$

It has two different roots

$$x_{1,2}=\frac{-b\pm\sqrt{\Delta}}{2a}.$$

R Alternatively, we can say that these two real roots are conjugate to each other:

$$\boxed{x_1=\frac{-b-\sqrt{\Delta}}{2a},\quad x_2=\frac{-b+\sqrt{\Delta}}{2a}.}$$

3. If $\Delta<0$,

$$Q(x)=ax^2+bx+c=a\left[\left(x+\frac{b}{2a}\right)^2+\overbrace{\frac{-\Delta}{(2a)^2}}^{\text{positive}}\right].$$

Clearly, $Q(x)$ is the sum of a non-negative number and a positive number cannot be equal to 0. Thus $ax^2+bx+c=0$ has no real root.

The following table summarizes the preceding section.

$\Delta=b^2-4ac$	Factoring form	Roots
$\Delta<0$	No factoring	No real roots
$\Delta=0$	$ax^2+bx+c=a\left(x+\frac{b}{2a}\right)^2$	$x_1=x_2=\frac{-b}{2a}$
$\Delta>0$	$ax^2+bx+c=a(x-x_1)(x-x_2)$	$x_{1,2}=\frac{-b\pm\sqrt{\Delta}}{2a}$

■ **Example 1.7** Solve

$$z^2=z-1,\quad 4t^2-t-3=0,\quad \frac{2x^2}{5}-2x+1=0,\quad 4x^2+4x=-1.$$

(i) The equation is equivalent to $z^2 - z + 1 = 0$ with $a = 1$, $b = -1$, $c = 1$. Thus

$$\Delta = b^2 - 4ac = (-1)^2 - 4(1)(1) = -3 < 0 \Longrightarrow \boxed{S = \emptyset}.$$

(ii) $4t^2 - t - 3 = 0$, with $a = 4$, $b = -1$, $c = -3$. Thus

$$\Delta = b^2 - 4ac = (-1)^2 - 4(4)(-3) = 49 > 0 \Longrightarrow t_{1,2} = \frac{-(-1) \pm \sqrt{49}}{2(4)},$$

$$t_{1,2} = \frac{1 \pm 7}{8} \Longrightarrow \boxed{S = \left\{\frac{-3}{4}, 1\right\}}.$$

(iii) $\dfrac{2x^2}{5} - 2x + 1 = 0$ with $a = \dfrac{2}{5}$, $b = -2$, $c = 1$. Thus

$$\Delta = b^2 - 4ac = (-2)^2 - 4\frac{2}{5}(1) = \frac{12}{5} > 0 \Longrightarrow x_{1,2} = \frac{-(-2) \pm \sqrt{\frac{12}{5}}}{2\left(\frac{2}{5}\right)} = \frac{10 \pm \sqrt{5 \times 12}}{4}$$

$$\boxed{S = \left\{\frac{5 \pm \sqrt{15}}{2}\right\}}.$$

(iv) The equation is equivalent to $4x^2 + 4x + 1 = 0$, with $a = 4$, $b = 4$, $c = 1$. Thus

$$\Delta = b^2 - 4ac = (-4)^2 - 4(4)(1) = 0 \Longrightarrow x = \frac{-4}{2(4)} \Longrightarrow \boxed{S = \left\{\frac{-1}{2}\right\}}.$$

■

■ **Example 1.8** Factor each quadratic expression.

$$A(z) = z^2 - z + 1, \quad B(t) = 4t^2 - t - 3, \quad C(x) = \frac{2}{5}x^2 - 2x + 1, \quad D(x) = 4x^2 + 4x + 1.$$

(i) Since the discriminant $\Delta < 0$, $A(z)$ cannot be factored.

(ii) Since the roots of $B(t) = 0$ are $t_1 = \dfrac{-3}{4}$ and $t_2 = 1$,

$$B(t) = a(t - t_1)(t - t_2) = 4\left(t - \frac{-3}{4}\right)(t - 1) \Longrightarrow \boxed{B(t) = (4t + 3)(t - 1)}.$$

(iii) Since the roots of $C(x) = 0$ are $x_1 = \dfrac{5 - \sqrt{15}}{2}$, $x_2 = \dfrac{5 + \sqrt{15}}{2}$,

$$C(x) = a(x - x_1)(x - x_2) = \frac{2}{5}\left(x - \frac{5 - \sqrt{15}}{2}\right)\left(x - \frac{5 + \sqrt{15}}{2}\right),$$

$$\boxed{C(x) = \left(\frac{2}{5}x - \frac{5 - \sqrt{15}}{5}\right)\left(x - \frac{5 + \sqrt{15}}{2}\right)}.$$

(iv) Since $\Delta = 0$ and the root is $x = -\dfrac{1}{2}$,

$$D(x) = 4\left(x - \frac{-1}{2}\right)^2 = 2^2\left(x - \frac{-1}{2}\right)^2 \Longrightarrow \boxed{D(x) = (2x + 1)^2}.$$

■

1.4.3 Vieta's Rules

We may start by the following observations for the following

1. **quadratic** polynomials:

$$a(x-x_1)(x-x_2) = a\left(x^2-xx_1-xx_2+x_1x_2\right) = ax^2 - a(x_1+x_2)x + ax_1x_2,$$

and by identification with ax^2+bx+c, we obtain ($S=x_1+x_2,\ P=x_1x_2$).

$$\begin{cases} b = & -a(x_1+x_2) & = & -aS, \\ c = & ax_1x_2 & = & aP \end{cases} \implies S = \frac{-b}{a} \text{ and } P = \frac{c}{a}.$$

Thus for $2x^2-5x-7=0$ the sum of the roots $S=-b/a = 5/2$ and their product is $P=-7/2$.
If we consider $a=1$, we may conclude if $\Delta>0$ that

$$x^2+bx+c = x^2-Sx+P=0 \quad \text{has two roots, } x_1,\ x_2 \text{ with sum } S \text{ and product } P.$$

Therefore, if we know the sum of two numbers and their product, we may retrieve them by solving the quadratic equation:

$$x^2-Sx+P=0.$$

Example 1.9 Find two numbers knowing their sum (11) and their product (13).
It is sufficient to solve the quadratic equation: $x^2-11x+13=0$. Its discriminant $\Delta = 11^2-4(1)(13) = 121-52 = 69 > 0$, so the two numbers are

$$x_1 = \frac{-b-\sqrt{\Delta}}{2a} = \frac{11-\sqrt{69}}{2}, \quad x_2 = \frac{-b+\sqrt{\Delta}}{2a} = \frac{11+\sqrt{69}}{2}.$$

Let us check:

$$S = \frac{11-\sqrt{69}+11+\sqrt{69}}{2} = \frac{2(11)}{2} = 11, \quad P = \frac{(11-\sqrt{69})(11+\sqrt{69})}{4} = \frac{11^2-69}{4} = \frac{52}{4} = 13.$$

■

2. for **cubic** polynomials:

$$\begin{aligned} ax^3+bx^2+cx+d &= a(x-x_1)(x-x_2)(x-x_3), \\ &= a\left([x^2-(x_1+x_2)x+x_1x_2](x-x_3)\right), \\ &= a\left(x^3-(x_1+x_2)x^2+x_1x_2x - x^2x_3+(x_1+x_2)x_3x - x_1x_2x_3\right), \\ &= a\left(x^3-(x_1+x_2+x_3)x^2+[x_1x_2+x_2x_3+x_3x_1]x - x_1x_2x_3\right), \end{aligned}$$

and by identification we obtain

$$S_1 = x_1+x_2+x_3 = -b/a, \quad S_2 = x_1x_2+x_2x_3+x_3x_1 = c/a, \quad P = x_1x_2x_3 = -d/a.$$

Example 1.10 (Genuine Proof for an Identity)
Prove that for any real numbers a,b,c, we have[1]:

$$a^3+b^3+c^3-3abc = (a+b+c)\left[a^2+b^2+c^2-ab-bc-ca\right].$$

To this end, let us assume that $a,b,$ and c are the real roots of a cubic polynomial; by using Vieta's rule, we can write the following:

$$Q(x) = (x-a)(x-b)(x-c) = x^3-S_1x^2+S_2x-P,$$

[1] One can establish this by a direct proof, utilizing the formulas for the sum of cubes and the cube of a sum of two numbers to derive the correct factorization.

where $P=abc$, $S_1=a+b+c$, and $S_2=ab+bc+ca$. Now, since $Q(a)=Q(b)=Q(c)=0$, we have

$$\begin{cases} Q(a)=a^3-S_1a^2+S_2a-P &= 0,\\ Q(b)=b^3-S_1b^2+S_2b-P &= 0,\\ Q(c)=c^3-S_1c^2+S_2c-P &= 0. \end{cases}$$

By summing, we get

$$\begin{aligned} & a^3+b^3+c^3-S_1(a^2+b^2+c^2)+S_2(a+b+c)-3P=0,\\ \Longleftrightarrow\quad & a^3+b^3+c^3-3P=S_1(a^2+b^2+c^2)-S_2(a+b+c),\\ \Longleftrightarrow\quad & a^3+b^3+c^3-3P=S_1(a^2+b^2+c^2)-S_2S_1,\\ \Longleftrightarrow\quad & a^3+b^3+c^3-3P=S_1\left[a^2+b^2+c^2-S_2\right]. \end{aligned}$$

Thus we obtain the following

$$\boxed{a^3+b^3+c^3-3abc=(a+b+c)\left(a^2+b^2+c^2-ab-bc-ca\right)}.$$

■

(R) Using the fact that $a^2+b^2-2ab=(a-b)^2$ and similarly for the other squares, we obtain

$$\begin{aligned} a^2+b^2+c^2-ab-bc-ca &= \frac{2(a^2+b^2+c^2-ab-bc-ca)}{2},\\ &= \frac{(a^2+b^2-2ab)+(b^2+c^2-2bc)+(c^2+a^2-2ca)}{2},\\ &= \frac{(a-b)^2+(b-c)^2+(c-a)^2}{2}. \end{aligned}$$

Thus we obtain:

$$\boxed{a^3+b^3+c^3-3abc=\frac{a+b+c}{2}\left[(a-b)^2+(b-c)^2+(c-a)^2\right]}$$

1.4.4 Parabola $y=ax^2+bx+c,\ a\neq 0$

- If $a>0$ the parabola opens up, we say it is **Convex** or **Concave up**.
- If $a<0$ the parabola opens down, we say it is **Concave** or **Concave down**.
- The **vertex** is given by $\boxed{V\left(\alpha,\ \frac{-\Delta}{4a}\right),\quad \alpha=\frac{-b}{2a}}$.
- $x=\alpha$ is the **symmetry axis**.
- If $\Delta=0$, the parabola touches the x-axis at $x=\alpha=\dfrac{-b}{2a}$.
- If $\Delta<0$, the parabola does not cross the x-axis at all.
- If $\Delta>0$, the parabola crosses the x-axis at $x_{1,2}=\dfrac{-b\pm\sqrt{\Delta}}{2a}$.

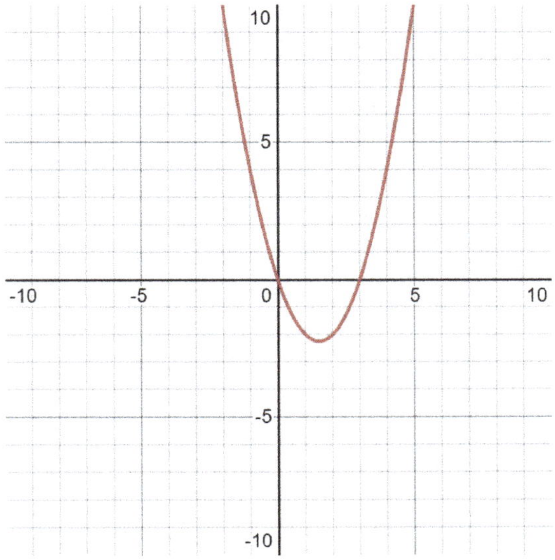

Figure 1.1: $y = x^2 - 3x$ with $a > 0$ and $\Delta > 0$

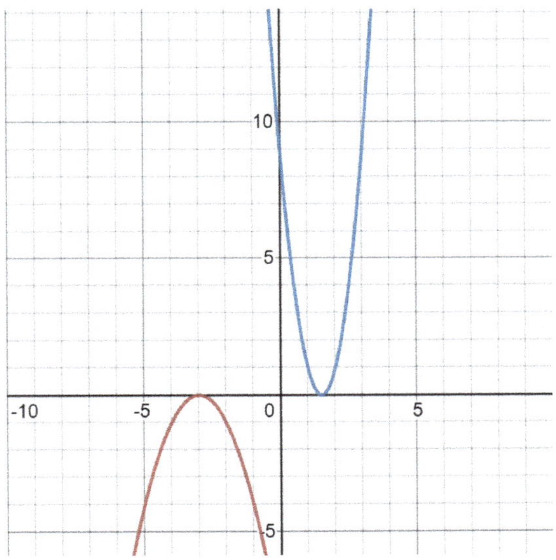

Figure 1.2: $\Delta = 0$ for both cases $a > 0$ (blue) and $a < 0$ (red)

1.4.5 Extrema of $ax^2 + bx + c$

$$a < 0 \Longrightarrow cx^2 + bx + c \leq \frac{-\Delta}{4a} = \beta = \frac{4ac - b^2}{4a}, \quad \forall x \in \mathbb{R}. \tag{1.2}$$

$$a > 0 \Longrightarrow cx^2 + bx + c \geq \frac{-\Delta}{4a} = \beta = \frac{4ac - b^2}{4a}, \quad \forall x \in \mathbb{R}. \tag{1.3}$$

Proof. It is obvious after completing the square (1.1) and the sign of a. Indeed,

$$ax^2 + bx + c = a(x - \alpha)^2 + \beta, \quad \alpha = \frac{-b}{2a}, \quad \beta = \frac{4ac - b^2}{4a} = \frac{-\Delta}{4a}.$$

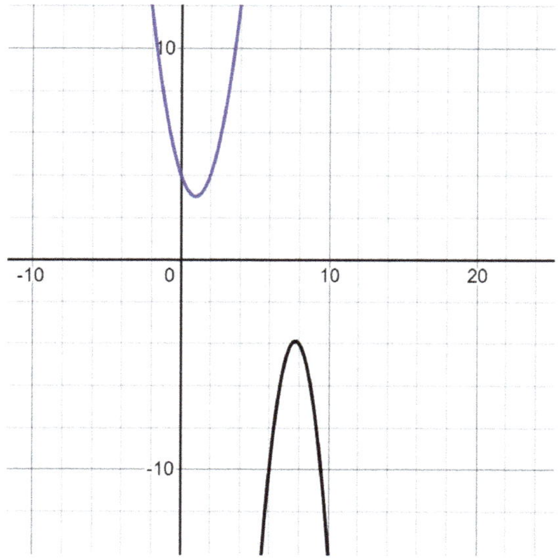

Figure 1.3: $\Delta = 0$ for both cases. $a > 0$ (purple) and $a < 0$ (black)

- If $a < 0$, then we have:

$$a(x-\alpha)^2 \leq 0 \Longrightarrow a(x-\alpha)^2 + \beta \leq \beta.$$

- Similarly, if $a > 0$, then we have:

$$a(x-\alpha)^2 \geq 0 \Longrightarrow a(x-\alpha)^2 + \beta \geq \beta.$$

■ **Example 1.11** We want to show that for all $a, b, c, d \in \mathbb{R}$, the inequality

$$(a+b+c+d)^2 \geq 4(ab+bc+cd+da) \quad \text{holds.}$$

Let f be a quadratic polynomial defined by setting $a = x$:

$$f(x) = (x+b+c+d)^2 - 4(xb+bc+cd+dx).$$

We need to rewrite $f(x)$ in standard form:

$$\begin{aligned} f(x) &= x^2 + 2x(b+c+d) + (b+c+d)^2 - 4x(b+d) - 4(bc+cd) \\ &= x^2 + 2x(c-b-d) + [(b+d)+c]^2 - 4(b+d)c \\ &= x^2 + 2x(c-b-d) + [(b+d)-c]^2. \end{aligned}$$

Now, we can calculate its discriminant:

$$\begin{aligned} \Delta &= 4[c-b-d]^2 - 4(b+d-c)^2) \\ &= 4[c-b-d]^2 - 4(c-b-d)^2) \\ &= 0. \end{aligned}$$

Finally, by using $\Delta = 0$, we obtain:

$$f(x) = (x+b+c+d)^2 - 4(xb+bc+cd+dx) \geq -\frac{\Delta}{4} = 0,$$

which completes the proof. ■

R *It is easy to see that for $P(x) = ax^2 + bx + c$ and $a \neq 0$, we have:*

$$a > 0 \Longrightarrow P(x) \leq \max\{P(u),\ P(v)\},\ \ \forall x \in [u,\ v]. \tag{1.4}$$

$$a < 0 \Longrightarrow P(x) \geq \min\{P(u),\ P(v)\},\ \ \forall x \in [u,\ v]. \tag{1.5}$$

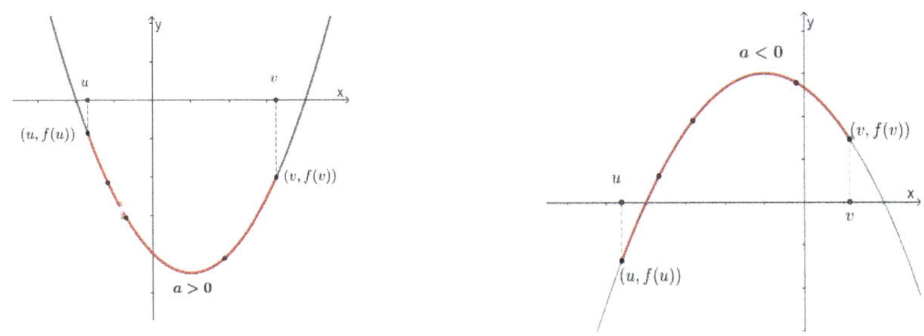

Figure 1.4: Max and Min on closed Interval

We will use the above concept to solve a non easy inequality that can be cast as a nice optimization problem.

■ **Example 1.12** Prove that

$$x^2 + y^2 + z^2 \leq 2 + xyz,\ \ \forall\, x,\, y,\, z \in [0,\ 1].$$

Let f be the quadratic polynomial in x:

$$f(x) = x^2 - (yz)x + y^2 + z^2 - 2.$$

It suffices to prove that $f(x) \leq 0$ for all $x,\, y,\, z \in [0,\ 1]$. Since we have $x \in [0,\ 1]$ and the parabola opens up, we can apply the result (1.4) above as follows:

$$f(x) \leq \max\{f(0), f(1)\} = \max\{y^2 + z^2 - 2,\ y^2 + z^2 - yz - 1\}.$$

If we set $g(y) = y^2 - yz + z^2 - 1$ and $h(y) = y^2 + z^2 - 2$ with $y \in [0,\ 1]$, by applying the same technique for the two parabolas, we get:

$$g(y) \leq \max\{g(0),\ g(1)\}\ =\ \max\{z^2 - 1,\ z^2 - z\} = z^2 - z.$$

The last part of the inequality comes from:

$$z \leq 1 \Longleftrightarrow -z \geq -1 \Longleftrightarrow z^2 - z \geq z^2 - 1.$$

Similarly, we get:

$$h(y) \leq \max\{h(0),\ h(1)\}\ =\ \max\{z^2 - 2,\ z^2 - 1\} = z^2 - 1.$$

Finally, using the fact that $z \in [0,\ 1]$, we obtain :

$$f(x)\ \leq\ \max\{z^2-1,\ z^2-z\}\ =\ z^2-z \leq 0.$$

i.e., $f(x)\ \leq\ 0$ for all $x \in [0,\ 1]$ which completes the proof.

■

1.4.6 Sign of $\mathbf{ax^2+bx+c}$

In this part, we resume the study of the sign of ax^2+bx+c in general, where $a \neq 0,\ a,\ b,\ c \in \mathbb{R}$. The general method is to first find the roots (if any) and then conclude as it is shown in the table below.

$\Delta = \mathbf{b^2-4ac}$	Sign of $\mathbf{ax^2+bx+c}$
$\Delta < 0$	Sign of **a** for any $x \in \mathbb{R}$
$\Delta = 0$	Sign of **a** for any $x \in \mathbb{R},\ x \neq -b/(2a)$
$\Delta > 0$	Sign of **a** outside the roots

The sign results in this table could easily be drawn graphically.

■ **Example 1.13** Solve

$$z^2-z+1>0, \quad -4t^2+t+3 \leq 0, \quad \frac{2}{5}x^2-2x+1<0, \quad 4x^2+4x+1>0.$$

(i) Since the discriminant $\Delta < 0$, $sign(z^2-z+1) = sign(a) = sign(1) > 0$. Thus

$$z^2-z+1>0\ \forall z \in \mathbb{R} \Longrightarrow \boxed{S=(-\infty,\ \infty)}.$$

(ii) Since the discriminant $\Delta > 0$, we have $sign(-4t^2+t+3) = sign(a) = sign(-4) < 0$ outside the roots $t_1 = \dfrac{-3}{4}$ and $t_2 = 1$. Finally,

$$\boxed{S=\left(-\infty,\ \frac{-3}{4}\right] \bigcup [1,\ \infty)}$$

(iii) Since the roots of $\frac{2}{5}x^2-2x+1=0$ are $x_1 = \dfrac{5-\sqrt{15}}{2},\ x_2 = \dfrac{5+\sqrt{15}}{2}$, $sign(\frac{2}{5}x^2-2x+1) = sign(2/5) > 0$ outside the roots, and then

$$\frac{2}{5}x^2-2x+1\ <\ 0\ \text{ if } x \text{ is between the roots.}$$

Thus $\boxed{S=\left(\dfrac{5-\sqrt{15}}{2},\ \dfrac{5+\sqrt{15}}{2}\right)}$.

(iv) Since $\Delta = 0$ and the root is $x = -1/2$,

$$sign(4x^2+4x+1) = sign(4) > 0 \Longrightarrow 4x^2+4x+1>0 \text{ if } x \neq -\frac{1}{2}.$$

Finally, $\boxed{S=\mathbb{R} \setminus \left\{-\dfrac{1}{2}\right\}}$

Ⓡ Since the quadratic expression $4x^2+4x+1$ vanishes at $x=-1/2$, this value has to be rejected from the solution set.

■

■ **Example 1.14** Show that

$$x^2+y^2+z^2+xy+yz+zx+x+y+z+3/8 \geq 0.$$

Let us apply again the same approach, where the coefficient of x^2 is positive

$$h(x) = x^2+(1+y+z)x+y^2+z^2+yz+y+z+3/8 \geq 0.$$

It suffices to show the discriminant $\Delta_x \leq 0$

$$\Delta_x = (1+y+z)^2 - 4(1)(y^2+z^2+yz+y+z+3/8) \leq 0, \ \forall y,\ z.$$

$$\Delta_x = 1+y^2+z^2+2y+2yz+2z-4y^2-4z^2-4yz-4y-4z-3/2 = -3y^2-3z^2-2yz-2y-2z-1/2.$$

again, we will consider

$$-\Delta_x = g(y) = 3y^2+2(z+1)y+3z^2+2z+1/2 \geq 0.$$

Now, we need to show that the discriminant corresponding to $g(y)$ is negative since the coefficient of y is $3>0$.

$$\Delta_y = 4(z+1)^2 - 4(3)(3z^2+2z+1/2) = -4\left[8z^2+4z+1/2\right] = -32\left[z^2+\frac{z}{2}+\frac{1}{16}\right],$$

thus

$$\Delta_y = -32\left(z+\frac{1}{4}\right)^2 \leq 0 \Longrightarrow g(y) \geq 0 \Longrightarrow \Delta_x \leq 0 \Longrightarrow h(x) \geq 0.$$

■

1. If $a>0$ and there exists x_0 such that $f(x_0)<0 \Longrightarrow \Delta>0$.
2. If $a<0$ and there exists x_0 such that $f(x_0)>0 \Longrightarrow \Delta>0$.

and

1. If $a>0$ and there exists x_0 such that $f(x_0)\leq 0 \Longrightarrow \Delta \geq 0$.
2. If $a<0$ and there exists x_0 such that $f(x_0)\geq 0 \Longrightarrow \Delta \geq 0$.

Proof. It is easy to see that when $a>0$ the parabola opens up and if at x_0 we have $f(x_0)<0$ it implies that for sure the parabola will cross the x-axis in two different values. Thus, $\Delta>0$. The same argument is still valid when $a<0$ since the parabola opens down.

In the next example we introduce a `genuine original idea called the hidden` Δ.

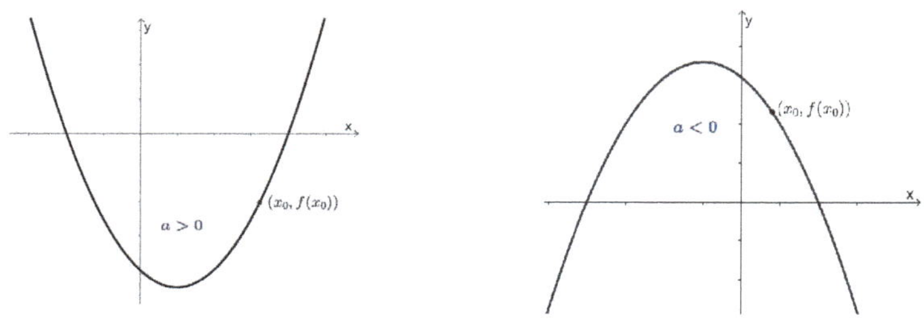

Figure 1.5: Δ principle

■ **Example 1.15 — Hidden Δ.** If $a \le b \le c \le d$ are four real numbers, prove that

$$(a+b+c+d)^2 \ge 8(ac+bd).$$

We encourage readers to start working on this question before studying the solution. The given inequality

$$(a+b+c+d)^2 - 8(ac+bd) \ge 0 \Longleftrightarrow (a+b+c+d)^2 - 4(2)(ac+bd) \ge 0 \twoheadrightarrow \Delta = B^2 - 4AC \ge 0,$$

where $A = 2,\ B = (a+b+c+d),\ C = (ac+bd)$. This discriminant corresponds to the following quadratic function

$$f(x) = 2x^2 - (a+b+c+d)x + (ac+bd) \hookrightarrow \Delta \ge 0 \Rightarrow f(x) = 0 \text{ has at least one root.}$$

According to the products ac and bd, we may re-write $f(x)$ as follows:

$$f(x) = x^2 - (a+c)x + ac + x^2 - (b+d)x + ad = (x-a)(x-c) + (x-b)(x-d).$$

Now, we need to use the assumption $a \le b \le c \le d$. For instance

$$f(b) = (b-a)(b-c) \le 0.$$

Finally, there exists b such that the quadratic function with a positive coefficient of x^2 satisfies: $f(b) \le 0$, then we conclude that $\Delta \ge 0$ which completes the proof. ■

(R) The Intermediate Value Theorem (IVT) can also be applied to the continuous polynomial function $f(x) = (x-a)(x-c) + (x-b)(x-d)$ on $[b,\ d]$ as follows:

$$\begin{array}{l} f(b) = (b-a)(b-c) < 0 \\ f(d) = (d-a)(d-c) > 0 \end{array},$$

then there exists $t \in (b,\ d)\ :\ f(t) = 0 \Longrightarrow \Delta \ge 0$ which completes the proof.

In the following example, we introduce a novel application of the discriminant Δ to address a significant question regarding the range of a specific function.

■ **Example 1.16** Find the range of the following function $y = \dfrac{x^2+4}{x^2+x+4}$.

Since the quadratic equation $x^2+x+4=0$ has no real roots (as indicated by its discriminant $\Delta = -15 < 0$), the given expression is defined for all real numbers $(-\infty,\ \infty)$ and it defines a function, then

$$y = (x^2+4)/(x^2+x+4) \Longleftrightarrow y(x^2+x+4) = x^2+4 \Longleftrightarrow (y-1)x^2+yx+4y-4=0,$$

where the last quadratic equation in x must have at least one solution. Therefore, its discriminant $\Delta_x = y^2-4(y-1)(4)(y-1) \geq 0$. i.e.,

$$\Delta_x = y^2-16(y-1)^2 = [y-4(y-1)][y+4(y-1)] \ = \ (4-3y)(5y-4) \ \geq 0.$$

The obtained quadratic expression in y: $-15y^2+32y-16$, with $a=-15<0$, has opposite sign of a on $[4/5,\ 4/3]$. Consequently, the range of y is $[4/5,\ 4/3]$.

■

1.4.7 Variations of ax^2+bx+c

According to the canonical form

$$P(x) = ax^2+bx+c = a(x-\alpha)^2+\beta,$$

we conclude that if $\boxed{a<0}$ we have:

x	$-\infty$		α		$+\infty$
$P(x)$	$-\infty$	$\nearrow$	β	$\searrow$	$-\infty$

And if $\boxed{a>0}$

x	$+\infty$		α		$-\infty$
$P(x)$	$+\infty$	$\searrow$	β	$\nearrow$	$+\infty$

■ **Example 1.17** Determine

$$\max_{x\in\mathbb{R}} \{x-x^2\}.$$

Note that $x \to x-x^2$ is quadratic with $a=-1<0,\ b=1$, and $c=0$.
Let us consider its variations, where

$$\alpha = \frac{-1}{-2} = \frac{1}{2},\ \beta = \frac{-\Delta}{4a} = \frac{-1}{-4} = \frac{1}{4}.$$

x	$-\infty$	$\frac{1}{2}$	$+\infty$
$x-x^2$	$-\infty$ ↗	$\frac{1}{4}$	↘ $-\infty$

Therefore,

$$\boxed{\max_{x\in\mathbb{R}}\{x-x^2\} = \frac{1}{4} \text{ at } x=\frac{1}{2}}.$$

■

■ **Example 1.18** Consider the following functions:

$$f(x)=x^2+x-12, \qquad g(x)=\left|x^2+x-12\right|, \qquad h(x)=x^2+|x|-12.$$

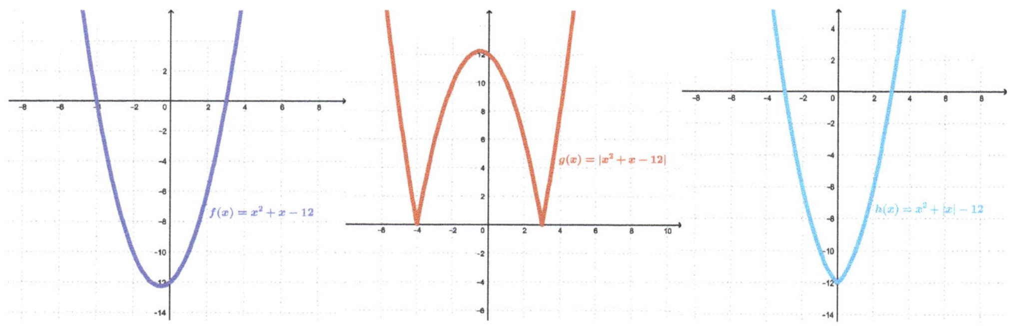

Figure 1.6: Example 1.13

The graphs of functions f, g, and h are shown in Figure 1.6 from left to right. We observe that the graph of function f demonstrates a (classical) quadratic function that opens upward. The graph of function g illustrates an absolute value function, which is always positive and above the x-axis. Interestingly, the graph of h depicts that it is symmetric on the y-axis and that h is an even function.

■

(R) The preceding question in Example 1.17, can be solved directly

- by completing the square:

$$x-x^2=-(x^2-x)=-\left[(x-1/2)^2-\frac{1}{4}\right]=\frac{1}{4}-\underbrace{(x-1/2)^2}_{\geq 0}\leq\frac{1}{4},$$

  and equality holds when $x=1/2$.
- by using the derivative (Calculus 1) if we denote $f(x)=x-x^2$:

$$f'(x)=1-2x=0\Longrightarrow x=1/2,\ f''(x)=-2<0 \Longrightarrow \max_x f(x)=f(1/2)=1/4.$$

■ **Example 1.19** Let $x\in[0,\ 1]$, $y\in[1,\ 2]$, and $z\in[2,\ 3]$. Our goal is to demonstrate that

$$x^2+y^2+z^2-xy-yz-zx\geq\frac{3}{4}.$$

We start by considering a quadratic function in x with coefficients:
$a = 1,\ b = -(y+z)$, and $c = y^2+z^2-yz$.

$$f(x) = x^2-(y+z)x+y^2+z^2-yz, \quad x \in [0,\ 1] \Longrightarrow \alpha = \frac{-b}{2a} = \frac{y+z}{2}.$$

Since $y \in [1,\ 2]$ and $z \in [2,\ 3]$, we can deduce that $\alpha \in [3/2, 5/2]$, which implies that for all $x \leq 1$, f is a decreasing function on the interval $[0,\ 1]$. Consequently, for all $x \leq 1$, we have

$$f(x) \geq f(1) \Longrightarrow f(x) \geq 1-(y+z)+y^2+z^2-yz \Longrightarrow f(x) \geq z^2-z(1+y)+y^2-y+1 = h(z).$$

Now, let us analyze for $z \in [2,\ 3]$ the quadratic function $h(z) = z^2-z(1+y)+y^2-y+1$ with coefficients:
$a = 1,\ b = -(1+y)$, and $c = y^2-y+1$ it has $\alpha = \dfrac{-b}{2a} = \dfrac{y+1}{2} \in [1,\ 3/2]$. Consequently, for $z \in [2,\ 3]$, we find that $z > \alpha$, indicating that h is an increasing function on the interval $[2,\ 3]$. This yields:

$$2 \leq z \Longrightarrow h(2) \leq h(z) \Longrightarrow 2^2-2(1+y)+y^2-y+1 = y^2-3y+3 \leq h(z).$$

Now, using inequality (1), we obtain

$$f(x) \geq h(z) \geq y^2-3y+3 \geq -\frac{\Delta}{4} = -\frac{9-12}{4} = \frac{3}{4},$$

which establishes the desired inequality. Since f is decreasing and h is increasing, equality case is achieved at $x = 1$, $z = 2$, and $y = \dfrac{-b}{2a} = \dfrac{3}{2}$. Let us verify this:

$$x^2+y^2+z^2-xy-yz-zx = 1+\frac{9}{4}+4-\frac{3}{2}-3 = \frac{9}{4}-\frac{6}{4} = \frac{3}{4}.$$

■

1.5 Chapter 1 Practice Problems

Question 1 Solve for x:

1. $\big|x-|3x+1|\big| = 4$.
2. $\Big|\big||x|-2\big|-1\Big| = 3$.
3. $|x-1|+2|x|-3|x+1|-|x+2| = x$.

Question 2 If $\Big||4z+1|-\alpha\Big| = 6$ has three distinct solutions, find the value of α.

Question 3 If $|x+1|-|x-3| = a$ is an equation in x, and it has infinitely many solutions, find the value of a.

Question 4 Let p be an integer satisfying $|2p+7|+|2p-1| = 8$. Find the number of solutions.

Question 5 If the equation $||x-2|-1| = a$ has exactly 3 solutions for x, then find the value of a.

Question 6 Given that the equation $|x| = ax+1$ has exactly one negative solution and has no positive solution, find the range of a.

Question 7 Solve the following equations by factorisation (if possible):

1. $3x^2+4x-1=0$.
2. $x^2-(p+q)x+pq=0$.
3. $8x^2-22x=21$.
4. $2x^2=a^2-ax$.
5. $6x^2-x=2$.
6. $2x+3=5x^2$.

Question 8 For what value of k will the following equations have real roots?

1. $2x^2+kx+2=0$.
2. $kx^2=4-3x$.
3. $3x^2-12x+6k=0$.
4. $200kx^2-300x+400=0$.

Question 9 Evaluate $4x^3-4x^2-66x+1$ for $x=\frac{3+\sqrt{2}}{2}$.

Question 10 Sketch the following graphs:

1. $y=2x^2+x+2$.
2. $y=-x^2+3x$.
3. $y=-2x^2+2x+1$.
4. $y=\sqrt{2-8x+8x^2}$.
5. $y=x^2-5x+6$.

Question 11 Let $y=f(x)=x^2+x-2$.

1. Sketch $y=f(x)$.
2. Explain how can we conclude the graph of $y=g(x)=|f(x)|$ from the graph of f.
3. Sketch $y=x^2+|x|-2$. (Explain all the steps.)

Question 12 Solve:

1. $x^2+\dfrac{5\sqrt{3}}{2}+3=0$.
2. $x^2+7x+7=0$.
3. $6\sqrt{x}=x+5$.
4. $4-8x^2+12=0$.
5. $x^4-8x^2+4\sqrt{3}=0$.

Question 13 Solve:

1. $x^2-\dfrac{a^2+b^2}{2}x+\dfrac{a^2b^2}{4}=0$.
2. $abx^2+(b^2-ac)x=bc$.
3. $k^2x^2+(a^2-b^2)x=b^2$.

Question 14 Solve:

1. $\dfrac{1}{a+b+x} = \dfrac{1}{a} + \dfrac{1}{b} + \dfrac{1}{ax}$, $abx \neq 0$.
2. $abx^2\sqrt{6} + x\sqrt{2} - 5x\sqrt{3} = 5$.
3. $(a+b)^2x^2 + 8(a^2-b^2)x + 16(a-b)^2 = 0$, $a \neq \pm b$.

Question 15 Solve:

1. If $P(\sqrt{2}) = 0$, find the other zeros if $P(x) = 2x^3 - x^2 - 4x + 2$.
2. If $Q(2 \pm \sqrt{3}) = 0$, find the other zeros if $Q(x) = 2x^4 - 9x^3 + 5x^2 + 3x - 1$.

Question 16 If one root of $2x^2 - 5x + a$ is reciprocal of the other, find the value of a.

Question 17 Solve for x

$$2x^2 - 3x + \frac{3}{x} + \frac{2}{x^2} = 1.$$

Question 18 Solve for x:

$$2\sqrt{\frac{x}{x+1}} - 5\sqrt{\frac{1+x}{x}} = 3.$$

Question 19 Find the domain of each expression:

1. $A(x) = \sqrt{2-x} - \sqrt{2x+10}$.
2. $B(x) = \sqrt{x^2 - 5x + 6}$.
3. $C(x) = \dfrac{1}{2x^3 - x^6 - 1}$.
4. $D(x) = \dfrac{1}{\sqrt{8x^2 - 8x - 20}}$.
5. $E(x) = \sqrt{x^2 + 4x - 21} + \sqrt{x^2 - x - 6} + \sqrt{6x^2 - 5x - 39}$.
6. $F(x) = \sqrt[4]{x^3 - 2x^2 + 1}$.
7. $G(x) = \ln(5 - 2x - 2x^2)$.
8. $H(x) = \sqrt{6x - 9x^2 - 1}$.

Question 20 Discuss the domain of f in terms of the real parameter t if

$$f_t(x) = \sqrt{2\sqrt{tx} - x - t}.$$

Question 21 Show that

$$(1+a+b)^2 \geq 4(a+b)^2, \ \forall a,\ b \in [0,\ 1/2].$$

Question 22 Find the relation between the roots of the following quadratic equations:

$$ax^2 + bx + c = 0 \quad \text{and} \quad cx^2 + bx + a = 0.$$

Question 23 Propose a condition on a, b, c, so that we always have

$$a^3+b^3+c^3 \geq 3abc.$$

Question 24 Prove that

$$a^2+b^2+c^2+3-2(a+b+c) \geq 0, \quad \forall a, b, c \in \mathbb{R}.$$

Question 25 Prove that

$$a^2+1+b^2-ab-a-b \geq 0, \quad \forall a, b \in \mathbb{R}.$$

Question 26 Show that

$$x^2+y^2+z^2 \leq 2+xyz, \quad \forall x, y, z \in [0, 1].$$

Question 27 If $0 \leq a < b$ and $n \geq 1 \in \mathbb{N}$, show that

$$na^{n-1} < \frac{b^n-a^n}{b-a} < nb^{n-1}.$$

Question 28 Show that for any triangle ABC with side lengths a, b, and c, we have

$$a^2+b^2+c^2 < 2(ab+bc+ca).$$

Question 29 Show that if $a+b+c=0$ then $ax^2+bx+c=0$ has at least one real root.

2. Logic

Chapter Key Terms

(a) *statement*
(b) *logical connectors*
(c) *disjunction, conjunction, and negation*
(d) *quantifiers*
(e) *implication and bi-conditional*
(f) *converse, inverse, contrapositive, and contradiction.*

2.1 Logic

Logic is a sensible way of understanding the situation presented to one that allows one to deduce an outcome from a given input. Logic is commonly used in our daily routine tasks, and many of us employ logic in our everyday lives unknowingly. Routine patterns developed in performing daily living tasks are suggestive and may not reflect mathematical logic precision. Logic is an essential tool for problem-solving. It enables us to think more critically and rationally when faced with complex problems. It can also help us make better decisions by providing a method of reasoning that is rigorous and systematic. This chapter introduces logic in a mathematical way, as well as the rules for developing an argument in a formal way.

2.1.1 Statements

In mathematics, the concept of a [/bf statement] refers to a declarative statement that is either true or false (but not both). The truth value of a statement indicates whether it is true or false. Statements are often used in mathematical proofs, as well as in everyday life. They are the building blocks of logic and reasoning. A statement can be simple, such as "2 + 2 = 4", or complex, such as "If A, then B". So, statements are an important tool in mathematics, both for proving and understanding logic and reasoning.

■ **Example 2.1** Consider the following sentences:

1. An odd number is not divisible by 2.
2. Area of a circle with radius r is πr^2.
3. A number is irrational if it is not rational.
4. The set of prime numbers is infinite.
5. Every integer is a rational number.

Note that all of the above sentences are true statements. ■

■ **Example 2.2** The following are some examples of false statements:

1. All prime numbers are odd.
2. The circumference of a circle with radius r is $2\pi r^2$.
3. A number is irrational if it is not a real number.
4. A subset is always a finite set.
5. A right triangle is isosceles.

■

■ **Example 2.3** The following are some sentences that are not statements:

1. Divide both sides by 3.
2. A circle is round.
3. What is the real solution of $2x = 3$?
4. y is a prime number.
5. Find a finite subset.

■

■ **Example 2.4** The following are few more examples of the sentences or expressions that are not statements, but small tweaks (or similar sentences) make them statements:

1. Divide both sides by 3 of $3x = 15$.
2. A circle is a polygon.
3. The real solution of $2x = 3$ is $x = 1.5$.
4. $y = 3$ is a prime number.
5. $A = \{1,2,3\}$ is a finite subset of $\mathbb{Z}$.

■

In general, we use the letters P, Q, R, and S to name a specific statement. For the cases when more statements are involved and more letters are needed we can use subscripts. Example 2.5 shows how these letters are used; we leave it to the readers to find the truth value of each statement.

■ **Example 2.5**

1. P: If it is a triangle, then it has three sides.
2. Q: An empty set is a subset of every set.
3. R: The real solution of $x^2 + 1 = 0$ is $x = \pm 1$.
4. S_1: If a number is prime, then it is an odd number.
5. S_2: If zero is even, then one is a prime number.

■

With logic, it is easy to validate the presented information when proving that a statement is true. Logic also helps us understand the combined pieces of information contained in a statement. In the next section we study basic methods of joining simple statements to form a new combined statement.

2.2 Logical Connectives

Statements can be combined into compound statements using connecting words called **logical connectives**. We begin this section with the connectors called **sentential connectives** (and, or, not, if ... then, and if and only if) in mathematical logic, which may not be identical to their everyday use. One should carefully review their specific mathematical meanings before concluding the truth value of a combined statement.

Definition 2.2.1 If P and Q are statements, then $P \wedge Q$ ("P and Q") is also a statement, and it is true precisely when both P and Q are both true. The statement $P \wedge Q$ called the **conjunction** of P and Q. The conjunction is also referred as the **AND** statement.

■ **Example 2.6** Let $P(n)$: $3n$ is odd and $Q(n)$: $2n$ is even be two statements where n is an integer. Then the conjunction, $P(n) \wedge Q(n)$: $3n$ is odd and $2n$ is even is true for all odd integers n. ■

Definition 2.2.2 If P and Q are statements, then $P \vee Q$ ("P or Q") is also a statement, and it is true when at least one of P and Q is true. The statement $P \vee Q$ called the **disjunction** of P and Q. The disjunction is also referred as the **OR** statement.

■ **Example 2.7** Let $P(n)$ and $Q(n)$ be the statements from Example 2.6. Then, $P(n) \vee Q(n)$: $3n$ is odd or $2n$ is even . We note that the disjunction, $P(n) \vee Q(n)$ is true for all integers n. ■

Definition 2.2.3 If P is a statements, then $\neg P$ (negation of P) is also a statement. We write $\neg P$ as **negation** of P and note that $\neg P$ is true if and only if P is false.

The logical connectors: conjunction ($\wedge$), disjunction ($\vee$), and negation ($\neg$) can be summarized in Tables 2.1, 2.2, and 2.3 respectively.

P	Q	$P \wedge Q$
T	T	T
T	F	F
F	T	F
F	F	F

Table 2.1: Conjunction

P	Q	$Q \vee P$
T	T	T
T	F	T
F	T	T
F	F	F

Table 2.2: Disjunction

P	$\neg P$
T	F
F	T

Table 2.3: Negation

(R) When we have 2 statements, the corresponding truth table contains 4 rows, which is $2^{\#\text{ of statements}}$. Thus if we will have 3 statements, then we will have to fill a truth table with $2^3 = 8$ rows, and in general for n statements, we will have a truth table with 2^n rows.

Definition 2.2.4 If P and Q are statements, then P nor Q (P exclusive or Q) is a statement that is true when P is true, and Q is false or P is false and Q is true. The statement Pnor$Q$, denoted by $P \triangledown Q$ is called the **exclusive disjunction** of P and Q. The exclusive disjunction is also referred as the **nor** statement. Table 2.4 is the corresponding truth table.

■ **Example 2.8** Let $P(n)$ and $Q(n)$ be the statements from Example 2.6. Then, $P(n) \triangledown Q(n)$: $3n$ is odd nor $2n$ is even. We note that $P(n) \triangledown Q(n)$ is true for all even integers, n. ■

Definition 2.2.5 Two statements P and Q are **equivalent** if they have same truth values or have the same truth table. The equivalence of P and Q is denoted as $P \equiv Q$.

P	Q	$P \triangledown Q$
T	T	F
T	F	T
F	T	T
F	F	F

Table 2.4: Exclusive disjunction

■ **Example 2.9** Let n be an integer. Consider the following statements:

1. n is either positive or odd.
2. n is odd or positive or both.
3. n is positive and odd.

Note that the first two items say the same thing, but the third does not. For example, the first two are true for $n = 2$, while the third is false. ■

Theorem 2.2.1 Let P, Q, and R be three statements. Then the following are true:

1. $P \vee Q \equiv Q \vee P$, (commutative law)
2. $P \wedge Q \equiv Q \wedge P$, (commutative law)
3. $P \vee (Q \vee R) \equiv (P \vee Q) \vee R$, (associative law)
4. $P \wedge (Q \wedge R) \equiv (P \wedge Q) \wedge R$, (associative law)
5. $P \vee (Q \wedge R) \equiv (P \vee Q) \wedge (P \vee R)$, (distributive law)
6. $P \wedge (Q \vee R) \equiv (P \wedge Q) \vee (P \wedge R)$, (distributive law)

Proof.

1. Obvious by the definition of $\vee$.
2. Obvious by the definition of $\wedge$.
3. $P \vee (Q \vee R) \equiv (P \vee Q) \vee R$,

P	Q	R	$(Q \vee R)$	$\mathbf{P} \vee (\mathbf{Q} \vee \mathbf{R})$	$(P \vee Q)$	$(\mathbf{P} \vee \mathbf{Q}) \vee \mathbf{R}$
T	T	T	T	T	T	T
T	T	F	T	T	T	T
T	F	T	T	T	T	T
T	F	F	F	T	T	T
F	T	T	T	T	T	T
F	T	F	T	T	T	T
F	F	T	T	T	F	T
F	F	F	F	F	F	F

4. Proof in Section 2.9, Question 2.
5. $P \vee (Q \wedge R) \equiv (P \vee Q) \wedge (P \vee R)$.

P	Q	R	$Q \wedge R$	$\mathbf{P} \vee (\mathbf{Q} \wedge \mathbf{R})$	$P \vee Q$	$P \vee R$	$(\mathbf{P} \vee \mathbf{Q}) \wedge (\mathbf{P} \vee \mathbf{R})$
T	T	T	T	T	T	T	T
T	T	F	F	T	T	T	T
T	F	T	F	T	T	T	T
T	F	F	F	T	T	T	T
F	T	T	T	T	T	T	T
F	T	F	F	F	T	F	F
F	F	T	F	F	F	T	F
F	F	F	F	F	F	F	F

6. Proof in Section 2.9, Question 2.

■

■ **Example 2.10** Let P be the following statement n is a positive integer. We can express its negation, $\neg P$, in several ways:

1. $\neg P$: n is not a positive integer.
2. $\neg P$: It is not true that n is a positive integer.
3. $\neg P$: It is false that n is a positive integer.
4. $\neg P$: Either n is zero or a negative integer.

■

We note that negating a statement is not (usually) the same thing as stating its opposite, usually the case in a daily routine. In the previous example, $\neg P$ stated in (4): *Either n is zero or a negative integer*, rather than n is a negative integer. In literature, we may find other notations used for $\neg P$ are $\sim P$ and $\overline{P}$.

■ **Example 2.11** Suppose we want to obtain that only one of P or Q is true, but not both P and Q are true. In symbols, we have

$$(P \vee Q) \wedge \neg (P \wedge Q)$$

This entire statement can be expressed by a truth table, consisting of possible four combinations of true, false values of P and Q. We illustrate the truth table in Table 2.5. The truth values of the individual statements $(P \vee Q)$, $(P \wedge Q)$, and $\neg(P \wedge Q)$ are recorded in columns 3, 4, and 5 respectively. Column 5 is just the negation of column 4, containing its opposite values. Finally, the sixth column is the conjunction of columns 3 and 5. The column 6 is also called the resultant column. Table 2.5 clearly indicates that $(P \vee Q) \wedge \neg(P \wedge Q)$ is true when either P or Q is true but not both. ■

P	Q	$P \vee Q$	$P \wedge Q$	$\neg(P \wedge Q)$	$(P \vee Q) \wedge \neg(P \wedge Q)$
T	T	T	T	F	F
T	F	T	F	T	T
F	T	T	F	T	T
F	F	F	F	T	F

Table 2.5: Truth Table of Example 2.11

2.3 Quantifiers

2.3.1 Universal and Existential Quantifiers

With the help of logical connectors, $\wedge, \vee, \neg, \Longrightarrow$, and, $\Longleftrightarrow$, we can deconstruct many English sentences into their symbolic equivalents. By using this symbol, we are able to understand the logical structure and meaning of sentences as well as their logical equivalence. Despite this, these five symbols are not sufficient to capture every mathematical statement's full meaning. We now introduce two new symbols that correspond to two common English phrases in mathematics that can help overcome this defect. If $P(x)$ is a statement over the domain D, the truth set of $P(x)$ is the set of all elements of D that make $P(x)$ true. **Quantifiers** are words that describe quantities such as *some* or *all* and tell how many elements a given statement is true for. The symbol $\forall$ denotes *for all* or *for every* and is called the **universal quantifier** . The symbol $\exists$ stands for the phrase *there exists a* or *there is a* and is called the **existential quantifier**. Statements that contain them are called quantified statements. A statement beginning with $\forall$ is called a universally quantified statement, and one beginning with $\exists$ is called an existentially quantified statement.

■ **Example 2.12** The following are some examples of quantified statements and their corresponding symbolic representations:

a. For every non-negative real number x, there is a real number y such that $y^2 = x$, can be expressed in symbolic form: $\forall x \in [0,\ \infty),\ \exists\, y \in \mathbb{R}:\ y^2 = x.$
b. Given any two irrational numbers a and b, it follows that *ab is not rational*, can be expressed in symbolic form: $\forall\, a, b \notin \mathbb{Q},\ ab \notin \mathbb{Q}.$
c. *No real numbers have squares equal to* -2, can be expressed as: $\forall\, x \in \mathbb{R},\ x^2 \neq -2.$
d. For every real number, if its absolute value is greater than 1, then it is either less than -1 or greater than 1. In symbols, $\forall\, x \in \mathbb{R},\ |x| > 1 \Longrightarrow (x < -1)\ \vee\ (x > 1).$
e. For every natural number there is another natural number larger than the first. The symbolic form is: $\forall\, n \in \mathbb{N},\ \exists\, m \in \mathbb{N}:\ m > n.$

■

2.3.2 Negating Quantified Statements

In general it is easy to figure out that the quantified statement $P:\ \forall x \in \mathbb{R}\ x^3 > 0$ is not true because we can provide a real number $x = -2$ that makes the statement false. In other terms, the negation of the statement can be expressed as follows:

$$\exists x \in \mathbb{R}:\ x^3 \leq 0$$

It is important to keep in mind the main rules for negating quantifiers[1]:

- $\forall x:\ P(x)$; its negation is $\exists\, x:\ \neg P(x)$.
- $\exists x:\ P(x)$; its negation is $\forall\, x:\ \neg P(x)$.

For example, the following two statements are equivalent:

$$\neg\Big(\forall x \in \mathbb{R}:\ x^3 > 0\Big)\ \equiv\ \exists x \in \mathbb{R}:\ x^3 \leq 0.$$

Note that the universal set (see the example above) in any statement should not be changed when negating the statement. In addition, one should be careful to respect the negation rules.

R Generally, in mathematics we cross expressions with many quantifiers, which we call *nested quantifiers*. Negating such expressions follows the same preceding approach as we can see in the following examples:

$$\neg\Big((\forall x)(\forall y)(\exists z):\ P(x,y,z)\Big)\ \equiv\ (\exists x)(\exists y)(\forall z):\ \neg P(x,y,z).$$

■ **Example 2.13** Express the negation to the following quantified expressions:

$$\forall x \in A,\ \exists y \in B,\ P(x,y), \qquad \exists x \in A,\ \forall y \in B,\ P(x,y),$$

$$\neg\Big(\forall x \in A,\ \exists y \in B,\ P(x,y)\Big) \equiv \exists x \in A,\ \forall y \in B,\ \neg P(x,y),$$

$$\neg\Big(\exists x \in A,\ \forall y \in B,\ P(x,y)\Big) \equiv \forall x \in A,\ \exists y \in B,\ \neg P(x,y).$$

■

[1] The negation of the statement: *There exists an x such that* $P(x)$ *holds* is *For all x the statement* $P(x)$ *does not hold.*

■ **Example 2.14** Express the negation to the following:

$$(\forall x \in A)\ (\exists y \in B),\ \ [P(x,y) \Longrightarrow Q(x,y)],$$

$$\neg\Big(\ (\forall x \in A)\ (\exists y \in B),\ \ [P(x,y) \Longrightarrow Q(x,y)]\ \Big) \equiv R(x,y),$$

where

$$R(x,y) \equiv (\exists x \in A)\ (\forall y \in B),\ \neg[P(x,y) \Longrightarrow Q(x,y)].$$

The last negation will be seen in the next section.

■

2.4 Implications

There is another logical connective which denotes the relationship between two predicates in a sentence of the form *if, then*. Given any two predicates P and Q, a new statement *If* **P** then **Q** is formed and symbolically denoted as $P \Longrightarrow Q$. A statement of the form $P \Longrightarrow Q$ is called a **conditional statement** because it means Q will be true under the condition that P is true. The statement $P \Longrightarrow Q$, also read as P implies Q, has a very specific meaning. The statement $P \Longrightarrow Q$ asserts that if P is true, then Q must also be true. In other words, we can think of $P \Longrightarrow Q$ as being a reward (Q) and a promise made for performing a good deed (P) so, whenever a good deed P is performed, the reward Q is awarded; therefore the promise will be true also. There is only one way this promise can be broken (or be false), which is if P is true but Q is false. Thus the truth table for the promise $P \Longrightarrow Q$ is given in Table 2.6. In other words, we say that the condition P being true implies Q to be true. Such a statement is called an **implication**.

In logic, the construction of *If* **P** *then* **Q** means exactly what the truth values for conditional statement expresses in Table 2.6.

The following list presents several grammatical constructions that mean *If* **P** *then* **Q**.

- P implies Q.
- Q if P.
- Q whenever P.
- Q, provided that P.
- When P, then Q.
- P is a sufficient condition for Q.
- For P, it is necessary that Q happens.
- Q is a necessary condition for P.

In other words, if P is true then Q must also be true. This implication says nothing about the truth of Q when P is false, only that if P is true, Q must also be true. Therefore, when P is false, the statement cannot be false. If P is true, Q must also be true, otherwise the statement is incorrect. We note that the statement *If* **P** *then* **Q** is not equivalent to *If* **Q** *then* **P**. The truth table for the statement *If* **Q** *then* **P** is shown in Table 2.7.

P	Q	$P \Longrightarrow Q$
T	T	T
T	F	F
F	T	T
F	F	T

Table 2.6: Conditional Truth Table $P \Longrightarrow Q$.

P	Q	$Q \Longrightarrow P$
T	T	T
T	F	T
F	T	F
F	F	T

Table 2.7: Conditional Truth Table $Q \Longrightarrow P$.

Example 2.15 Negation of an implication
The truth table shown in Table 2.9 shows that the two statements *If* **P** *then* **Q** and *not* **P** *or* **Q** are equivalent, despite the fact that you might need a little more thought to make sense of this.

P	Q	$\neg P$	$P \Longrightarrow Q$	$\neg P \vee Q$
T	T	F	T	T
T	F	F	F	F
F	T	T	T	T
F	F	T	T	T

Table 2.8: Implication Truth Table.

P	Q	$\neg Q$	$P \Longrightarrow Q$	$\neg(P \Longrightarrow Q)$	$P \wedge \neg Q$
T	T	F	T	F	F
T	F	T	F	T	T
F	T	F	T	F	F
F	F	T	T	F	F

Table 2.9: Negation of an implication

This is true because for a given implication, *If* **P** *then* **Q**, there are two possibilities. We could have P and therefore Q. Likewise, we would have **not** P, or we could have Q. Thus P implies Q is equivalent to **Q** *or not* **P**, which is usually written as *not* **P** or **Q**. The equivalence is shown in Table 2.8. Consequently, the statements: negation of (*if* **P**, *then* **Q**) and **P** *and not* **Q** are equivalent, see Table 2.9. ■

2.5 Bi-conditional Statements

It is notable that the implication *If* **P** *then* **Q** is not same as the implication *If* **Q** *then* **P**. Suppose that x is some integer, and consider the following statements:
$P \Longrightarrow Q$: If x is divisible by 4, then x is divisible by 2.
$Q \Longrightarrow P$: If x is divisible by 2, then x is divisible by 4.

- The statement $P \Longrightarrow Q$ is true, whereas $Q \Longrightarrow P$ is false ($x = 6$ is an even number, and it is not divisible by 4).
- The statement $Q \Longrightarrow P$ is called the **converse** of $P \Longrightarrow Q$. A conditional statement and its converse, in general, express entirely different things.
- When two predicates P and Q have same truth value, either both are true or else both false; P and Q are **equivalent** or **bi-conditional**. The equivalence of P and Q is denoted as $P \Longleftrightarrow Q$, and its truth table is shown in Table 2.10.

It is important to know that the bi-conditional statement $P \Longleftrightarrow Q$ is a conjunction of the two implications, $P \Longrightarrow Q$ and $Q \Longrightarrow P$. That is, $P \Longleftrightarrow Q \equiv (P \Longrightarrow Q) \wedge (Q \Longrightarrow P)$, shown in Table 2.11. Additionally, we note that $P \Longleftrightarrow Q \equiv Q \Longleftrightarrow P$.

2.6 Logical Equivalence

Two statements P and Q are logically equivalent if for any instance, both P and Q are true or else both P and Q are false.

P	Q	$P \Longleftrightarrow Q$
T	T	T
T	F	F
F	T	F
F	F	T

Table 2.10: Bi-conditional Truth Table.

P	Q	$\mathbf{P \Longleftrightarrow Q}$	$P \Longrightarrow Q$	$Q \Longrightarrow P$	$(\mathbf{P \Longrightarrow Q}) \wedge (\mathbf{Q \Longrightarrow P})$
T	T	T	T	T	T
T	F	F	F	T	F
F	T	F	T	F	F
F	F	T	T	T	T

Table 2.11: Example of bi-conditional equivalence

■ **Example 2.16** Let us construct a truth table (Table 2.12) to show the following equivalence:

$$P \Longrightarrow Q \equiv (P \wedge \neg Q) \Longrightarrow (Q \wedge \neg Q)$$

P	Q	$\neg Q$	$\mathbf{P \Longrightarrow Q}$	$P \wedge \neg Q$	$Q \wedge \neg Q$	$(\mathbf{P} \wedge \neg \mathbf{Q}) \Longrightarrow (\mathbf{Q} \wedge \neg \mathbf{Q})$
T	T	F	T	F	F	T
T	F	T	F	T	F	F
F	T	F	T	F	F	T
F	F	T	T	F	F	T

Table 2.12: Equivalence in Example 2.16

■

Some logically equivalent statements are natural and intuitive. There is a pair of logically equivalent statements called De Morgan's laws; the equivalence of the statements in De Morgan's laws can be extremely useful to quickly obtain additional pieces of information in terms of equivalence.

Definition 2.6.1 — De Morgan's Laws.

1. $\neg(P \wedge Q) \equiv (\neg P) \vee (\neg Q)$.

2. $\neg(P \vee Q) \equiv (\neg P) \wedge (\neg Q)$.

P	Q	$\neg P$	$\neg Q$	$\neg P \vee \neg Q$	$\neg(P \wedge Q)$	$P \wedge Q$
T	T	F	F	F	F	T
T	F	F	T	T	T	F
F	T	T	F	T	T	F
F	F	T	T	T	T	F

Table 2.13: De Morgan's Law $\neg P \vee \neg Q \equiv \neg(P \wedge Q)$

2.6.1 More on Implications

Let $P \Longrightarrow Q$ be an implication. We define following related statements based on the implication:

P	Q	$\neg P$	$\neg Q$	$\neg P \wedge \neg Q$	$\neg(P \vee Q)$	$P \vee Q$
T	T	F	F	F	F	T
T	F	F	T	F	F	T
F	T	T	F	F	F	T
F	F	T	T	T	T	F

Table 2.14: De Morgan's Law $\neg P \wedge \neg Q \equiv \neg(P \vee Q)$

- The **converse** is a statement obtained by interchanging the hypothesis and conclusion of the given implication. Therefore, the converse of P implies Q is the implication

$$Q \Longrightarrow P.$$

- The **inverse** of an implication is described as a statement obtained by negating the hypothesis and conclusion. The statement

$$\neg P \Longrightarrow \neg Q$$

is the inverse of the original implication

$$P \Longrightarrow Q.$$

- **Contrapositive** of the implication as described as

$$\neg Q \Longrightarrow \neg P$$

obtained by negating and interchanging the hypothesis and conclusion.

■ **Example 2.17** Consider the valid implication: *If it is raining, then there are clouds.*

- The converse would be *If there are clouds, then it is raining.* This is not always true.
- The inverse is *If it is not raining, then there are no clouds*, which is not always true.
- The contrapositive of the implication is the following *If there are no clouds, then it is not raining.* This statement is equivalent to our original implication.

■

We note the following equivalences:
- $P \Longrightarrow Q \equiv \neg Q \Longrightarrow \neg P.$
- $Q \Longrightarrow P \equiv \neg P \Longrightarrow \neg Q.$

■ **Example 2.18** We write the contrapositive, converse, and inverse along with their symbolic representations of the following statement: If a real number is greater than 2, then its square is greater than 4.

$$\forall x \in \mathbb{R},\ x > 2 \implies x^2 > 4, \text{ which is equivalent to } x > 2 \implies x^2 > 4:$$

- **converse**: If the square of a real number is greater than 4, then the number is greater than 2.

$$x^2 > 4 \implies x > 2.$$

- **inverse**: If a real number is less than or equal to 2, then the square of this number is less than or equal to 4.

$$x \leq 2 \implies x^2 \leq 4$$

- **contrapositive**: If a square of a real number is less than or equal to 4, then this number is less than or equal to 2.

$$x^2 \leq 4 \implies x \leq 2.$$

■

(R) It is important to know when a condition (or statement) is sufficient for another statement and when it is necessary:

- **Sufficient condition** P is a sufficient condition for Q means that: $P \Longrightarrow Q$.
- **Necessary condition** P is a necessary condition for Q means that: If $\neg P \Longrightarrow \neg Q$.

For example, if Q: "f is differentiable at $x = a$" and P: " f is continuous at $x = a$", then we have the following

- P is a necessary condition for Q because if f is not continuous at $x = a$, then f is not differentiable at $x = q$.
- Continuity is a necessary condition but not sufficient for differentiability.
- Q is a sufficient condition for P because if f is differentiable at $x = a$, then f is continuous at $x = a$.
- Studying is a necessary condition, but it is not sufficient to get "an A grade".
- Having four sides for a polygon is a necessary condition, but it not sufficient to be a square.

2.7 Tautologies and Contradictions

A statement is a **logical tautology** if it is always true (regardless of the truth values of its component statements). Similarly, a statement is a **logical contradiction** (also called an **absurdity**) if it is always false (regardless of the truth values of its component statements).

■ **Example 2.19** The following truth tables illustrate some interesting examples of logical tautology and contradiction:

- logical tautology: $P \vee \neg P$
- logical contradiction: $Q \wedge \neg Q$
- logical contradiction: $(P \vee \neg P) \Longrightarrow (Q \wedge \neg Q)$

P	$\neg P$	$P \vee \neg P$
T	F	T
F	T	T

Table 2.15: Tautology

Q	$\neg Q$	$Q \wedge \neg Q$
T	F	F
F	T	F

Table 2.16: Contradiction

■

$P \vee \neg P$	$Q \wedge \neg Q$	$(P \vee \neg P) \Longrightarrow (Q \wedge \neg Q)$
T	F	F
T	F	F
T	F	F
T	F	F

Table 2.17: Contradiction

Finally, we prove the equivalence of the following statements **without using truth tables** and close this section.

■ **Example 2.20** $(P\vee\neg P)\Longrightarrow(Q\wedge\neg Q)$ is a contradiction (or absurdity).
Let us denote **T** for the tautology and **F** for the contradiction. Then,

$$\begin{aligned}(P\vee\neg P)\Longrightarrow(Q\wedge\neg Q) &\equiv \neg(P\vee\neg P)\ \vee\ (Q\wedge\neg Q),\\ &\equiv \neg(\mathbf{T})\vee\ \mathbf{F},\\ &\equiv \mathbf{F}\vee\ \mathbf{F},\\ &\equiv \mathbf{F}.\end{aligned}$$

Thus $(P\vee\neg P)\Longrightarrow(Q\wedge\neg Q)$ is a contradiction. ■

■ **Example 2.21** $P\Longrightarrow Q\equiv(P\wedge\neg Q)\Longrightarrow(Q\wedge\neg Q)$.
Let us start from the RHS:

$$\begin{aligned}RHS\equiv(P\wedge\neg Q)\Longrightarrow(Q\wedge\neg Q) &\equiv \neg(P\wedge\neg Q)\ \vee(Q\wedge\neg Q),\\ &\equiv (\neg P\vee\neg(\neg Q))\ \vee\mathbf{F}, && \text{(De Morgan's rule)}\\ &\equiv \neg P\vee Q,\\ &\equiv P\Longrightarrow Q.\end{aligned}$$

■

■ **Example 2.22** $(P\vee Q)\Longrightarrow R\equiv[(P\Longrightarrow R)\wedge(Q\Longrightarrow R)]$.
Let us start from the LHS:

$$\begin{aligned}(P\vee Q)\Longrightarrow R &\equiv \neg(P\vee Q)\ \vee R,\\ &\equiv (\neg P\wedge\neg Q)\ \vee R, && \text{(De Morgan's rule)}\\ &\equiv R\vee(\neg P\wedge\neg Q), && \text{(commutativity)}\\ &\equiv (R\vee\neg P)\wedge(R\vee\neg Q), && \text{(distributive)}\\ &\equiv (\neg P\vee R)\wedge(\neg Q\vee R), && \text{(commutativity)}\\ &\equiv (P\Longrightarrow R)\wedge(Q\Longrightarrow R).\end{aligned}$$

■

■ **Example 2.23** $Q\Longrightarrow P\ \equiv\ (Q\wedge\neg P)\Longrightarrow(P\wedge\neg Q)$
Let us start from the RHS.

$$\begin{aligned}(Q\wedge\neg P)\Longrightarrow(P\wedge\neg Q) &\equiv \neg(Q\wedge\neg P)\vee(P\wedge\neg Q)\\ &\equiv (\neg Q\vee P)\ \vee(P\wedge\neg Q), && \text{(De Morgan's rule)}\\ &\equiv [(\neg Q\vee P)\ \vee P]\wedge[(\neg Q\vee P)\ \vee\neg Q], && \text{(distribution)}\\ &\equiv [(\neg Q\vee(P\ \vee P)]\wedge[(\neg Q\vee\neg Q)\ \vee P], && \text{(associativity)}\\ &\equiv (\neg Q\vee P)\wedge(\neg Q\ \vee P),\\ &\equiv \neg Q\vee P,\\ &\equiv Q\Longrightarrow P.\end{aligned}$$

■

1	$P\wedge Q\ \equiv\ Q\wedge P$	$P\vee Q\ \equiv\ Q\vee P$	commutative law
2	$P\wedge P\equiv P$	$P\vee P\equiv P$	idempotent law
3	$P\wedge\mathbf{T}\equiv P$	$P\vee\mathbf{T}\equiv\mathbf{T}$	identity law(1)
4	$P\wedge\mathbf{F}\equiv\mathbf{F}$	$P\vee\mathbf{F}\equiv P$	identity law(2)
5	$P\wedge(Q\wedge R)\equiv(P\wedge Q)\wedge R$	$P\vee(Q\vee R)\equiv(P\vee Q)\vee R$	associative laws
6	$P\wedge(Q\vee R)\equiv(P\wedge Q)\vee(P\wedge R)$	$P\vee(Q\wedge R)\equiv(P\vee Q)\wedge(P\vee R)$	distributive laws
7	$\neg(\neg P)=P$	$\neg(P\wedge P)=\neg P$	involution law
8	$P\vee\neg P\equiv\mathbf{T},P\wedge\neg P\equiv\mathbf{F}$	$\neg\mathbf{T}\equiv\mathbf{F},\quad\neg\mathbf{F}\equiv\mathbf{T}$	complement laws
9	$\neg(P\vee Q)\equiv\neg P\wedge\neg Q$	$\neg(P\wedge Q)\equiv\neg P\vee\neg Q$	De Morgan laws
10	$P\ \Longrightarrow\ Q\ \equiv\ \neg P\ \vee\ Q$	$Q\ \Longrightarrow\ P\ \equiv\ P\ \vee\ \neg Q$	implication
11	$P\ \Longrightarrow\ Q\ \equiv\ \neg Q\ \Longrightarrow\ \neg P$	$\neg P\ \vee\ Q\equiv\ \neg Q\ \Longrightarrow\ \neg P$	contraposition
12	$P\iff Q\equiv(P\Longrightarrow Q)\wedge(Q\Longrightarrow P)$	$Q\iff P\equiv P\iff Q$	bi-conditional

2.8 Chapter 2 Practice Problems

Question 1 Given three statements P, Q, and R, prove the following

1. $P \wedge (Q \wedge R) \equiv (P \wedge Q) \wedge R$, (associativity),
2. $P \wedge (Q \vee R) \equiv (P \wedge Q) \vee (P \wedge R)$. (distribution).

Question 2 Write the truth tables to the logical statements

$$P \vee (Q \Longrightarrow R), \text{ and } P \wedge (Q \vee \neg R).$$

Question 3 For the statements P, Q, and R show (by making a truth table) that:

$$((P \wedge Q) \Longrightarrow R) \equiv (Q \wedge (\sim R)) \Longrightarrow (\sim P).$$

Question 4 Construct a truth table for

$$(\neg P \vee R) \vee [P \wedge (\neg R \vee Q)].$$

Question 5 Express the negation of the following statements:

(i) $\exists\, y \in \mathbb{Z}:\ x < y.$
(ii) $\exists\, x \in \mathbb{R}\ \exists\, y \in \mathbb{Z} \ni\ x = y + 2.$
(iii) $\forall\, x \in \mathbb{R}\ \exists\, n \in \mathbb{Z}:\ x^2 \geq n.$
(iv) $\forall\, x \in \mathbb{R} \exists\, y \in \mathbb{R}:\ x - y = 3.$
(v) $\forall x \in \mathbb{R}\ \sqrt{x^2} = x \vee \sqrt{x^2} = -x.$
(vi) $\forall x \in \mathbb{R}\ \exists\, y \in \mathbb{Z}:\ x^2 - 2y = 0.$

Question 6 True/ False (Justify your answer).

(i) $\forall\, x \in \mathbb{R}\ \exists\, n \in \mathbb{Z}:\ x^2 \geq n.$
(ii) $\forall\, x \in \mathbb{R}\ \exists\, y \in \mathbb{R}:\ x - y = 3.$

Question 7 True/ False (Justify your answers):

(a) $\forall x \in \mathbb{R},\ (\sqrt{x^2} = x \vee \sqrt{x^2} = -x).$
(b) $\forall x \in \mathbb{R}\ \exists\, y \in \mathbb{Z}:\ x^2 - 2y = 0.$

Question 8 True/ False (Justify your answers):

(a) $\forall z \in \mathbb{R}\ \forall\, n \in \mathbb{Z}:\ z^2 - n^3 = 0.$
(b) $\exists t \in \mathbb{R}\ \ni \forall\, m \in \mathbb{Z},\ t^2 + m = 0.$

Question 9 State each statement in *If* **P**, *then* **Q**" form.

1. An integer is divisible by 10 only if it is divisible by 5.
2. The number 27 is prime only if 3 is not.
3. If a function is integrable, then it is necessary that it is continuous.
4. Whenever a function is integrable, it is continuous.
5. For a number x to be positive it is sufficient that $x > 0$.
6. A function is rational if it is a polynomial.
7. Every perfect square is a square of an integer.
8. A perfect square is an integer that is the square of an integer.
9. For a perfect square, it is necessary that it is a square of an integer.
10. An integer to be a square of an integer, it is sufficient that it is a perfect square.
11. Whenever a matrix A is invertible, it is non-singular.
12. Good company is a necessary for a memorable journey.

Question 10 Prove the following statements without using truth tables:

a. $(P \Longrightarrow Q) \Longrightarrow R \equiv R \vee (\neg Q \wedge P)$.

b. $P \Longrightarrow Q \equiv (P \wedge \neg Q) \Longrightarrow (Q \wedge \neg Q)$.
c. $P \Longrightarrow (Q \wedge R) \equiv (\neg Q \vee \neg R) \Longrightarrow \neg P$.

Question 11 Assume $P: \sqrt{2}$ is rational, $Q: 2/3$ is rational, and $R: \sqrt{5}$ is irrational. Determine the truth value of each statement:
a. $\neg P \wedge Q \Longrightarrow R$.
b. $P \vee Q \Longrightarrow \neg R$.

Question 12 Assume $P: \sqrt{17}$ is irrational, $Q: \frac{12}{4}$ is rational, and $R: \pi$ is a real number. Determine the truth value of each statement:
a. $(R \Longrightarrow Q) \Longleftrightarrow P$.
b. $(R \Longleftrightarrow P) \wedge (R \Longleftrightarrow Q)$.

Question 13 $P(n): 3^{n-1}$ is prime and $Q(n): n^2$ is odd.
Determine all $n \in S = \{1,2,3,4,5\}$ for which $P(n) \Longleftrightarrow Q(n)$ is a true bi-conditional statement.

Question 14 Establish the following equivalence using a truth table:

$$((P \wedge Q) \Longrightarrow R) \equiv (Q \wedge (\sim R)) \Longrightarrow (\sim P).$$

Question 15 For the statements
- $P(n)$: if n is prime, then $\sqrt{n}$ is irrational;
- $Q(n)$: if n is odd integer, then n^3 is odd;
- $R(n)$: if n is a prime number, then $2n$ is even;

determine the truth values for each of the following implication
a. $(\neg P \wedge \neg Q) \Longrightarrow \neg R$.
b. $\neg(P \wedge Q) \Longrightarrow \neg R$.

Question 16 For the statements P, Q, and R, without using a truth table, show that:

$$((P \wedge Q) \Longrightarrow R) \equiv (Q \wedge (\sim R)) \Longrightarrow (\sim P).$$

Question 17 Check whether the following statement is a contradiction by using a truth table:

$$(\neg Q \vee R) \wedge (P \Longrightarrow (\neg R \wedge \neg(\neg Q))) \wedge (P \vee R).$$

Question 18 Let $S: [\neg q \wedge (p \vee r)] \Longrightarrow p$ and $T: [p \vee (q \vee \neg r)]$. Prove that $S \Longleftrightarrow T$.

Question 19 For the statements P, Q, and R, determine (by a truth table) if the following equivalence is true:

$$((P \wedge (\neg R)) \Longleftrightarrow Q \equiv (Q \vee (\neg P)) \Longleftrightarrow \neg R.$$

Question 20 Rewrite each of the following implications using sufficient conditions and determine their truth values:
- If a triangle is equilateral, then it is isosceles.
- A triangle is isosceles if it has two equal sides.
- If a triangle has two equal angles, then it is isosceles.
- A triangle is isosceles if it is equilateral.
- A triangle is isosceles if it has two sides and two angles equal.

Question 21 Rewrite each of the following implications using sufficient conditions and determine their truth values:
- If a triangle is equilateral, then it is isosceles.

- A triangle is isosceles if it has two equal sides.
- If a triangle has two equal angles, then it is isosceles.
- A triangle is isosceles if it is equilateral.
- A triangle is isosceles if it has two sides and two angles equal.

Question 22 Rewrite each of the following implications using sufficient conditions and determine their truth values:

- If a triangle is a right triangle, then all of its angles are equal.
- A triangle is a right triangle if any two of its angles are complementary.
- A triangle is a right triangle if the square of the length of its longest side is equal to sum of the squares of the lengths of two smallest sides.
- If a triangle is a right triangle, then twice of the area of the triangles equals the area of a rectangle.
- A triangle is a right triangle if sum of two of its angles is equal to the third angle.

Question 23 Decide whether the following pairs of statements are logically equivalent:

a. $P \wedge Q$ and $\neg(\neg P \vee \neg Q)$.

b. $(P \Longrightarrow Q) \vee R$ and $\neg((P \wedge \neg Q) \wedge \neg R)$.

c. $(\neg P) \wedge (P \Longrightarrow Q)$ and $\neg(Q \Longrightarrow P)$.

d. $\neg(P \Longrightarrow Q)$ and $P \wedge \neg Q$.

Question 24 Establish the following equivalence.

a. $P \Longrightarrow Q \equiv [\neg Q \Longrightarrow \neg P] \equiv \neg P \vee Q$.

b. $P \Longleftrightarrow Q \equiv (P \wedge Q) \vee (\neg P \wedge \neg Q)$.

Question 25 Check whether the following statement is a tautology, without using a truth table:

$$(\neg Q \vee R) \wedge (P \Longrightarrow (\neg R \wedge \neg(\neg Q))) \wedge (P \vee R).$$

Question 26 Check whether the following statement is a tautology or contradiction or neither with and without the use of a truth table:

$$(A \wedge B) \vee (\neg A \wedge B) \vee (\neg A \wedge \neg B) \vee (A \wedge \neg B).$$

Question 27 Using truth tables, prove that the following are tautologies:

a. $[P \wedge (P \Longrightarrow Q)] \Longrightarrow P$ (modus ponens).

b. $[\neg Q \wedge (P \Longrightarrow Q)] \Longrightarrow \neg(P \wedge Q)$ (modus tollens).

Question 28 Using truth tables, prove that the following are tautologies:

a. $[(P \implies Q) \wedge (Q \implies R)] \implies (P \implies R)$.

b. $[(P \vee Q) \implies R] \iff (P \implies R) \wedge (Q \implies R)$.

2.9 Chapter 2 Additional Practice Problems with Solutions

Question 1 Construct a truth table to verify if the following is a tautology:

$$(\neg P \vee R) \vee [P \wedge (\neg R \vee Q)].$$

Solution.

P	Q	R	$\neg P \vee R$	$\neg R \vee Q$	$P \wedge (\neg R \vee Q)$	$(\neg P \vee R) \vee (P \wedge (\neg R \vee Q))$
T	T	T	T	T	T	T
T	T	F	F	T	T	T
T	F	T	T	F	F	T
T	F	F	F	T	T	T
F	T	T	T	T	F	T
F	T	F	T	T	F	T
F	F	T	T	F	F	T
F	F	F	T	T	F	T

Question 2 Let P, Q, and R be three statements. Using truth tables, prove the following

1. $P \wedge (Q \wedge R) \equiv (P \wedge Q) \wedge R$, (associative law).

2. $P \wedge (Q \vee R) \equiv (P \wedge Q) \vee (P \wedge R)$. (distributive law).

Solution.

1. $P \wedge (Q \wedge R) \equiv (P \wedge Q) \wedge R$ (associative law).

P	Q	R	$Q \wedge R$	$P \wedge (Q \wedge R)$	$P \wedge Q$	$(P \wedge Q) \wedge R$
T	T	T	T	T	T	T
T	T	F	F	F	T	F
T	F	T	F	F	F	F
T	F	F	F	F	F	F
F	T	T	T	F	F	F
F	T	F	F	F	F	F
F	F	T	F	F	F	F
F	F	F	F	F	F	F

2. $P \wedge (Q \vee R) \equiv (P \wedge Q) \vee (P \wedge R)$ (distributive law).

P	Q	R	$P \wedge Q$	$P \wedge R$	$(P \wedge Q) \vee (P \wedge R)$	$Q \vee R$	$P \wedge (Q \vee R)$
T	T	T	T	T	T	T	T
T	T	F	T	F	T	T	T
T	F	T	F	T	T	T	T
T	F	F	F	F	F	F	F
F	T	T	F	F	F	T	F
F	T	F	F	F	F	F	F
F	F	T	F	F	F	T	F
F	F	F	F	F	F	F	F

Question 3 Which of these statements are true for all integers n?

a. $n > 4 \Longrightarrow n \geqslant 1$.

b. $n \geqslant 1 \Longrightarrow n < -1$.

c. $n^3 = 8 \Longrightarrow n = 2$.

d. $n^2 = 9 \Longrightarrow n = 3$.

e. $n - 1 = 5 \Longrightarrow n \geqslant 6$.

Solution.

a. Let us denote p : "$n > 4$", q : "$n \geq 1$".
   - If $n < 1$, then we have p and q are false, and then $p \Longrightarrow q$ is true.
   - If $1 \leq n \leq 4$, then we have p false and q true; therefore, $p \Longrightarrow q$ is true.
   - If $n > 4$, then we have p and q are true, and then $p \Longrightarrow q$ is true. Hence, the statement is true for any integer n.

b. Let us denote p : "$n \geq 1$", q : "$n < 1$". The case where p is true, and q is false makes $p \Longrightarrow q$ false. Hence, the statement is not true for any integer n.

c. Let us denote p : "$n^3 = 8$", q : "$n = 2$".
   - If $n \neq 2$, then we have p and q are false, and then $p \Longrightarrow q$ is true.
   - If $n = 2$, then we have p and q are true, and then $p \Longrightarrow q$ is true. Hence, the statement is true for any integer n.

d. Let us denote p : "$n^2 = 9$", q : "$n = 3$".
   - If $n = 3$, then we have p and q are true, and then $p \Longrightarrow q$ is true.
   - If $n \neq 3$ and $n \neq -3$, then we have p and q are false, and then $p \Longrightarrow q$ is true.
   - If $n = -3$, then we have p true and q false, and then $p \Longrightarrow q$ is false. Hence, the statement is not true for any integer n.

e. Let us denote p : "$n - 1 = 5$", q : "$n \geqslant 6$".
   - If $n = 6$, then we have p and q are true, and then $p \Longrightarrow q$ is true.
   - If $n > 6$, then we have p false and q true, and then $p \Longrightarrow q$ is true.
   - If $n < 6$, then we have p and q are false, and then $p \Longrightarrow q$ is true. Hence, the statement is true for any integer n.

Thus (a), (c), and (e) are true statements for all integers.

Question 4 Construct a truth table showing that the following three statements are equivalent:

(a) $\neg P \Longrightarrow Q$.

(b) $Q \vee P$.

(c) $\neg Q \Longrightarrow P$.

Solution.
In the following truth table, the boldface columns indicate that the statements (a), (b), and (c) are equivalent.

P	Q	$\neg P$	$\neg Q$	$\neg P \Longrightarrow Q$	$Q \vee P$	$\neg Q \Longrightarrow P$
T	T	F	F	T	T	T
T	F	F	T	T	T	T
F	T	T	F	T	T	T
F	F	T	T	F	F	F

Question 5 Construct a truth table for the following logical statements:

$$P \vee (Q \Longrightarrow R) \text{ and } P \wedge (Q \vee \neg R).$$

Solution.

P	Q	R	$Q \Longrightarrow R$	$\mathbf{P} \vee (\mathbf{Q} \Longrightarrow \mathbf{R})$	$\neg R$	$Q \vee \neg R$	$\mathbf{P} \wedge (\mathbf{Q} \vee \neg \mathbf{R})$
T	T	T	T	T	F	T	T
T	T	F	F	T	T	T	T
T	F	T	T	T	F	F	F
T	F	F	T	T	T	T	T
F	T	T	T	T	F	T	F
F	T	F	F	F	T	T	F
F	F	T	T	T	F	F	F
F	F	F	T	T	T	T	F

Question 6 Using a truth table, prove that the following logical statements are equivalent:

$$((P \wedge Q) \Longrightarrow R) \text{ and } (Q \wedge \neg R) \Longrightarrow \neg P.$$

Solution. In the following truth table, the boldface columns indicate that the statements are equivalent.

P	Q	R	$P \wedge Q$	$(\mathbf{P} \wedge \mathbf{Q}) \Longrightarrow \mathbf{R}$	$Q \wedge \neg R$	$\neg P$	$(\mathbf{Q} \wedge \neg \mathbf{R}) \Longrightarrow \neg \mathbf{P}$
T	T	T	T	T	F	F	T
T	T	F	T	F	T	F	F
T	F	T	F	T	F	F	T
T	F	F	F	T	F	F	T
F	T	T	F	T	F	T	T
F	T	F	F	T	F	T	T
F	F	T	F	T	F	T	T
F	F	F	F	T	F	T	T

Question 29 Check whether the following statement is a contradiction by using a truth table:

$$(\neg Q \vee R) \wedge (P \Longrightarrow (\neg R \wedge \neg(\neg Q))) \wedge (P \vee R).$$

P	Q	R	$\neg Q \vee R$	$\neg R \wedge Q$	$P \Longrightarrow (\neg R \wedge Q)$	(4)$\wedge$ (5)	$P \vee R$	(7)$\wedge$ (8)
T	T	T	T	F	F	F	T	F
T	T	F	F	T	T	F	T	F
T	F	T	T	F	F	F	T	F
T	F	F	T	F	F	F	T	F
F	T	T	T	F	T	T	T	T
F	T	F	F	T	T	F	F	F
F	F	T	T	F	T	T	T	T
F	F	F	T	F	T	T	F	F
(1)	(2)	(3)	(4)	(5)	(6)	(7)	(8)	(9)

Question 7 Without using a truth table and by constructing one, determine whether the following statement is a tautology, contradiction, or neither:

$$(A \wedge B) \vee (\neg A \wedge B) \vee (\neg A \wedge \neg B) \vee (A \wedge \neg B).$$

Solution. Let S denote the given statement, that is

$$S : (A \wedge B) \vee (\neg A \wedge B) \vee (\neg A \wedge \neg B) \vee (A \wedge \neg B).$$

1. Without using a truth table, we proceed as follows:

$$\begin{aligned}
S &\equiv (A \wedge B) \vee (\neg A \wedge B) \vee (\neg A \wedge \neg B) \vee (A \wedge \neg B), \\
&\equiv (B \wedge A) \vee (B \wedge \neg A) \vee (\neg B \wedge \neg A) \vee (\neg B \wedge A), \\
&\equiv (B \wedge (A \vee \neg A) \vee (\neg B \wedge (\neg A \vee A), \\
&\equiv (B \wedge \mathbf{T}) \vee (\neg B \wedge \mathbf{T}), \quad (\mathbf{T} \text{ denotes the tautology}) \\
&\equiv B \vee \neg B, \\
&\equiv \mathbf{T}.
\end{aligned}$$

   Thus the statement S is a tautology.

2. Let us construct a truth table for S:

A	B	$\neg A$	$\neg B$	$A \wedge B$	$\neg A \wedge B$	$(\neg A \wedge \neg B$	$A \wedge \neg B$	S
T	T	F	F	T	F	F	F	T
T	F	F	T	F	F	F	T	T
F	T	T	F	F	T	F	F	T
F	F	T	T	F	F	T	F	T

   Thus the statement S is a tautology.

3. Sets

Chapter Key Terms

(*a*) *sets*
(*b*) *operations on sets*
(*c*) *set properties*
(*d*) *difference of sets and complements*
(*e*) *indexed families and Cartesian products*

3.1 Introduction to Sets

It is natural for us to classify items in a sorted way. For example, think of the clothes and accessories you wear, types of food you eat, or even the music you listen to. A set is simply a collection of objects, meaning it is just things grouped together with a certain property in common. For example, a set can be the collection of items you carry in your backpack: laptop, pens, pencils, a notebook, a textbook, a calculator, and so on. We may call this as a set of your *school supplies*, or any other name that describes the set and its contents. We can think of many other sets, such as, your school's soccer team, fiction books in the public library, apps in your cell phone, and the fingers of a human hand.

Definition 3.1.1 — sets. A **set** is a collection of objects. The objects that make up a set are called its **elements** (or members). For example, the players on a soccer team represent the elements of the team as a set. The fingers on a hand represent the elements of a hand as a set of fingers.

3.1.1 Set Notation

There are two common methods for defining sets: set description notation (or list notation), and set builder notation.

Definition 3.1.2 — The set description notation. A set with a small numbers is often expressed by listing its elements separated by commas and enclosed in curly brackets.

For example, the collection $\{1,3,5,7\}$ is a set containing four elements: the numbers $1,3,5$ and 7.

(R)

(a) In set description notation, the order in which the elements are given is irrelevant. For example, $\{2,5,6\}$ and $\{5,2,6\}$ are the same set.
(b) Repetitions of elements don't matter. For example, the sets: $\{1,7,11\}$, $\{11,1,7,11\}$, and $\{1,1,1,7,11\}$ all represent the same set $\{1,7,11\}$. Note that usually we avoid repeated elements in the same set.

Some sets contain infinitely many elements (e.g., the set of all positive integers, $\{1,2,3,4,5,...\}$). A set is called an **infinite set** if it contains infinitely many elements. If a set is not an infinite set, then it is called a **finite** set.

It is quite common to use uppercase letters to name a set. For example, we can name the aforementioned set as $A=\{2,3,5,7\}$. To express that an element is in the set, we use the notation $\in$ which means *belong to*. For example, $5\in A$, but $6\notin A$. The symbol $\notin$ used here is for something doesn't belong to the set. Some sets are so significant that they reserve special symbols as their names. Take for example, the set of natural numbers is denoted as

$$\mathbb{N}=\{1,2,3,4,5,...\}.$$

The set of all integers

$$\mathbb{Z}=\{...,-3,-2,-1,0,1,2,3,...\}$$

is another fundamental set. The set of rational numbers $\mathbb{Q}$ and the set of real numbers $\mathbb{R}$ are commonly used sets by their names. Later on we will provide a list of special sets that have symbolic names.

Definition 3.1.3 — The set builder notation. A set may be denoted by the expression $\{x:P(x)\}$, where P is the defining condition called predicate. Such notation is called set builder notation. The set contains all elements of the type x for which the predicate $P(x)$ is true.

■ **Example 3.1** List the elements of each set:

$$A=\{z\in\mathbb{N}:-5\leqslant z\leqslant 5\},\ B=\{x\in\mathbb{R}:x^3=8\}.$$

Note that

- $z\in A$ such that $z\in\mathbb{N}$ and $z\in[-5,\ 5]$, therefore, $A=\{1,2,3,4,5\}$
- $x\in B$ such that $x\in\mathbb{R}$ and $x^3=8$. Since $x=\sqrt[3]{8}=2$, therefore, $B=\{2\}$

■

■ **Example 3.2** List the elements of each set:

$$C=\{z\in\mathbb{Z}:|z|\leqslant 3\},\ D=\{z\in\mathbb{Z}:|2z|>8\}$$

Note that

- $z\in C$ such that $z\in\mathbb{Z}$ and $|z|\leqslant 3$, or, $z\in[-3,\ 3]\cap\mathbb{Z}$. That is, $C=\{-3,-2,-1,0,1,2,3\}$
- $z\in D$ such that $z\in\mathbb{Z}$ and $|z|>4$, or $\{z<-4\ \vee\ z>4\}\cap\mathbb{Z}$. That is, $D=\mathbb{Z}\setminus[-4,\ 4]$

■

3.1.2 Cardinality of Sets

Sets may have elements others than numbers. For example, $S = \{Yes, No\}$ is set of two words, perhaps representing the answer choices for some questions. The set $S = \{\{1\}, \{2\}\}$, is a set consists of two elements $\{1\}$ and $\{2\}$ which are sets as well. The cardinality of a finite set is the number of elements in the set. For example, the cardinality of $S = \{1, 2, 3, ..., m\}$ is m. Another set worth mentioning is the **empty set**, a set with no element and denoted by $\{\ \}$ or $\emptyset$. We will adopt the notation $\emptyset$ for an empty set, but note that the set $\{\emptyset\}$ is not an empty set. It has one element which is $\emptyset$, so its cardinality is one.

The **cardinality**, or **size**, of a set is the number of elements in the set. If the cardinality of an empty set is zero, we write $n(\emptyset) = 0$. Consider $T = \{1, \{2\}\}$, which has two elements, 1 and $\{2\}$, so the cardinality of T, $n(T) = 2$. We observe that $S \neq T$.

Two sets are **equal** if they contain exactly same elements. For example, $A = \{2, 3, 5, 7\}$ and $B = \{3, 5, 2, 7\}$ are equal sets. We will use the notation $|A|$ for the cardinality of a set A. In literature, one may find the following notations:

$$n(A) \ = \ card(A) \ = \ |A|.$$

If set A and set B have the same number of elements, we say they have **same cardinality**, meaning there is a one-to-one correspondence from set A to set B.

As noted earlier, the set of natural numbers $\mathbb{N}$ is an infinite set. We denote the cardinality of $\mathbb{N}$ as $\aleph_0$ aleph null.

Definition 3.1.4 — denumerable set. A set A is called a **denumerable set** if the set A has the same cardinality as the set of natural numbers $\mathbb{N}$. That is, $|A| \ = \aleph_0$. For example, $\mathbb{Z}$ and $\mathbb{Q}$ are both denumerable sets. Sometimes denumerable sets are called **countably infinite**.

Definition 3.1.5 — countable set. We say a set A is **countable** if it is either finite or denumerable.
A set is called **uncountable** if it is not a countable set. For example, the set $\mathbb{Q}$ is countable, and the set $\mathbb{R}$ is uncountable (see sections 9.5.1 - 9.5.3).

(R) The cardinality of a countable set is denoted as $\aleph_0$. For example, $|\mathbb{N}| = |\mathbb{Z}| = |\mathbb{Q}| = \aleph_0$.
The cardinality of an uncountable set is denoted as c or $\aleph_1$.
For example, $|\mathbb{R}| = |(0,1)| = \aleph_1 = c$.

3.1.3 Special Set Notations

There are some sets that are used so often that they have special names. Here we include a list of special sets with their names and symbols:

1. the empty set: $\emptyset = \{\}$.
2. the natural numbers: $\mathbb{N} = \{1, 2, 3, 4, 5, ...\}$.
3. the integers: $\mathbb{Z} = \{..., -3, -2, -1, 0, 1, 2, 3, 4, 5, ...\}$.
4. the rational numbers: $\mathbb{Q} = \{x : x = \frac{p}{q}$, where $p, q \in \mathbb{Z}$ and $q \neq 0\}$.
5. the real numbers: $\mathbb{R} = (-\infty, \infty)$ (the set of all real numbers on the number line).
6. the irrational numbers: $\mathbb{I} = \{x \in \mathbb{R} : x \notin \mathbb{Q}\}$.

Notice $\mathbb{R}$ is the set of all numbers that can be expressed on a real line. Recall from your study of calculus that $(-\infty, \infty)$ is an **open interval**.

Given two numbers $a, b \in \mathbb{R}$ with $a \leqslant b$, we extend our list of special sets that are formed from various intervals on the real line:

1. closed interval: $[a, b] = \{x \in \mathbb{R} : a \leqslant x \leqslant b\}$

2. half open interval: $[a,b) = \{x \in \mathbb{R} : a \leqslant x < b\}$
3. half open interval: $(a,b] = \{x \in \mathbb{R} : a < x \leqslant b\}$
4. open interval:$(a,b) = \{x \in \mathbb{R} : a < x < b\}$

3.2 Subsets

A set A is called a **subset** of a set B if every element of A is also an element of B. We say A is contained in B (or A is a subset of B) and denote this by writing $A \subseteq B$. If B has an element that is not an element of the subset A, we say A is a **proper subset** of B and write it as $A \subset B$. If A, B, and C are sets such that $A \subseteq B$ and $B \subseteq C$, then $A \subseteq C$. This definition tells us that to prove $A \subseteq B$, we must show that *if* $x \in A$, *then* $x \in B$ is a true statement. Note that every set is a subset of itself. While working with sets, we consider a **universal set U**, in which all sets under consideration are contained in **U**. The universal set can be $\mathbb{Z}$ if we are dealing only with integers, or $\mathbb{R}$ if we are dealing only with real numbers. On the other hand, the universal set being considered may be neither $\mathbb{Z}$ nor $\mathbb{R}$. In fact, $\mathbb{U}$ may not even be a set of numbers.

■ **Example 3.3** The set of natural numbers is a subset of the set of integers; symbolically, we write $\mathbb{N} \subset \mathbb{Z}$. In addition, $\mathbb{Q} \subset \mathbb{R}$ and $\mathbb{R} \subset \mathbb{C}$. Therefore, it follows that $\mathbb{Q} \subset \mathbb{C}$. ■

R

- Every set is a subset of itself: $A \subseteq A$.
- $\emptyset$ is a subset of any set A : $\emptyset \subset A$. Indeed, $\forall x \in \emptyset \implies x \in A$. The implication is true since the first statement is always false.

A convenient way to represent sets is with diagrams called **Venn diagrams**. Figure 3.1 shows Venn diagrams for sets $A = \{1,2,3,4,5\}$, $B = \{2,4,6,8,10\}$, $C = \{3,6,9\}$, and $D = \{1,3,5,7,9\}$. The rectangle of a Venn diagram represents $U = \{1,2,3,4,5,6,7,8,9,10\}$.

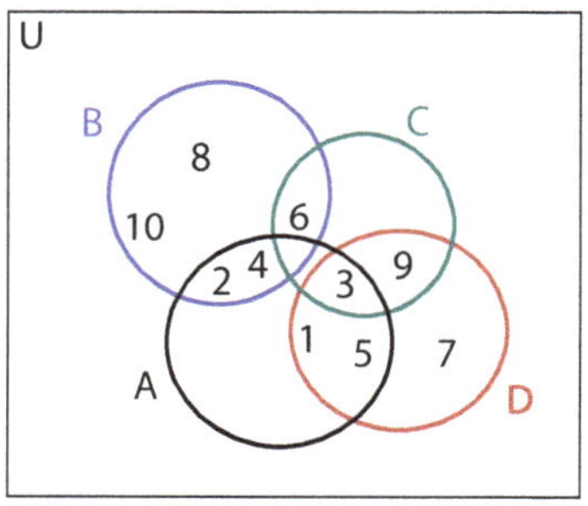

Figure 3.1: An example of a Venn diagram

Let A and B be two sets. We say that A is equal to B, which is written as $A = B$, if $A \subseteq B$ and $B \subseteq A$. When proving two sets A and B are equal, we show that the following conjunction is a true statement:
if $x \in A$, then $x \in B$, and if $x \in A$, then $x \in B$.

■ **Example 3.4** Let

$$A = \{2,3\},$$
$$B = \{1,2,3\},$$
$$C = \{2,4,6\},$$
$$D = \{x \in \mathbb{Z} : x^2 - 5x + 6 = 0\}.$$

The following statements are all true:

- $A \subseteq B$, more accurately $A \subset B$,
- $A = D$,
- $C = 2 \times B = 2B$,
- $B \supseteq D$, more accurately $B \supset D$.

■

R The set of the multiples of p can be expressed as $p\mathbb{Z}$. Take the following, for instance:
- the multiples of 2 or the set of even numbers: $2\mathbb{Z} = \{..., -4, -2, 0, 2, 4, 6, ...\}$.
- the set of multiples of 3 is $3\mathbb{Z} = \{..., -6, -3, 0, 3, 6, 9, ...\}$.
- the set of positive multiples of 11 is $11\mathbb{N} = \{11, 22, 33, ...\}$.

Proposition 3.2.1 Let A be a given set. Then we have

$$a \in A \iff \{a\} \subset A.$$

Theorem 3.2.2 Let A, B, and C be sets, then

$$\begin{cases} A \subseteq B \\ B \subseteq C \end{cases} \implies A \subseteq C.$$

Proof. By the definition of inclusions, we need to prove that for all x in A, x is in C. To this end, let us consider

$$x \in A \underbrace{\implies}_{A \subseteq B} x \in B \underbrace{\implies}_{B \subseteq C} x \in C.$$

Thus we have proven that $A \subseteq C$. ■

3.3 Power Sets

Definition 3.3.1 — power set. The set consisting of all subsets of a given set A is called the **power set** of A and is denoted by $\mathscr{P}(A)$.

$$\mathscr{P}(A) = \{X : X \subseteq A\}.$$

■ **Example 3.5** Let $A = \{1, a\}$ and $B = \{\emptyset\}$:

$$\mathscr{P}(A) = \{\{1\}, \{a\}, \{1, a\}, \emptyset\} = \{\{1\}, \{a\}, A, \emptyset\}$$

.

$$\mathscr{P}(B) = \{\{\emptyset\}, \emptyset\}$$

. ■

Notice that $n(\mathscr{P}(A)) = 2^{n(A)} = 4$ and $n(\mathscr{P}(B)) = 2^{n(B)} = 2$. In fact, for any finite set S with n elements, the $\mathscr{P}(S)$ has 2^n elements; that is, $n(\mathscr{P}(S)) = 2^{n(S)}$.

■ **Example 3.6** Let $C = \{1, a, b\}$.

$$\mathscr{P}(C) = \{\{1\}, \{a\}, \{b\}, \{1, a\}, \{1, b\}, \{a, b\}, \{1, a, b\}, \emptyset\}.$$

■

Notice that $n(\mathscr{P}(C)) = 2^{n(C)} = 2^3 = 8$.

■ **Example 3.7** Let $A = \{\emptyset\}$. Then $\mathscr{P}(A) = \{\emptyset, \{\emptyset\}\}$.
Note that it is always the case that $n(A) < n(\mathscr{P}(A))$ for any finite set A. ■

Theorem 3.3.1 *If A is a given set with a finite number of elements: $n(A)$ is bounded, then*

$$n(\mathscr{P}(A)) = 2^{n(A)}.$$

Proof. See Chapter 7, Question 21. ∎

(R) The power set captures the idea of completeness when dealing with subsets. By including all possible subsets, it ensures that no subset of the original set A is left out. This is important in mathematical reasoning and proof.

3.4 Set Operations

There are many ways to combine two sets to produce another set. The set operations are similar to the basic operations used in arithmetic (addition, subtraction, multiplication, and division).

Definition 3.4.1 — union set. The **union** of two sets A and B, denoted by $A \cup B$, is the set of all elements belonging to either A or B (or both); that is,

$$A \cup B = \{x : x \in A \text{ or } x \in B\} = \{x : (x \in A) \vee (x \in B)\}.$$

Example 3.8 Let $A = \{a,b,c\}$ and $B = \{b,c,d,e\}$. The union set of A and B is

$$A \cup B = \{a,b,c,d,e\}.$$

∎

Proposition 3.4.1 — properties of union of sets. *For any sets A, B, and C, it follows that*

1. $A \cup B = B \cup A$ (commutative Property).
2. $A \cup (B \cup C) = (A \cup B) \cup C$ (associative Property).
3. $A \subseteq A \cup B$ and $B \subseteq A \cup B$.
4. $A \cup \emptyset = A$.
5. $A \subseteq B$ iff $A \cup B = B$.
6. If $A \subseteq B$, then $A \cup C \subseteq B \cup C$.

Proof.

1. $A \cup B = B \cup A$
   If x is an element in $A \cup B$, then, by definition of union, we have $x \in A$ or $x \in B$. But if x is in A or B, then it is in B or A, and by definition of union, $x \in B \cup A$. Therefore, $A \cup B \subseteq B \cup A$. For the reverse inclusion, let x be an element of $B \cup A$, then we know $x \in B$ or $x \in A$. But, $x \in B$ or $x \in A$ implies that x is in A or B; hence, $x \in A \cup B$. Therefore, $B \cup A \subseteq A \cup B$. Hence, $A \cup B = B \cup A$.

2. $A \cup (B \cup C) = (A \cup B) \cup C$
   First, let x be any element in $A \cup (B \cup C)$. This means that $x \in A$ or $x \in (B \cup C)$. If $x \in A$, then $x \in (A \cup B)$; hence, $x \in (A \cup B) \cup C$.
   On the other hand, if $x \notin A$, then $x \in (B \cup C)$. This means $x \in B$ or $x \in C$. If $x \in B$, then $x \in (A \cup B) \implies x \in (A \cup B) \cup C$. If $x \in C$, then $x \in (A \cup B) \cup C$. Hence, $A \cup (B \cup C) \subseteq (A \cup B) \cup C$. The other inclusion is identical. Let x be an element of $(A \cup B) \cup C$. Then $x \in (A \cup B)$ or $x \in C$. If $x \in (A \cup B)$, we know $x \in A$ or $x \in B$. If $x \in A$, then $x \in A \cup (B \cup C)$. If $x \in B$, then $x \in (B \cup C)$; hence, $x \in A \cup (B \cup C)$. On the other hand, if $x \in C$, then $x \in (B \cup C)$, $x \in A \cup (B \cup C)$. Therefore, $(A \cup B) \cup C \subseteq A \cup (B \cup C)$. Thus $A \cup (B \cup C) = (A \cup B) \cup C$.

3. $A \subseteq A \cup B$
   Let $x \in A$. Thus it is true that at least one of $x \in A$ or $x \in B$ is true. Since $x \in A$ or $x \in B$ is true, by definition of union, $x \in A \cup B$.
4. $B \subseteq A \cup B$
   Let $x \in B$. Thus it is true that at least one of $x \in A$ or $x \in B$ is true. Since $x \in A$ or $x \in B$ is true, by definition of union, $x \in A \cup B$.

5. $A \cup \emptyset = A$
   If $x \in A \cup \varnothing$, then $x \in A$ or $x \in \varnothing$ by definition of union. Since $x \notin \varnothing$ (by definition of $\varnothing$), we must have $x \in A$. Hence, $A \cup \varnothing \subseteq A$.
   On the other hand, if $x \in A$, then $x \in A \cup \varnothing$ by definition of union, so $A \subseteq A \cup \varnothing$. Therefore, $A \cup \varnothing = A$.

6. $A \subseteq B$ iff $A \cup B = B$
   First assume that $A \subseteq B$. If $x \in A \cup B$, then $x \in A$ or $x \in B$ by definition, so in particular, $x \in B$. This proves $A \cup B \subseteq B$. Now if $x \in B$, then $x \in A \cup B$. This proves $B \subseteq A \cup B$. Together, this implies $A \cup B = B$.
   On the other hand, assume that $A \cup B = B$. If $x \in A$, then $x \in A \cup B$, and, by assumption, $x \in B$. This proves $A \subseteq B$.

7. $A \subseteq B$, then $A \cup C \subseteq B \cup C$
   Let $A \subseteq B$. If $x \in A \cup C$, then $x \in A$ or $x \in C$ by definition. If $x \in A$, then, by assumption, $x \in B$, too, so $x \in B \cup C$. This proves that $A \cup C \subseteq B \cup C$.

■

Definition 3.4.2 — intersection set. The **intersection** of two sets A and B, denoted by $A \cap B$, is the set of all elements belonging to both A and B, that is,

$$A \cap B = \{x : x \in A \text{ and } x \in B\} = \{x : (x \in A) \wedge (x \in B)\}.$$

■ **Example 3.9** Let $A = \{a, b, c\}$ and $B = \{b, c, d, e\}$. The intersection set of A and B is

$$A \cap B = \{b, c\}.$$

■

Figure 3.2 shows a variety of Venn diagrams for two sets A and B. The gray colored regions in Figure 3.2 show intersection, union, and complement of A and B, respectively. The complement set will be defined in section 3.5.

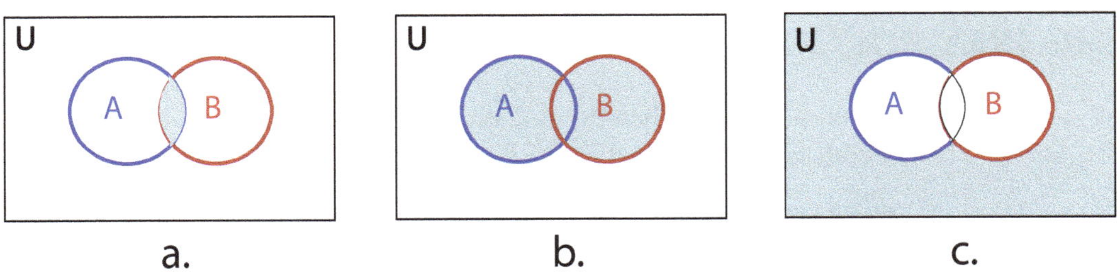

Figure 3.2: Venn Diagram of intersection, union, and complement

Proposition 3.4.2 — properties of intersection of sets. For any sets A, B, and C, it follows that:

1. $A \cap B = B \cap A$ (Commutative Property).
2. $A \cap (B \cap C) = (A \cap B) \cap C$ (Associative Property).
3. $A \cap B \subseteq A$ and $A \cap B \subseteq B$.
4. $A \cap \emptyset = \emptyset$.
5. $A \subseteq B$ iff $A \cap B = A$.
6. If $A \subseteq B$ then, $A \cap C \subseteq B \cap C$.

Proof.

1. $A \cap B = B \cap A$
   If x is any element in $A \cap B$, then we know, by definition of intersection, that $x \in A$ and $x \in B$. Hence, $x \in B$ and $x \in A$, so $x \in B \cap A$. Therefore, $A \cap B \subseteq B \cap A$. Conversely, if x is any element of $B \cap A$, then we know $x \in B$ and $x \in A$. Hence, $x \in A$ and $x \in B$. This implies $x \in A \cap B$. Hence, $B \cap A \subseteq A \cap B$. So, $A \cap B = B \cap A$.
2. $A \cap (B \cap C) = (A \cap B) \cap C$
   We will prove both inclusions at once. x is in $A \cap (B \cap C)$ if and only if $x \in A$ and $x \in (B \cap C)$; further, $x \in (B \cap C)$ if and only if $x \in B$ and $x \in C$. Similarly, $x \in (A \cap B) \cap C$ if and only if $x \in A$ and $x \in B$ and $x \in C$. Thus $A \cap (B \cap C)$ and $(A \cap B) \cap C$ have exactly the same elements (since $x \in A \cap (B \cap C)$ if and only if $x \in A$ and $x \in B$ and $x \in C$ if and only if $x \in (A \cap B) \cap C$). Thus $A \cap (B \cap C) = (A \cap B) \cap C$.
3. $A \cap B \subseteq A$
   Let $x \in A \cap B$. By definition of intersection, $x \in A$ and $x \in B$. Thus in particular, $x \in A$ is true. Similarly, $A \cap B \subseteq B$.
4. $A \cap \emptyset = \emptyset$
   If $x \in A \cap \varnothing$, then, by definition of intersection, $x \in A$ and $x \in \varnothing$. But by definition of $\varnothing$, there is no x such that $x \in \varnothing$. Hence, $A \cap \varnothing$ must be empty. By uniqueness of the empty set, we have $A \cap \varnothing = \varnothing$.
5. $A \subseteq B$ iff $A \cap B = A$
   First assume that $A \subseteq B$. If $x \in A \cap B$, then $x \in A$ and $x \in B$ by definition, so in particular, $x \in A$. This proves $A \cap B \subseteq A$. Now if $x \in A$, then by assumption $x \in B$ too, so $x \in A \cap B$. This proves $A \subseteq A \cap B$. Together this implies $A = A \cap B$.
   On the other hand, assume that $A \cap B = A$. If $x \in A$, then by assumption, $x \in A \cap B$, so $x \in A$ and $x \in B$. In particular, $x \in B$. This proves $A \subseteq B$.
6. $A \subseteq B$ then, $A \cap C \subseteq B \cap C$
   Let $A \subseteq B$. If $x \in A \cap C$, then $x \in A$ and $x \in C$ by definition. If $x \in A$, then by assumption, $x \in B$ too, so $x \in B \cap C$. This proves that $A \cap C \subseteq B \cap C$.

■

Corollary 3.4.3 $A \cap B \subseteq A \cup B$.

Proof. Since $A \cap B \subseteq A$ and $A \subseteq A \cup B$, then $A \cap B \subseteq A \cup B$. Or one could proceed as: $A \cap B \subseteq B$ and $B \subseteq A \cup B$. Thus in both cases it proves that $A \cap B \subseteq A \cup B$. ■

Proposition 3.4.4 — **distribution laws of union and intersection.** For any sets A, B, and C, it follows that

- $A \cap (B \cup C) = (A \cap B) \cup (A \cap C)$.
- $A \cup (B \cap C) = (A \cup B) \cap (A \cup C)$.

Proof. We proceed by using logical equivalences here. Let us assume that

$$p : x \in A, \quad q : x \in B \text{ and } r : x \in C.$$

- $A \cap (B \cup C) = (A \cap B) \cup (A \cap C)$.

$$\begin{aligned}
\text{Let } x \in A \cap (B \cup C) &\iff (x \in A) \text{ and } [x \in (B \cup C)] \\
&\iff (x \in A) \text{ and } [(x \in B) \text{ or } (x \in C)] \\
&\iff p \wedge (q \vee r) \\
&\iff (p \wedge q) \vee (p \wedge r) \\
&\iff [(x \in A) \text{ and } (x \in B)] \text{ or } [(x \in A) \text{ and } (x \in C)] \\
&\iff x \in (A \cap B) \cup (A \cap C).
\end{aligned}$$

- $A \cup (B \cap C) = (A \cup B) \cap (A \cup C)$

$$\begin{aligned}
\text{Let } x \in A \cup (B \cap C) &\iff (x \in A) \text{ or } [x \in (B \cap C)] \\
&\iff (x \in A) \text{ or } [(x \in B) \text{ and } (x \in C)] \\
&\iff p \vee (q \wedge r) \\
&\iff (p \vee q) \wedge (p \vee r) \\
&\iff [(x \in A) \text{ or } (x \in B)] \text{ and } [(x \in A) \text{ or } (x \in C)] \\
&\iff x \in (A \cup B) \cap (A \cup C).
\end{aligned}$$

■

R Power sets are intimately connected with set operations such as union, intersection, and complement. For example, when you take the union of all subsets in a power set, you get the original set A. The intersection of all subsets in a power set gives you the intersection of all possible combinations of subsets of A. Furthermore, power sets can serve as building blocks for generating increasingly larger sets, as exemplified by Cantor's Theorem.

Definition 3.4.3 — difference set. The **difference** of two sets A and B, denoted by $A - B$, is the set of all elements in A that are not in B; that is,

$$A - B = \{x : x \in A \text{ and } x \notin B\}.$$

■ **Example 3.10** Let $A = \{a, b, c\}$ and $B = \{b, c, d, e\}$. Then

$$A - B = \{a\}, \quad \text{but} \quad B - A = \{d, e\}.$$

■

■ **Example 3.11** Take the following:

$$A - A = \emptyset, \qquad A - \emptyset = A, \text{ and } \quad \emptyset - A = \emptyset.$$

- $A - A = \emptyset$
  Let $x \in A - A \implies (x \in A) \text{ and } (x \notin A) \implies x \in \emptyset$. Therefore, $A - A \subseteq \emptyset$. Since $\emptyset \subseteq A - A$, we have $A - A = \emptyset$.

- $A - \emptyset = A$
  Since $A - \emptyset \subseteq A$, we need only to prove the other inclusion. Let $x \in A \implies (x \in A)$ and $\underbrace{(x \notin \emptyset)}_{tautology} \implies$ $x \in A - \emptyset$. Therefore, $A \subseteq A - \emptyset$; thus $A - \emptyset = A$.
- $\emptyset - A = \emptyset$
  Let $x \in \emptyset - A \iff \underbrace{\underbrace{(x \in \emptyset)}_{contradiction} \text{ and } \underbrace{(x \in A)}_{p}}_{false} \iff x \in \emptyset$.

■

Definition 3.4.4 — disjoint set. Two sets A and B are said to be **disjoint** if $A \cap B = \emptyset$. One may notice that two sets may be distinct but not disjoint.

For example, $A = \{1,2,3\}$ and $B = \{1,5,3\}$ are certainly distinct ($2 \in A$ and $2 \notin B$). Clearly, A and B are not disjoint since $A \cap B = \{1,3\} \neq \emptyset$.

■ **Example 3.12** Let $A = \{a,b,c\}$ and $B = \{d,e\}$. Then A and B are disjoint as $A \cap B = \emptyset$. ■

(R) Here is a list of fundamental facts about the operation on number sets.

1. $\mathbb{N} \cup \mathbb{Z} = \mathbb{Z}$.
2. $\mathbb{Q} \cup \mathbb{I} = \mathbb{R}$.
3. $\mathbb{N} \cap \mathbb{Z} = \mathbb{N}$.
4. $\mathbb{Q} \cap \mathbb{R} = \mathbb{Q}$.
5. $\mathbb{R} - \mathbb{Q} = \mathbb{I}$.
6. $\mathbb{R} - \mathbb{I} = \mathbb{Q}$.
7. $\mathbb{Q} \cap \mathbb{I} = \emptyset$.

Theorem 3.4.5 *The main result of the cardinal of the union of two sets is given by*

$$\boxed{n(A \cup B) = n(A) + n(B) - n(A \cap B)}$$

Proof. Theorem 8.6.8 proves this result. ■

(R) Since $n(\emptyset) = 0$, we conclude that

$$\boxed{\text{If } A \cap B = \emptyset \implies n(A \cup B) = n(A) + n(B)}.$$

■ **Example 3.13** We already proved that for any sets E, F,

$$\boxed{E = (E-F) \cup (E \cap F) \quad \text{and} \quad (E-F) \cap (E \cap F) \ = \ \emptyset}.$$

Since, in general,

$$\begin{aligned} x \in E &\iff [(x \in E) \wedge (x \in F)] \vee [(x \in E) \wedge (x \notin F)] \\ &\iff [(x \in E \cap F)] \vee [x \in (E-F)] \\ &\iff x \in (E-F) \cup (E \cap F), \end{aligned}$$

the first part is proved. For the second part, let

$$x \in (E-F) \Longrightarrow (x \in E) \ \wedge \ (x \notin F) \Longrightarrow x \notin F \Longrightarrow x \notin \ E \cap F.$$

Thus $(E-F) \cap (E \cap F) \ = \ \emptyset$. ■

Theorem 3.4.6 *For any sets A, B, we have*

$$\boxed{n(A-B) \ = \ n(A) \ - \ n(A \cap B)}.$$

Proof. Using results,

$$A = (A-B)\cup(A\cap B) \text{ and } (A-B)\cap(A\cap B) = \emptyset \implies n(A) = n(A-B) + n(A\cap B).$$

Thus $n(A-B) \;=\; n(A) \;-\; n(A\cap B)$. ■

■ **Example 3.14** Let A, B, and C be three sets. If $n(A) = 32$, $n(B) = 31$, $n(A\cup B) = 47$, and $n(G-F) = 20$, find the following values:
(a) $n(B\cup C)$.
(b) $n(B-A)$.

(a) Since $n(B\cup C) = n(B) + \underbrace{n(C) - n(B\cap C)}_{n(C-B)} = n(B) + n(C-B) = 31 + 20 = 51$.
(b) Since $n(B-A) = n(B) - n(B\cap A) = 31 - n(B\cap A)$, it suffices to find $n(B\cap A)$.
Let us consider

$$n(A\cap B) = n(A) + n(B) - n(B\cup A) = 32 + 31 - n(A\cup B) = 63 - 47 = 16,$$

thus we have, $n(B-A) = 31 - 16 = 15$. Therefore, $n(B-A) = 15$.

■

3.5 Complement

Assume we are considering a certain universal set $\mathbf{U}$; that is, all sets being discussed are subsets of $\mathbf{U}$. For a set A, its **complement** is denoted by $\bar{A}$. In other words, the complement of a set A is the set $\mathbf{U} - A$:

$$\overline{A} = \mathbf{U} - A = \{x : x \in \mathbf{U} \text{ and } x \notin A\}.$$

Figure 3.2 (c) illustrates the complement of sets A and B.

■ **Example 3.15** Let $\mathbf{U} = \{1,2,3,4,5,6,7,8,9,10\}$ be the universal set.
For sets $A = \{1,2,3,4,5\}$, $B = \{2,4,6,8,10\}$, $C = \{3,6,9\}$, and $D = \{1,3,5,7,9\}$, we have the following complements:

- $\overline{A} = \{6,7,8,9,10\}$
- $\overline{B} = D$ and $\overline{D} = \overline{\overline{B}} = B$
- $\overline{C} = \{1,2,4,5,7,8,10\}$
- $\overline{\overline{A}} = \overline{\{6,7,8,9,10\}} = \{1,2,3,4,5\} = A$

■

■ **Example 3.16** Let $\mathbf{U} = \mathbb{R}$. The complement sets of $[0,1]$, $(-10,\ 10]$, $(-\infty,\ \pi)$, and $[0,\ \sqrt{5})$ are the following:

- $\overline{[0,1]} = (-\infty,\ 0)\cup(1,\ \infty)$.
- $\overline{(-10,\ 10]} = (-\infty,\ -10]\cup(10,\ \infty)$.
- $\overline{(-\infty,\ \pi)} = [\pi,\ \infty)$.
- $\overline{[0,\ \sqrt{5})} = (-\infty,\ 0)\cup[\sqrt{5},\ \infty)$.

■

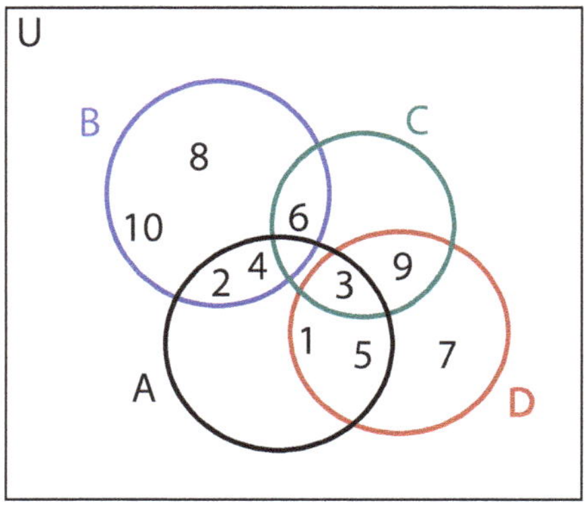

Figure 3.3: Venn diagram of Example 3.15

■ **Example 3.17** Let $A = \{11, 12, 13\}$.

- If $\mathbf{U} = \{1, 2, 3, ..., 13, 14, 15\}$, then $\overline{A} = \{1, 2, 3, 4, 5, 6, 7, 8, 9, 10, 14, 15\}$.
- If $\mathbf{U} = \mathbb{N}$, then $\overline{A} = \{1, 2, 3, 4, 5, 6, 7, 8, 9, 10, 14, 15, 16, ...\}$.
- If $\mathbf{U} = \mathbb{Z}$, then $\overline{A} = \{..., -3, -2, -1, 0, 1, 2, 3, 4, 5, 6, 7, 8, 9, 10, 14, 15, 16, ...\}$.

■

(R) Let A and B be two sets:

$$A - B = A \cap \overline{B}.$$

Indeed,

$$A - B = \{x \in A \text{ and } x \notin B\} = \{x \in A \text{ and } x \in \overline{B}\} = A \cap \overline{B}.$$

After these examples, we can state the following Proposition 3.5.1.

Proposition 3.5.1 Let $\mathbf{U}$ be the universal set, thus for any subset A, we have:

$$\overline{\overline{A}} = A, \quad \overline{A} \cup A = \mathbf{U} \quad \text{and} \quad \overline{A} \cap A = \emptyset.$$

■ **Example 3.18** Using the sets from Example 3.15, determine each of the following:

1. $\overline{A} \cup A$.
2. $\overline{B} \cap B$.
3. $(\overline{A} \cup \overline{B} \cup \overline{C} \cup \overline{D}) \cup \mathbf{U}$.
4. $(\overline{A} \cup \mathbf{U}) \cap (\overline{B} \cup \mathbf{U}) \cap (\overline{C} \cup \mathbf{U}) \cap (\overline{D} \cup \mathbf{U})$.

1. $\overline{A} \cup A = \mathbf{U}$.
2. $\overline{B} \cap B = \emptyset$.
3. $(\overline{A} \cup \overline{B} \cup \overline{C} \cup \overline{D}) \cup \mathbf{U} = \mathbf{U}$.
4. $(\overline{A} \cup \mathbf{U}) \cap (\overline{B} \cup \mathbf{U}) \cap (\overline{C} \cup \mathbf{U}) \cap (\overline{D} \cup \mathbf{U}) = \mathbf{U}$.

■

Theorem 3.5.2 *De Morgan's laws*

- *The complement of the union of two sets A and B is equal to the intersection of their complements.*

$$\overline{A \cup B} = \overline{A} \cap \overline{B}.$$

- *The complement of the intersection of two sets A and B is equal to the union of their complements.*

$$\overline{A \cap B} = \overline{A} \cup \overline{B}.$$

Proof. See Section 5.1 Example 5.6. ■

Corollary 3.5.3

$$x \notin (A \cup B) \iff x \notin A \text{ and } x \notin B,$$

and

$$x \notin (A \cap B) \iff x \notin A \text{ or } x \notin B.$$

Proof. It suffices to consider $p:\ x \in A,\ \ q:\ x \in B$ and use De Morgan's laws for logic.

Theorem 3.5.4 *Let $A,\ B$ be two subsets of a universal set U. Then we have:*

$$A \subseteq B \iff \overline{B} \subseteq \overline{A}.$$

Proof. See Section 5.1 Example 5.7. ■

Theorem 3.5.5 *If A be a subset of the universal set $\mathbf{U}$, then*

$$\boxed{n(\overline{A}) = n(\mathbf{U}) - n(A)}.$$

Proof. Since $A \cup \overline{A} = \mathbf{U}$ and $A \cap \overline{A} = \emptyset$,

$$n(\mathbf{U}) = n(A \cup \overline{A}) = n(A) + n(\overline{A}) \Longrightarrow n(\overline{A}) = n(\mathbf{U}) - n(A).$$

■

Definition 3.5.1 — symmetric difference. The **symmetric difference** between two sets A and B, denoted Δ, is defined as the following:

$$A \,\Delta\, B = (A - B) \cup (B - A).$$

Theorem 3.5.6 *The symmetric difference between sets A, B, C satisfies the following conditions:*

1. $A \,\Delta\, B = B \,\Delta\, A$.
2. $A \,\Delta\, A = \emptyset$.
3. $A \,\Delta\, \emptyset = A$.
4. $A \,\Delta\, B = (A \cup B) - (A \cap B)$.
5. $A \,\Delta\, (B \,\Delta\, C) = (A \,\Delta\, B) \,\Delta\, C$.

Proof.

1. $A \,\Delta\, B = B \,\Delta\, A$.

$$A \,\Delta\, B = (A - B) \cup (B - A) = (B - A) \cup (A - B) = B \,\Delta\, A.$$

2. $A \,\Delta\, A = \emptyset$.

$$A \,\Delta\, A = (A - A) \cup (A - A) = \emptyset \cup \emptyset = \emptyset.$$

3. $A \,\Delta\, \emptyset = A$.

$$A \,\Delta\, \emptyset = (A - \emptyset) \cup (\emptyset - A) = A \cup \emptyset = A.$$

4. Left as an exercise, see Question 15 (ii).
5. Left as an exercise, see Question 15 (iii).

■

R According to property 4, we can see that when $x \in A \,\Delta\, B$ it is equivalent that $x \in A$ or $x \in B$ but is not in their intersection. This corresponds to the **exclusive or**.

3.6 Indexed Sets and Partitions

More than two sets will often be combined using the set operations we discussed earlier. In the case of four sets A, B, C, and D described in Example 3.15, the standard Venn diagram is shown in Figure 3.3. The union $A \cup B \cup C \cup D$ and intersection $A \cap B \cap C \cap D$ are defined as following:

$$A \cup B \cup C \cup D = \{x : x \in A \text{ or } x \in B \text{ or } x \in C \text{ or } x \in D\},$$

$$A \cap B \cap C \cap D = \{x : x \in A \text{ and } x \in B \text{ and } x \in C \text{ and } x \in D\}.$$

The union or intersection of two or more sets is often a useful to consider and requires additional notation. The union and intersection of the $n \geqslant$ sets $A_1, A_2, ..., A_n$ are denoted and defined as the following:

$$\bigcup_{i=1}^{n} A_i = A_1 \cup A_2 \cup ... \cup A_n = \{x : x \in A_1, \text{ or } x \in A_2, \ ..., \text{ or } x \in A_n\}.$$

In other words,

$$x \in \bigcup_{i=1}^{n} A_i \iff \exists\, k \in \{1,2,...,n\} : \ x \in A_k.$$

$$\bigcap_{i=1}^{n} A_i = A_1 \cap A_2 \cap ... \cap A_n = \{x : x \in A_1, \text{ and } x \in A_2, \ ..., \text{ and } x \in A_n\}.$$

Differently from the union, we have

$$x \in \bigcap_{i=1}^{n} A_i \iff \forall\, k \in \{1,2,...,n\} : \ x \in A_k.$$

■ **Example 3.19** Let $A_0 = \{0,2\}$, $A_1 = \{1,3\}$, $A_n = \{n, n+2\}$ for $n \in \mathbb{N}$. Determine the following indexed sets:

$$(1) \bigcup_{i=2}^{5} A_i, \quad (2) \bigcap_{i=3}^{5} A_i, \quad (3) \bigcup_{i=0}^{n} A_i, \quad (4) \bigcap_{i=0}^{n} A_i.$$

1. $\bigcup_{i=2}^{5} A_i = A_2 \cup A_3 \cup A_4 \cup A_5 = \{2,3,4,5,6,7\}.$
2. $\bigcap_{i=3}^{5} A_i = A_3 \cap A_4 \cap A_5 = \emptyset.$
3. $\bigcup_{i=0}^{n} A_i = A_0 \cup A_1 \cup ... \cup A_n = \{0,1,2,3,...,n+1,n+2\}.$
4. $\bigcap_{i=0}^{n} A_i = A_1 \cap A_2 \cap ... \cap A_n = \emptyset.$

■

■ **Example 3.20** Let $B_n = [0,\ n)$ for $n \in \mathbb{N}$. Determine $\bigcup_{k=1}^{N} B_k$ and $\bigcap_{k=1}^{N} B_k$. Directly, we may claim that

$$\bigcup_{k=1}^{N} B_k = [0,\ N) \quad \text{and} \quad \bigcap_{k=1}^{N} B_k = [0,\ 1).$$

Indeed, using the fact that $[0,\ 1)$ is contained in all sets $[0,k) = B_k$ with $k \geq 1$, we conclude that

$$[0,\ 1) \subset \bigcap_{k=0}^{N} B_k.$$

Now, let us take

$$x \in \bigcap_{k=1}^{N} B_k \implies \forall k \in \{1,2,\ldots,N\} : x \in B_k = [0,\ k) \implies x \in [0,\ 1).$$

Finally, $\bigcap_{k=1}^{N} B_k = [0,\ 1)$.

Let

$$x \in \bigcup_{k=1}^{N} B_k \implies \exists j \in \{1,2,\ldots,N\} : x \in B_j = [0,\ j) \implies x \in [0,\ N);$$

thus $\bigcup_{k=1}^{N} B_k \subset [0,\ N)$.

Now, let us consider

$$x \in [0,\ N) \implies \exists\, k \in \{1,2,3,\ldots,(N-1),N\} : x \in [0,\ k) \implies x \in \bigcup_{k=1}^{N} B_k.$$

Finally, $\bigcup_{k=1}^{N} B_k = [0,\ N)$. ■

Theorem 3.6.1 *Let $(A_i)_{i\in I}$ be an indexed family of subsets of a universal set* **U** *and let B also be a subset of* **U**. *Then we have the following:*

1. $$B \cap \left(\bigcup_{i\in I} A_i\right) = \bigcup_{i\in I} (A_i \cap B).$$
2. $$B \cup \left(\bigcap_{i\in I} A_i\right) = \bigcap_{i\in I} (A_i \cup B).$$
3. $$\overline{\bigcap_{i\in I} A_i} = \bigcup_{i\in I} \overline{A_i}.$$
4. $$\overline{\bigcup_{i\in I} A_i} = \bigcap_{i\in I} \overline{A_i}.$$

Proof. See Section 5.1 Example 5.8. ■

■ **Example 3.21** For any $m \in \mathbb{N}$, $m \geq 3$, we define $C_m = \left[\frac{1}{m},\, 2+\frac{1}{m}\right]$. Find $\bigcup_{k\geq 3} C_k$.

We claim that $\bigcup_{k\geq 3} C_k = \bigcup_{k\geq 3}\left[\frac{1}{k},\, 2+\frac{1}{k}\right] = \left(0,\, \frac{7}{3}\right]$. Indeed,

$$\text{let } x \in \bigcup_{k\geq 3} C_k \Longrightarrow \exists\, m \geq 3:\ x \in C_m \Longrightarrow \frac{1}{m} \leq x \leq 2+\frac{1}{m},$$

and since $m \geq 3$, we have $0 < \frac{1}{m} \leq x \leq 2+\frac{1}{m} \leq 2+\frac{1}{3} = \frac{7}{3}$. Therefore, $\bigcup_{k\geq 3} C_k \subseteq \left(0,\, \frac{7}{3}\right]$.

Now, let $x \in \left(0,\, \frac{7}{3}\right]$. We will consider two cases:

- If $x \in \left[\frac{1}{3}, \frac{7}{3}\right] \Longrightarrow x \in C_3 \Longrightarrow x \in \bigcup_{k\geq 3} C_k$.
- If $x \in \left(0, \frac{1}{3}\right)$: since $\lim_{m\to\infty} \frac{1}{m} = 0$ and $x > 0$, then there exists $m_1 \geq 3:\ x \geq \frac{1}{m_1}$, and then $x \in \left[\frac{1}{m_1},\, \frac{1}{3}\right) \Longrightarrow x \in \left[\frac{1}{m_1},\, 2\right]$ and therefore $x \in \left[\frac{1}{m_1},\, 2+\frac{1}{m_1}\right] = C_{m_1}$, and finally, $x \in \bigcup_{k\geq 3} C_k$.

Thus we conclude that $\bigcup_{k\geq 3} C_k = \left(0,\, \frac{7}{3}\right]$. ■

■ **Example 3.22** For any $m \in \mathbb{N}$, $m \geq 3$, we define $C_m = \left[\frac{1}{m},\, 2+\frac{1}{m}\right]$. Show that $\bigcap_{k\geq 3} C_k = \left[\frac{1}{3},\, 2\right]$.

Let $x \in \left[\frac{1}{3}, 2\right] \Longrightarrow \forall m \geq 3,\ 0 < \frac{1}{m} \leq \frac{1}{3} \leq x$ and $x \leq 2 < 2+\frac{1}{m}$. Thus we can affirm that $\forall m \geq 3,\ x \in \left[\frac{1}{m},\, 2+\frac{1}{m}\right] = C_m \Longrightarrow x \in \bigcap_{m\geq 3} C_m$.
Now, let us prove the other side of the inclusion and consider

$$x \in \bigcap_{m\geq 3} C_m \Longrightarrow x \in C_3 \Longrightarrow x \geq \frac{1}{3}.$$

For any $m \geq 3$, we have $x \in C_m \Longrightarrow x \leq 2+\frac{1}{m}$ and then $x \leq \lim_{m\to\infty} 2+\frac{1}{m} = 2 \Longrightarrow x \leq 2$. Thus $x \in \left[\frac{1}{3},\, 2\right]$. Since this is true for every x in $\bigcap_{m\geq 3} C_m$, we have $\bigcap_{m\geq 3} C_m \subset \left[\frac{1}{3},\, 2\right]$. Finally,

$$\bigcap_{m\geq 3} C_m = \left[\frac{1}{3},\, 2\right].$$

■

Definition 3.6.1 — partition. The subsets $X_1,\, X_2, \ldots, X_n$ of a given set X form a **partition** of X if and only if the following three conditions are fulfilled:

(i) $X_i \neq \emptyset,\quad \forall\, i:\ 1 \leq i \leq n.$

(ii) $X_i \cap X_j = \emptyset,\quad \forall\, i,\ j \in \{1,2,\ldots,n\}, i \neq j.$

(iii) $X = X_1 \cup X_2 \cup \ldots \cup X_n = \bigcup_{i=1}^{n} X_i$.

■ **Example 3.23** If $X = \mathbb{Z}$, it is clear that the family of sets:

$$X_1 = \{\ldots, -3, -2, -1\},\ X_2 = \{0\},\ X_3 = \mathbb{N} \text{ form a partition of the set of integers } \mathbb{Z}.$$

Indeed, it is easy to check that the three subsets of $\mathbb{Z}$ are not empty and are disjoint with each other. In addition

$$X_1 \cup X_2 \cup X_3 = \{\ldots, -3, -2, -1\} \cup \{0\} \cup \{1, 2, 3, 4, \ldots\} = \{\ldots, -3, -2, -1, 0, 1, 2, 3, \ldots\} = \mathbb{Z}.$$

■

■ **Example 3.24** If $X = \mathbb{R}$, it is clear that the family of sets:

$$X_1 = (-\infty,\ 0),\ X_2 = [0,\ \infty) \text{ form a partition of } \mathbb{R}.$$

Indeed, it is easy to check that the two subsets of $\mathbb{R}$ are not empty and are disjoint with each other. In addition,

$$X_1 \cup X_2 = (-\infty,\ 0) \cup [0,\ \infty) = (-\infty,\ \infty) = \mathbb{R}.$$

■

3.7 The Cartesian Product

A pair of sets A and B may be multiplied by each other to obtain a new set denoted as $A \times B$. This operation is called the **Cartesian product**. The elements of a Cartesian product set are called **ordered pairs**.

An ordered pair, denoted as (x, y), is a list of two numbers x and y, enclosed in parentheses and separated by a comma. The ordered pair (x, y) is defined as a single element consisting of a pair of elements in which x is the first element (or coordinate) of the ordered pair (x, y) and y is the second element (or coordinate). For two ordered pairs (x, y) and (w, z) to be equal, we must have corresponding elements equal. That is,

$$(x, y) = (w, z) \quad \text{if and only if} \quad x = w \quad \text{and} \quad y = z.$$

Moreover, $(x, y) \neq (y, x)$ if $x \neq y$.

Definition 3.7.1 — Cartesian set. The **Cartesian product** (or simply cross product) $A \times B$ of two sets A and B is the set consisting of all ordered pairs whose first coordinate belongs to A and whose second coordinate belongs to B. That is, $A \times B = \{(a, b) : a \in A \text{ and } b \in B\}$.

■ **Example 3.25** Let $A = \{0\}$ and $B = \{\alpha, \beta,\ \gamma,\ \eta\}$; then

$$A \times B = \{(0, \alpha), (0, \beta), (0, \gamma), (0, \eta)\}.$$

and

$$B \times A = \{(\alpha, 0), (\beta, 0), (\gamma, 0), (\eta, 0)\}.$$

Clearly, $A \times B \neq\ A \times B$. ■

■ **Example 3.26** Let $E = \{a, b\}$ and $F = \{x, y, z\}$; then

$$E \times F = \{(a, x), (a, y), (a, z), (b, x), (b, y), (b, z)\}.$$

■

R

- *If $A=\emptyset$ or $B=\emptyset$, then $A\times B=\emptyset$.*
- *The Cartesian product $\mathbb{R}\times\mathbb{R}$ is the set of all points in the Euclidean plane.*

Theorem 3.7.1 *Let A, B, C, and D be sets. Then*
1. $(A\cup B)\times C \;=\; (A\times C)\cup(B\times C)$.
2. $(A\cap B)\times C \;=\; (A\times C)\cap(B\times C)$.
3. $(A\times B)\cap(C\times D) \;=\; (A\cap C)\times(B\cap D)$.
4. $(A\times B)\cup(C\times D) \;\subseteq\; (A\cup C)\times(B\cup D)$.

Proof.
1. To prove $(A\cup B)\times C \;=\; (A\times C)\cup(B\times C)$, we will prove its equivalence of the following bi-conditional statement: $(A\cup B)\times C \iff (A\times C)\cup(B\times C)$.

   Let us consider

$$\begin{aligned}
(x,y)\in\;(A\cup B)\times C &\iff x\in(A\cup B)\text{ and } y\in C,\\
&\iff [(x\in A)\text{ or }(x\in B)]\text{ and }(y\in C),\\
&\iff ((x\in A)\text{ and }(y\in C))\text{ or}((x\in B)\text{ and }(y\in C)),\\
&\iff (x,y)\in(A\times C)\text{ or }(x,y)\in(B\times C),\\
&\iff (x,y)\in(A\times C)\cup(B\times C).
\end{aligned}$$

2. Let us consider

$$\begin{aligned}
(x,y)\in\;(A\cap B)\times C &\implies x\in A\cap B\text{ and } y\in C,\\
&\implies [(x\in A)\;\wedge\;(x\in B)]\;\wedge\;(y\in C).
\end{aligned}$$

   If we denote $p:x\in A,\;\; q:x\in B,\;\; r:y\in C$, we have the following equivalences:

$$\begin{aligned}
(x,y)\in\;(A\cap B)\times C &\iff x\in A\cap B\text{ and } y\in C,\\
&\iff [(x\in A)\;\wedge\;(x\in B)]\;\wedge\;(y\in C),\\
&\iff [(x\in A)\;\wedge(y\in C)\;\wedge\;(x\in B)]\;\wedge\;(y\in C),\\
&\iff [(x\in A)\;\wedge(y\in C)]\;\wedge\;[(x\in B)\;\wedge\;(y\in C)],\\
&\iff (p\wedge r)\wedge(q\wedge r),\\
&\iff ((x\in A)\wedge(y\in C))\wedge((x\in B)\wedge(y\in C)),\\
&\iff [(x,y)\in A\times C]\wedge[(x,y)\in B\times C],\\
&\iff (x,y)\in(A\times C)\cup(B\times C).
\end{aligned}$$

3. See Question 29.
4. See Question 20.

■

■ **Example 3.27** Let $F=\{1,2,3,4\}$. The subsets

$A=\{(m,n)\in F\times F:\; m<n\}$, $B=\{(m,n)\in F\times F:\; m=n\}$, and $C=\{(m,n)\in F\times F:\; m>n\}$

form a Partition of $F\times F$. Indeed, it is easy to check that

$$A=\{(1,2);(1,3);(1,4);(2,3);(2,4);(3,4)\},\quad B=\{(1,1);(2,2);(3,3);(4,4)\}\quad\text{and}$$

$$C=\{(2,1);(3,1);(3,2);(4,1);(4,2);(4,3)\}\quad\text{are not empty and}$$

$$A\cap B=\emptyset,\quad B\cap C=\emptyset,\quad A\cap C=\emptyset,\quad A\cup B\cup C=E\times E.$$

■

Corollary 3.7.2 For any sets A and B, we have $n(A \times B) = n(A) \times n(B)$.

3.8 Chapter 3 Practice Problems

Question 1 Write each of the following sets in set description notation (by listing their elements between braces):

1. $\{z \in \mathbb{Z} : -2 < z \leqslant 3\}$.
2. $\{z \in \mathbb{Z} : -5 \leqslant z \leqslant 5\}$.
3. $\{x \in \mathbb{R} : x^3 = 8\}$.
4. $\{z \in \mathbb{Z} : |z| \leqslant 3\}$.
5. $\{z \in \mathbb{Z} : |2z| > 6\}$.
6. $\{z \in \mathbb{Z} : |3z| - 1 = 8\}$.
7. $\{x \in \mathbb{R} : x^2 + 3x = -2\}$.
8. $\{x \in \mathbb{R} : x^3 + 3x^2 + 3x = -1\}$.
9. $\{x \in \mathbb{R} : 2x^2 + 3x = -2\}$.
10. $\{z \in \mathbb{R} : 4z^2 + 4z = -1\}$.
11. $\{z \in \mathbb{Z} : 4z^2 + 4z = -1\}$.

Question 2 Write each of the following sets in the set builder notation:

1. $\{4, 8, 12, 16, 20, \cdots\}$.
2. $\{1, 3, 5, 7, 9, \cdots\}$.
3. $\{\cdots, -20, -15, -10, -5, 0, 5, 10, 15, 20, \cdots\}$.
4. $\{-19, -17, -15, \cdots, 1, 3, 5, 7, 9\}$.
5. $\{\cdots, -24, -16, -8, 0, 8, 16, 24, \cdots\}$.
6. $\{1, \frac{1}{2}, \frac{1}{3}, \frac{1}{4}, \cdots\}$.
7. $\{\cdots, -\frac{3\pi}{2}, -\pi, -\frac{\pi}{2}, 0, \frac{\pi}{2}, \pi, \frac{3\pi}{2}, \cdots\}$.
8. $\{-1, -\frac{1}{2}, -\frac{1}{3}, \cdots, 0, \cdots, \frac{1}{3}, \frac{1}{2}, 1\}$.

Question 3 Write each of the following in set notation;

1. Let x be a real number that makes $x > 0$ and $x < 1$ true.
2. Let x be a rational number that makes $x > 0$ and $x < 1$ true.
3. Let x be an integer that makes $x > 0$ and $x < 1$ false.

Question 4 Determine whether the following statements are true or false:

1. If $\{a\} \in \mathscr{P}(A)$, then $a \in A$, but $\{a\} \notin A$.
2. If three sets A, B, and C are subsets of $\{a, b\}$ such that $|A| = |B| = |C| = 2$, then at least two of these sets are equal.
3. If $A = \{1, 2, \{3\}\}$, then $1 \in A$ and $2 \in A$, but $3 \notin A$.
4. If $A = \{1, 2, \{3\}\}$, then $\{1\} \subset A$, $\{2\} \subset A$ and $\{3\} \subset A$.

Question 5 Write the following sets by listing elements within braces:

1. $\{n \in \mathbb{N} : n^3 < 100\}$.
2. $\{m \in \mathbb{Z} : -10 \leqslant m^3 \leqslant 10\}$.
3. $\{r \in \mathbb{Q} : 4r^2 - 12 = 0\}$.
4. $\{n \in \mathbb{R} : x^2 - x = 0\}$.

Question 6 Find $A \cap B$, $A \cup B$, $A - B$, $B - A$ if the following are true:

(i) $A = \{4, 5, 7\}$, $B = \{6, 7, 10, 4\}$.
(ii) $A = \{3\}$, $B = \emptyset$.
(iii) $A = \{a, b, c, d, e\}$ and $B = \{a, b\}$.

Question 7 Let A, B, and C be sets. Prove that

$$A \subset B \implies A \cap C \subset B \cap C.$$

Question 8 Let A, B, and C be sets. Prove that

$$A \subset B \implies \overline{B} \cup C \subseteq \overline{A} \cup C.$$

Question 9 Let A and B be two sets. Prove that:

(i) $A = (A \cap B) \cup (A - B)$.

(ii) $A \cup B = (A - B) \cup (B - A) \cup (A \cap B)$.

Question 10 Let A and B be two sets. Prove the following

(i) $A \cap B \subseteq A$.

(ii) $A \cup B \supseteq A$.

Question 11 Let A and B be two sets. Prove the following

(i) $A \cup B = A \iff B \subseteq A$.

(ii) $A \cap B = A \iff A \subseteq B$.

Question 12 Let A and B be two sets. Prove the following

(i) $A - B = \emptyset \iff B \supseteq A$.

(ii) $A - B = A \iff A \cap B = \emptyset$.

Question 13 Show that for sets A, B, and C we have

$$(A \cup B) \cap C = (A \cap C) \cup (B \cap C).$$

Question 14 Let A, B, and C be subsets of $\mathbb{R}$. Complete each of the following sentences:

1. $x \notin A \cap B$ if and only if $\cdots$
2. $x \notin A \cup B$ if and only if $\cdots$
3. $x \notin A - B$ if and only if $\cdots$
4. $x \in \{(A \cap B) \cup (A \cap C)\}$ if and only if $\cdots$
5. $x \notin \{(A \cup B) \cap (A \cup C)\}$ if and only if $\cdots$

Question 15 The **symmetric difference** between two sets, denoted Δ, is defined as follows:

$$A \,\Delta\, B = (A - B) \cup (B - A).$$

(i) Determine $A \Delta B$ if $A = \{0,1,2,3,4,5,6,7,8\}$ and $B = \{-2,0,4,6,8,10,3\}$.

(ii) Show that for any sets A, B that $A \,\Delta\, B = (A \cup B) - (A \cap B)$.

(iii) Show that for any sets A, B, and C we have: $A \,\Delta\, (B \,\Delta\, C) = (A \,\Delta\, B) \,\Delta\, C$.

Question 16 Let $U = \{a,b,c,d,e,g,h,i,k,l,m,n\}$ be the universal set, and let $A = \{b,c,a,h,k\}$, $B = \{b,c,d,k\}$, and $C = \{b,d,e,g,m\}$.

(a) Find $A \bigcup B$ and $\overline{A} \bigcap \overline{B}$.

(b) Find $C - B$ and $A - (B - C)$.

Question 17 Prove that $(A - B) \cup (B - A) = (A \cup B) - (A \cap B)$.

Question 18 Prove or disprove the following:

(a) $A - (B \cup C) = (A - B) \cup (A - C)$.

(b) $A - (B \cap C) = (A - B) \cap (A - C)$.

Question 19 Let A be a set. Prove the following

1. $\emptyset \subseteq A$.
2. $A \subseteq A$.

Question 20 Prove that

$$A \subseteq B \iff \overline{B} \subseteq \overline{A}.$$

Question 21 Suppose $C = \{-2,0,4,3,6,7,1,9\}$, and $D = \{-1,5,6,8,4\}$, and let $U = \{n \in \mathbb{Z} : -3 \leq n \leq 10\}$ be the universal set. Find

$$C \cup D, \ \overline{C \cap D}, \ \overline{\overline{C} \cup D}.$$

Question 22 Let $(A_i)_{i \in I}$ be a family of subsets of a universal set E, and let B also be a subset of E. Then we have

$$B \cup \left(\bigcap_{i \in I} A_i \right) = \bigcap_{i \in I} (A_i \cup B).$$

Question 23 Let $(A_i)_{i \in I}$ be a family of subsets of a universal set E, and let B also be a subset of E. Then we have

$$\overline{\bigcup_{i \in I} A_i} = \bigcap_{i \in I} \overline{A_i}.$$

Question 24 For the set $\mathbf{S} = \{2,3,4,5,6,7,8\}$, consider the four subsets:

$$A_1 = \{4,5\},\ A_2 = \{2,3\},\ A_3 = \{4,5,6\},\ A_4 = \{5,7,8\}.$$

Is $P = \{B_1, B_2, B_3\}$ a partition of $\mathbf{S}$, where $B_i = A_{i+1} - A_i,\ i = 1,2,3$?

Question 25 Find $A \times B,\ B \times A$ if the following are true:
(i) $A = \{4,5,7\},\quad B = \{6,7,10,4\}$.
(ii) $A = \{3\},\quad B = \emptyset$.
(iii) $A = \{a,b,c,d,e\}$ and $B = \{a,b\}$.

Question 26 Let $A = \{1,3,5\}$, $B = \{2,3,5\}$, and $C = \{1,2,3\}$. Determine if the following are true:
(a) $A \times (B \cap C) = (A \times B) \cap (A \times C)$.
(b) $A \times (B - C) = (A \times B) - (A \times C)$.

Question 27 Let $A = \{2,3,5\}$, $B = \{1,3,4\}$, and $C = \{1,2,3,4\}$. Determine if the following are true:
(a) $B \times (C \cap A) = (B \times C) \cup (B \times A)$.
(b) $A \times (C - B) = (A \times C) - (A \times B)$.
(c) $A - (B \cup C) = (A - B) \cap (A - C)$.
(b) $A - (B \cap C) = (A - B) \cup (A - C)$.

Question 28 For $A = \{0, \{0\}\}$ and $B = \{\emptyset, \{\emptyset\}\}$, determine $A \times B$ and $\mathscr{P}(A) \times \mathscr{P}(B)$.

Question 29 Let $A,\ B,\ C$, and D be three sets. Then

$$(A \times B) \cap (C \times D) = (A \cap C) \times (B \cap D).$$

Question 30 Let $A,\ B,\ C$, and D be three sets. Then

$$(A \times B) \cup (C \times D) \subseteq (A \cup C) \times (B \cup D).$$

Question 31 Determine which of the following is true by first rewriting the sets A and B in the set-description notation:

$$A = \{m \in \mathbb{Z} : -27 \leqslant m^3 \leqslant 27\} \text{ and } B = \{r \in \mathbb{Q} : 3r^3 - 9 = 0\}.$$

(a) $n(A \times A) \leqslant n(A)$.
(b) $n(A \times A) = n(A)$.
(c) $n(A) \leqslant n(A \times A)$.

(d) $n(A \times B) \leqslant n(B)$.
(e) $n(B \times B) = n(B)$.
(f) $n(B) \leqslant n(B \times B)$.
(g) $n(A \times B) \leqslant n(A)$.
(h) $n(A \times B) \leqslant n(B)$.
(i) $n(B \times A) = n(B)$.
(j) $n(A \times B) = n(A)$.
(k) $n(A) \leqslant n(A \times B)$.
(l) $n(B) \leqslant n(A \times B)$.
(m) $n(A \times B) = n(B \times A)$.
(n) $n(B \times A) = n(A \times B)$.

Question 32 Determine the cardinality of the following sets:

1. $\{x \in \mathbb{R} : x^2 + 1 = 0\}$.
2. $\{\emptyset, \{\emptyset\}, \{\emptyset, \{\emptyset\}\}\}$.

Question 33 Find two sets A and B such that A is both an element of and a subset of B.

Question 34 Suppose $A = \{b, c, d\}$, $B = \{a, b\}$. Find

$$(A \times B) \cap (A \times A), \qquad \mathscr{P}(A) - \mathscr{P}(B).$$

Question 35 Find $\mathscr{P}(A)$ and $n(\mathscr{P}(A))$ for $A = \{\emptyset, \{\emptyset\}, \{\emptyset, \{\emptyset\}\}\}$.

Question 36 Give an example of a set A such that the following are true:

1. $A \subseteq \mathscr{P}(A)$.
2. $A \in \mathscr{P}(A)$.
3. $A \subseteq \mathscr{P}(A)$ and $n(A) = 5$.
4. $A \in \mathscr{P}(A)$ and $n(A) = 5$.

3.9 Chapter 3 Additional Practice Problems with Solutions

Question 1 Write each of the following sets in set description notation:

Solution.

1. $\{z \in \mathbb{Z} : -2 < z \leqslant 3\} = \{-1, 0, 1, 2, 3\}$.
2. $\{z \in \mathbb{Z} : -5 \leqslant z \leqslant 5\} = \{-5, -4, -3, -2, -1, 0, 1, 2, 3, 4, 5\}$.
3. $\{x \in \mathbb{R} : x^3 = 8\} = \{2\}$.
4. $\{z \in \mathbb{Z} : |z| \leqslant 3\} = \{-3, -2, -1, 0, 1, 2, 3\}$.
5. $\{z \in \mathbb{Z} : |2z| > 6\} = \{z \in \mathbb{Z} : |z| > 3\} = \{..., -5, -4, 4, 5, 6, ...\}$.
6. $\{z \in \mathbb{Z} : |3z| - 1 = 8\} = \{-3, +3\}$.
7. $\{x \in \mathbb{R} : x^2 + 3x = -2\} = \{-2, -1\}$.
8. $\{x \in \mathbb{R} : x^3 + 3x^2 + 3x = -1\} = \{-1\}$.
9. $\{x \in \mathbb{R} : 2x^2 + 3x = -2\} = \emptyset$.

10. $\{z\in\mathbb{R}: 4z^2+4z=-1\} = \{-1/2\}$.

Question 2 Write the following sets in the set builder notation:

Solution.

1. $\{4,8,12,16,20,\cdots\}=\{4k:\ k\in\mathbb{N}\}$.
2. $\{1,3,5,7,9,\cdots\} = \{2k+1:\ k\in\mathbb{Z},\ k\geq 0\} = \{2k-1:\ k\in\mathbb{N}\}$.
3. $\{\cdots,-20,\ -15,\ -10,\ -5,\ 0,\ 5,\ 10,\ 15,\ 20,\ \cdots\} = \{5k:\ k\in\mathbb{Z}\}$.
4. $\{-19,\ -17,\ -15,\ \cdots,\ 1,\ 3,\ 5,\ 7,\ 9\} = \{2k+1:\ k\in\mathbb{Z},\ -10\leq k\leq 4\}$.
5. $\{-3,\ +3\}=\{z\in\mathbb{Z}: |2z|-1=5\}$.
6. $\{-2,\ -1,\ 0,\ 1,2,\ \cdots\}=\{x\in\mathbb{Z}:\ x\geqslant -2\}$.
7. $\{\cdots,-24,\ 16,\ -8,\ 0,\ 8,\ 16,\ 24,\ ...\} = \{8k:\ k\in\mathbb{Z}\}$.
8. $\{1,\frac{1}{2},\ \frac{1}{3},\ \frac{1}{4},\ \cdots\} = \left\{\frac{1}{n}:\ n\in\mathbb{N}\right\}$.
9. $\{\cdots,\ -\frac{3\pi}{2},\ -\pi,\ -\frac{\pi}{2},\ 0,\ \frac{\pi}{2},\ \pi,\ \frac{3\pi}{2},\ \cdots\} = \left\{\frac{k\pi}{2}:\ k\in\mathbb{Z}\right\}$.
10. $\{-1,\ -\frac{1}{2},\ -\frac{1}{3},\ \cdots,\ 0,\ \cdots,\ \frac{1}{3},\ \frac{1}{2},\ 1\} = \left\{\frac{1}{n}:\ n\in\mathbb{Z},\ n\neq 0\right\}\cup\{0\}$.

Question 3 Write the following statements in set notations:

Solution.

1. Let x be a real number that makes $x>0$ and $x<1$ true.
   $\{x\in\mathbb{R}:\ 0<x<1\} = (0,\ 1)$.
2. Let x be a rational number that makes $x>0$ and $x<1$ true.
   $\left\{\frac{p}{q}:\ 0<\frac{p}{q}<1,\ p,q\in\mathbb{Z},\ pq\neq 0\right\} = (0,\ 1)\cap\mathbb{Q}$.
3. Let x be an integer that makes $x>0$ and $x<1$ false.
   $\{x\leq 0:\ x\in\mathbb{Z}\}\cup\{x\geq 1:\ x\in\mathbb{N}\}$.

Question 4 Determine whether the following statements are true or false:

Solution.

1. If $\{a\}\in\mathscr{P}(A)$, then $a\in A$ but $\{a\}\notin A$.
   False. Take $A=\{a,\ \{a\}\}$; we have $\{a\}\in\mathscr{P}(A)$, and a and $\{a\}$ are in A.

2. If three sets A, B, and C are subsets of $\{a,b\}$ such that $|A|=|B|=|C|=2$, then at least two of these sets are equal.
   True. $A\subseteq B,\ n(A)=n(B) \implies A=B$, and then all three sets should be equal to $\{a,b\}$.

3. If $A=\{1,2,\{3\}\}$, then $1\in A$ and $2\in A$, but $3\notin A$.
   True. Because $\{3\}\in A$ not 3.

4. If $A=\{1,2,\{3\}\}$, then $\{1\}\subset A$, $\{2\}\subset A$ and $\{3\}\subset A$.
   False. Because $\{\{3\}\}\subset A$.

Question 5 For the following sets, find $A\cap B$, $A\cup B$, $A-B$, and $B-A$:

(i) $A=\{4,5,7\},\ B=\{6,7,10,4\}$.
(ii) $A=\{3\},\ B=\emptyset$.
(iii) $A=\{a,b,c,d,e\}$ and $B=\{a,b\}$.

Solution.

(i) $A=\{4,5,7\},\ B=\{6,7,10,4\}$,

$$A\cap B=\{4,7\},\quad A\cup B=\{4,5,7,6,10\},\quad A-B=\{5\},\quad B-A=\{6,10\}.$$

(ii) $A = \{3\}$, $B = \emptyset$,

$$A \cap B = B, \quad A \cup B = A, \quad A - B = A, \quad B - A = B.$$

(iii) $A = \{a, b, c, d, e\}$ and $B = \{a, b\}$,

$$A \cap B = B, \quad A \cup B = A, \quad A - B = \{c, d, e\}, \quad B - A = \emptyset.$$

Question 6 Let A, B, and C be sets. Prove that

$$A \subset B \implies A \cap C \subset B \cap C.$$

Solution. See Section 5.1 Example 5.9.

Question 7 For sets A and B, prove the following:
(i) $A = (A \cap B) \cup (A - B)$.
(ii) $A \cup B = (A - B) \cup (B - A) \cup (A \cap B)$.

Solution. See Section 5.1 Example 5.10.

Question 8 Let A, B, and C be sets. Prove the following
(i) $A \cup B = A \iff B \subseteq A$.
(ii) $A \cap B = A \iff A \subseteq B$.

Proof. See Section 5.1 Example 5.11. ■

Question 9 Show that for each set A, B, and C we have:

$$(A \cup B) \cap C = (A \cap C) \cup (B \cap C).$$

Solution. See Section 5.1 Example 5.12.

Question 10 Let A, B, C be subsets of $\mathbb{R}$. Complete each of the following sentences:
1. $x \notin A \cap B$ if and only if $\cdots$
2. $x \notin A \cup B$ if and only if $\cdots$
3. $x \notin A - B$ if and only if $\cdots$
4. $x \in \{(A \cap B) \cup (A \cap C)\}$ if and only if $\cdots$
5. $x \notin \{(A \cup B) \cap (A \cup C)\}$ if and only if $\cdots$

Solution.
1. $x \in \overline{A \cap B} = \overline{A} \cup \overline{B} \iff x \notin A$ or $x \notin B$.
2. $x \notin A \cup B \iff x \in \overline{A \cup B} = \overline{A} \cap \overline{B} \iff x \notin A$ and $x \notin B$.
3. $x \notin A - B \iff x \in \overline{A - B} = \overline{A \cap \overline{B}} = \overline{A} \cup \overline{\overline{B}} = \overline{A} \cup B \iff x \notin A$ or $x \in B$.
4. $x \in [(A \cap B) \cup (A \cap C)] \iff x \in [A \cap (B \cup C)]$.
5. Let

$$\begin{aligned} x \notin [(A \cup B) \cap (A \cup C)] &\iff x \notin [A \cup (B \cap C)] \\ &\iff x \in \overline{A \cup (B \cap C)} \\ &\iff x \in \overline{A} \cap \overline{B \cap C} \\ &\iff x \in \overline{A} \cap (\overline{B} \cup \overline{C}). \end{aligned}$$

Question 11 The symmetric difference between two sets, denoted by Δ, is defined as follows:

$$A \,\Delta\, B = (A - B) \cup (B - A).$$

(i) Determine $A \Delta B$ if $A = \{0, 1, 2, 3, 4, 5, 6, 7, 8\}$ and $B = \{-2, 0, 4, 6, 8, 10, 3\}$.

(ii) Show that for any sets A, B that $A \Delta B = (A \cup B) - (A \cap B)$.

Solution.

(i) $A \Delta B = \{1,2,5,7\} \cup \{-2,10\} = \{-2,1,2,5,7,10\}$.

(ii) Let us denote $p : x \in A$, $q : x \in B$. Consider

$$\begin{aligned} x \in (A \cup B) - (A \cap B) &\iff x \in (A \cup B) \text{ and } x \notin (A \cap B), \\ &\iff [(x \in A) \text{ or } (x \in B)] \text{ and } x \in \overline{A \cap B}, \\ &\iff [(x \in A) \text{ or } (x \in B)] \text{ and } \left[(x \in \overline{A}) \text{ or} (x \in \overline{B})\right], \\ &\iff (p \vee q) \wedge (\neg p \vee \neg q), \\ &\iff [(p \vee q) \wedge \neg p] \vee [(p \vee q) \wedge \neg q)], \\ &\iff [(p \wedge \neg p) \vee (q \wedge \neg p)] \vee [(p \wedge \neg q) \vee (q \wedge \neg q)], \\ &\iff (q \wedge \neg p) \vee (p \wedge \neg q), \\ &\iff (p \wedge \neg q) \vee (q \wedge \neg p), \\ &\iff x \in (A \cap \overline{B}) \text{ or } x \in (B \cap \overline{A}), \\ &\iff x \in (A - B) \cup (B - A). \end{aligned}$$

Question 12 Let $U = \{a,b,c,d,e,g,h,i,k,l,m,n\}$ be the universal set, and let $A = \{b,c,a,h,k\}$, $B = \{b,c,d,k\}$, $C = \{b,d,e,g,m\}$.

(a) Find $A \bigcup B$ and $\overline{A} \bigcap \overline{B}$.

(b) Find $B - C$ and $A - (B - C)$.

Solution.

(a) $A \bigcup B = \{b,c,a,h,k,d\}$; $\overline{A} = \{d,e,g,i,l,m,n\}$, and $\overline{B} = \{a,e,g,h,i,l,m,n\}$. Thus $\overline{A} \cap \overline{B} = \{e,g,i,l,m,n\}$.

(b) $B - C = \{x \in U : x \in B \text{ and } x \notin C\} = \{c,k\}$ and $A - (B - C) = \{x \in A \text{ and } x \notin (B - C)\} = \{b,a,h\}$.

Question 13 Prove that $M = (M \bigcap \overline{N}) \bigcup (M \bigcap N)$ for any sets M and N.

Solution. Let us prove this directly. To this end, let $N \subset U$, and let us use the fact that

$$M = M \cap U = M \cap \left(N \cup \overline{N}\right) = (M \cap N) \cup (M \cap \overline{N}).$$

Question 14 Let A be a set. Prove that the following:

1. $\emptyset \subseteq A$.
2. $A \subseteq A$.

Solution.

1. Let $p : x \in \emptyset$ and $q : x \in A$. Our question is equivalent to prove that $p \Longrightarrow q$ is true. Since p is always false, we conclude that the implication $p \Longrightarrow q$ is always true for any truth value of q. Thus $\emptyset \subseteq A$.
2. Obvious.

Question 15 Prove that

$$A \subseteq B \iff \overline{B} \subseteq \overline{A}.$$

Solution. See Section 5.1 Example 5.13.

Question 16 Give an example of a set A such that the following are true:

1. $A \subseteq \mathscr{P}(A)$.
2. $A \in \mathscr{P}(A)$.
3. $A \subseteq \mathscr{P}(A)$ and $n(A) = 5$.
4. $A \in \mathscr{P}(A)$ and $n(A) = 5$.

Solution.

1. $A \subseteq \mathscr{P}(A)$.
   Let $A = \{\emptyset\}$
   $\mathscr{P}(A) = \{\emptyset, \{\emptyset\}\}$, therefore, $A \subseteq \mathscr{P}(A)$.
2. $A \in \mathscr{P}(A)$.
   Let $A = \{\emptyset\}$
   $\mathscr{P}(A) = \{\emptyset, \{\emptyset\}\}$, therefore, $A \in \mathscr{P}(A)$.
3. $A \subseteq \mathscr{P}(A)$ and $n(A) = 5$.
   Let $A = \{\emptyset, \{\emptyset\}, \{\{\emptyset\}\}, \{\{\{\emptyset\}\}\}, \{\{\{\{\emptyset\}\}\}\}\}$
   $\mathscr{P}(A) = \{\emptyset, \{\emptyset\}, \{\{\emptyset\}\}, \{\{\{\emptyset\}\}\}, \{\{\{\{\{\emptyset\}\}\}\}\}, \cdots, A\}$, therefore, $A \subseteq \mathscr{P}(A)$.
4. $A \in \mathscr{P}(A)$ and $n(A) = 5$.
   Let $A = \{\emptyset, \{\emptyset\}, \{\{\emptyset\}\}, \{\{\{\emptyset\}\}\}, \{\{\{\{\emptyset\}\}\}\}\}$,
   Clearly, $A \in \mathscr{P}(A)$.

Question 17 Let $(A_i)_{i \in I}$ be a family of subsets of a universal set E, and let B also be a subset of E. Then we have

$$B \cup \left(\bigcap_{i \in I} A_i \right) = \bigcap_{i \in I} (A_i \cup B).$$

Solution. See Section 5.1 Example 5.14.

Question 18 Let $(A_i)_{i \in I}$ be a family of subsets of a universal set E. Then we have:

$$\overline{\bigcup_{i \in I} A_i} = \bigcap_{i \in I} \overline{A_i}.$$

Solution. See Section 5.1 Example 5.15.

Question 19 Let $A = \{1,3,5\}$, $B = \{2,3,5\}$, and $C = \{1,2,3\}$. Determine if the following are true:

(a) $A \times (B \cap C) = (A \times B) \cap (A \times C)$.
(b) $A \times (B - C) = (A \times B) - (A \times C)$.

Solution.

(a) $A \times (B \cap C) = \{1,3,5\} \times \{2,3\} = \{(1,2), (1,3), (3,2), (3,3), (5,2), (5,3)\}$.
$A \times B = \{(1,2), (1,3), (1,5), (3,2), (3,3), (3,5), (5,2), (5,3), (5,5)\}$.
$A \times C = \{(1,1), (1,2), (1,3), (3,1), (3,2), (3,3), (5,1), (5,2), (5,3)\}$.
$(A \times B) \cap (A \times C) = \{(1,2), (1,3), (3,2), (3,3), (5,2), (5,3)\}$.
Then equality holds.

(b) $A \times (B - C) = (A \times B) - (A \times C)$.
$B - C = \{5\} \Longrightarrow A \times (B - C) = \{(1,5), (3,5), (5,5)\}$.
Since $A \times B = \{(1,2), (1,3), (1,5), (3,2), (3,3), (3,5), (5,2), (5,3), (5,5)\}$, and
$A \times C = \{(1,1), (1,2), (1,3), (3,1), (3,2), (3,3), (5,1), (5,2), (5,3)\}$,
$(A \times B) - (A \times C) = \{(1,5), (3,5), (5,5)\}$, the equality holds.

Question 20 Let A, B, C, and D be four sets, then

$$(A \times B) \cup (C \times D) \subseteq (A \cup C) \times (B \cup D).$$

Solution. See Section 5.1 Example 5.16.

Question 21 Determine the cardinality of the following sets:

1. $A = \{x \in \mathbb{R} : x^2 + 1 = 0\} = \{x \in \mathbb{R} : x^2 = -1\} = \emptyset \Longrightarrow n(A) = 0$.

2. $B=\{\emptyset,\{\emptyset\},\{\emptyset,\{\emptyset\}\}\} \implies n(B)=3$.

Question 22 Suppose $A=\{b,c,d\}$, and $B=\{a,b\}$. Find the following $(A\times B)\cap(A\times A)$, and $\mathscr{P}(A)-\mathscr{P}(B)$.

Solution. Directly, we have $A\times B=\{(b,a),(b,b),(c,a),(c,b),(d,a),(d,b)\}$ and $A\times A=\{(b,b),(b,c),(b,d),(c,b),(c,c),(c,d),(d,b),(d,c),(d,d)\}$, then

$$(A\times B)\cap(A\times A)=\{(b,b),(c,b),(d,b)\}.$$

$\mathscr{P}(A)=\{\emptyset,A,\{b\},\{c\},\{d\},\{b,c\},\{b,d\},\{c,d\}\}$ and $\mathscr{P}(B)=\{\emptyset,B,\{a\},\{b\}\}$, then

$$\mathscr{P}(A)-\mathscr{P}(B)=\{A,\{c\},\{d\},\{b,c\},\{b,d\},\{c,d\}\}.$$

Question 23 If we assume that $n(E)=32$, $n(F)=31$, $n(E\bigcup F)=47$, $n(E\bigcap G)=11$, $n(F-G)=20$ and $n(G-F)=22$. Find the following

(a) $n(E-F)$
(b) $n(F-E)$
(c) $n(F\bigcup G)$
(d) $n(E\bigcup G)$.

Solution.

(a) Since $E=(E-F)\cup(E\cap F)$ and $(E-F)\cap(E\cap F)=\emptyset$, $n(E)=n(E-F)+n(E\cap F)$. Using the fact that

$$n(E\cap F)=n(E)+n(F)-n(E\cup F)=32+31-47=16,$$

we conclude $n(E-F)=n(E)-n(E\cap F)=32-16=16 \implies n(E-F)=16$.

(b) Similarly to part (a), we have

$$n(F)=n(F-E)+n(F\cap E)\Longrightarrow n(F-E)=n(F)-n(E\cap F)=31-16=15.$$

Thus $n(F-E)=15$.

(c) $n(F\cup G)=n(F)+n(G)-n(F\cap G)=n(F)+\underbrace{n(G)-n(F\cap G)}_{n(G-F)}=31+22=53$.

Thus $n(F\cup G)=53$.

(d) Since $n(E\cup G)=n(E)+n(G)-n(E\cap G)=32+n(G)-11=n(G)+21$, it suffices to evaluate $n(G)$.

$$n(F\cup G)=n(F)+n(G)-n(F\cap G)=n(G)+\underbrace{n(F)-n(F\cap G)}_{n(F-G)=20}\Longrightarrow n(G)=53-20=33.$$

Finally, $n(E\cup G)=n(G)+21=33+21=54$.

Question 24 For the set $\mathbf{S}=\{1,2,3,4,5,6,7\}$, consider the following five subsets:

$$A_1=\{1,2\},\ A_2=\{3,4,5\},\ A_3=\{2,5,6\},\ A_4=\{2,7\},\ A_5=\{1,2\}.$$

Is $P=\{B_1,B_2,B_3,B_4\}$ a partition of S, where $B_i=A_i-A_{i+1}$, $i=1,2,3,4$.

Solution. Let us determine first the sets B_i, $i=1,2,3,4$:

$B_1=A_1-A_2=\{1,2\}$, $B_2=A_2-A_3=\{3,4\}$, $B_3=A_3-A_4=\{5,6\}$, and $B_4=A_4-A_5=\{7\}$.

$P=\{B_1,B_2,B_3,B_4\}$ is a partition of $\mathbf{S}$ if and only if all B_i are disjoint and their union is equal to S.

It is clear that all B_i, $i=1,2,3,4$ do not share any elements and $\bigcup_{i=1}^{4}B_i=\{1,2,3,4,5,6,7\}=S$.

Finally, $P=\{B_1,B_2,B_3,B_4\}$ is a partition of S.

4. Techniques of Proofs

Chapter Key Terms

(a) *trivial and vacuous proofs*
(b) *direct proof*
(c) *proof by contrapositive*
(d) *proof by contradiction*
(e) *proof by cases*

In mathematics, we often hear about *proofs*, which may seem daunting and complex at first glance. However, the true beauty of mathematics lies in its powerful language of proofs, built upon simple and successive logical rules of inference. What sets mathematics apart from other disciplines is its reliance on proofs. Once something is rigorously proven, we can be absolutely certain of its truth, both now and in perpetuity. Unlike many other sciences, mathematics isn't merely a theoretical framework subject to revision based on new observations; it's a realm where truths endure.

In this chapter, we will leverage the logical principles introduced in Chapter 2 to cultivate the art of constructing rigorous proofs. A proof, at its core, consists of a sequence of logically connected statements, mixed with definitions, previously established results, algebraic manipulations, and axioms.

Within this chapter, we will explore various techniques for proving implications, the classic "if P, then Q" statements. We will start with straightforward methods commonly employed to establish such implications. Additionally, we will occasionally introduce alternative techniques for proving the same implications. Our aim is to guide students in developing their proof-writing skills, progressing from the elementary to the more meticulous.

In this chapter, student's are expected to

- know previously introduced notions and logical results,
- understand why a specific technique is being used in the proof, and

- follow the logical sequence used to drive the conclusion.

Clearly, an implication *if P, then Q*, written as $P \implies Q$ involves two simple statements, P and Q, where the statement P is called the *hypothesis*, and Q is called the *conclusion*. Usually theorems in mathematics are about proving implications and or equivalences. The first section of the chapter discusses trivial and vacuous proofs, followed by other techniques to demonstrate implications. Recall that if P implies Q, then P is called the *antecedent* and Q is called the *consequent*.

Definition 4.0.1 — trivial and vacuous proofs. An implication is **trivially true** when its consequent is true. Since an implication *if P then Q* is true whenever Q is true regardless of the truth value of P, we say that $P \implies Q$ is trivially true. Such a proof is called a trivial proof.
On the other hand, an implication if P then Q is **vacuously true** when its antecedent (P) is always false. That is, if it can be shown that P is false regardless of the truth value of Q, then the implication is true vacuously. Such a proof is called a **vacuous proof**.

$$\text{trivial proof : antecedent} \implies \text{consequent(TRUE)}.$$

$$\text{vacuous proof : antecedent (FALSE)} \implies \text{consequent}.$$

■ **Example 4.1 Trivial Proof**

- Let $x \in \mathbb{R}$. If $x > 0$, then $x^2 + 2 > 0$.
  Proof- Since $x^2 + 2 > 0$ is true regardless of $x > 0$, so the statement is trivially true.
- If a polygon has 3 sides, it must be a triangle.
  Proof- By definition, a polygon with 3 sides is a triangle.
- All real numbers are non-negative when squared.
  Proof- By definition, a squared number is always non-negative.
- If the sun is a star then the moon is not.
  Proof- By definition, as a star, the sun produces its own energy through fusion reactions of gasses, while the moon doesn't.

■

■ **Example 4.2 Vacuous Proof**

- Let $x \in \mathbb{R}$. If $x^2 + 2 < 0$ then $x^3 > 5$.
  Proof- Since $x^2 + 2 < 0$ is false, the statement, $x^2 + 2 < 0$ then $x^3 > 5$, is true vacuously.
- If a polygon has less than 3 sides, then it is a circle.
  Proof- By definition, a polygon with less than 3 sides doesn't exist; the statement is true.
- If the square of a number is negative, then the number is prime.
  Proof- Since *the square of a number is negative* is a false, the statement is vacuously true.
- If the moon is a star, then the sun is a planet.
  Proof- The statement *the moon is a star* is false and prevents inferring the truth or false value of the consequent. Thus the statement is true vacuously.

■

4.1 Direct Proofs

Direct proofs (or techniques) are commonly used in mathematics to prove simple statements. A direct proof is also a viable method for proving complex statements in an efficient manner.

4.1.1 Definitions and Introductory Examples

Direct proofs always assume a hypothesis is true and then logically deduces a conclusion. When proving a conditional statement, **if** P**, then** Q or $P \implies Q$ by the direct-proof method, we assume

antecedent (or hypothesis) P to be true, and using rules of inference, we conclude that *consequent* Q must also be true.
Two statements P and Q are **equivalent** if and only if they have same truth values. For the equivalence, we write $P \equiv Q$ which is equivalent to

$$(P \implies Q) \wedge (Q \implies P),$$

as it was defined in Chapter 2, Sec 2.5.

R Usually, in a direct-proof method of $LHS \implies RHS$, we assume LHS is true, and using the definitions, rules, theorems, and sometimes axioms, we reach the true value of the RHS.

Before we discuss examples of direct proofs, let us introduce a few more notions that can be useful when constructing direct proofs.

4.1.2 Parity and Divisibility

By **parity**, we are referring to whether a number is even or odd. The parity of a number is determined by its remainder after being divided by 2. In general, we say any two integers belong to the same parity if both are even or both are odd; otherwise, they belong to opposite parities.

Definition 4.1.1 — even number. An integer n is said to be **even** if there exists $k \in \mathbb{Z}$ such that $n = 2k$.

Definition 4.1.2 — odd number. An integer n is said to be **odd** if there exists $k \in \mathbb{Z}$ such that $n = 2k+1$.

Addition of Even and Odd Numbers

(a) The sum (or difference) of two *even* numbers is *even*:

$$even \pm even = even.$$

(b) The sum (or difference) of two *odd* numbers is *even*:

$$odd \pm odd = even.$$

(c) The sum (or difference) of an *odd* number and an *even* number is *odd*:

$$even \pm odd = odd.$$

Theorem 4.1.1 The following properties hold for even and odd integers:
(1) The sum of two even (or odd) integers is an even integer.
(2) The sum of an even and an odd integer is an odd integer.

Proof. Let m and n be two integers.
(1) (a) Assume that m and n are both even integers.
Then there exist $p, q \in \mathbb{Z}$ such that $m = 2p$ and $n = 2q$.
Now, the sum

$$m+n = 2(p+q) = 2r, \qquad \text{where } r \in \mathbb{Z}.$$

Thus $m+n$ is an even integer.

(b) Assume that m and n are both odd integers.
Then there exist $p,\ q \in \mathbb{Z}$ such that $m = 2p+1$ and $n = 2q+1$.
Now, the sum

$$m+n = 2(p+q+1) = 2r, \qquad \text{where } r \in \mathbb{Z}.$$

Thus $m+n$ is an even integer.

(2) With out loss of generality (WOLOG), let us assume that m is even and n is odd.
Then there exist $p,\ q \in \mathbb{Z}$ such that $m = 2p$ and $n = 2q+1$.
Now, the sum

$$m+n = 2(p+q)+1 = 2r+1 \qquad \text{where } r \in \mathbb{Z}.$$

Thus $m+n$ is an odd integer.

■

Multiplication of Even and Odd Numbers:

(a) The product of two *even* numbers is *even*:

$$\text{even} \times \text{even} = \text{even}$$

(b) The product of two *odd* numbers is *odd* number:

$$\text{odd} \times \text{odd} = \text{odd}$$

(c) The product of an *even* and an *odd* is *even* number:

$$\text{even} \times \text{odd} = \text{even}$$

The theorem below uses direct proof to prove the above properties as implications.

Theorem 4.1.2 The following properties hold for even and odd integers:
(1) The product of two even (odd) integers is an even (odd) integer.
(2) The product of an even and odd integer is an even integer.

Proof. Let m and n be any two integers.

(1) (a) Assume that m and n are both even integers.
Then there exist $p,\ q \in \mathbb{Z}$ such that $m = 2p$ and $n = 2q$.
Now the product,

$$mn = 2p2q) = 2(2pq) = 2r, \qquad \text{where } r = (2pq) \in \mathbb{Z}.$$

Thus mn is an even integer.

(b) Assume that m and n are both odd integers.
Then there exist $p,\ q \in \mathbb{Z}$ such that $m = 2p+1$ and $n = 2q+1$.
Now the product,

$$mn = 2p+1 2q+1 = 2\ (2pq+p+q)+1 = 2r+1, \qquad \text{where } r = (2pq+p+q) \in \mathbb{Z}.$$

Thus mn is an odd integer.

(2) WLOG, let us assume that m is even and n is odd.
Then there exist $p,\ q \in \mathbb{Z}$ such that $m = 2p$ and $n = 2q+1$.
Now, the product

$$mn = 2p(2q+1) = 2(2pq+p) = 2r \qquad \text{where } r = 2pq+p \in \mathbb{Z}.$$

Thus mn is an even integer.

■

Theorem 4.1.3 The product of two consecutive integers is an even integer.

Proof. Let n be an integer. We want to prove that $n(n+1)$ is even.
Since n is either even or odd, then $(n+1)$ is either odd or even, and we conclude directly that $n(n+1)$ is even as a product of an odd number by an even one. Thus the product of any two consecutive integers is an even integer. ■

(R) Note that sometimes useful result like the preceding may be hidden, as in the following situations:

$$n - n^2 = n(1-n) = -(n-1)n \ \text{ is even}, \quad n^2 - 23n + 132 = (n-12)(n-11) \ \text{ is even.}$$

Definition 4.1.3 — m divides n. Let $m \neq 0$ and n be two integers. We say m divides n if there exists an integer q such that $n = mq$. We frequently use the notation $\mathbf{m} \mid \mathbf{n}$ for *m divides n*, the integer q is called the quotient of n divided by m.

Let $m \neq 0$ and n be two integers such that $m \mid n$. Then following statements are equivalent:
(a) m divides n.
(b) n is divisible by m.
(c) n is a multiple of m.
(d) m is a factor of n.
(e) m is a divisor of n.

(R) **Caution about the notation**
We caution that the notation $m \mid n$ doesn't represent the rational number $\frac{m}{n}$.
The notation $m \mid n$ is a proposition (statement) that involves two integers, m and n, such that m divides n, and the proposition could be either true or false. On the other hand, the notion: *divided by* in $\frac{n}{m}$ is an *operation* that divides integer m by the integer n.

■ **Example 4.3**
- $7 \mid 35$, since $35 = (5)(7)$.
- $7 \mid -21$, since $-21 = (-3)(7)$.
- $-7 \mid 63$, since $63 = (-7)(-9)$.

■

Definition 4.1.4 — m | n. Let $m \neq 0$ and n be two integers. We say m does not divide n whenever $n \neq mq$ for all $q \in Z$. We frequently use the notation $m \nmid n$ meaning *m does not divide n*.

For $m \neq 0$, let $m \nmid n$, then the following statements are equivalent:
(a) m doesn't divide n.
(b) n is not divisible by m.
(c) n is not a multiple of m.
(d) m is not a factor of n.
(e) m is not a divisor of n.

Example 4.4 $3 \nmid 5$.
One can argue that $5 = (3)(\frac{5}{3})$.
However, it is certainly not true in this case because $q = \frac{5}{3}$ is not an integer. ■

Definition 4.1.5 — greatest common divisor. For any two nonzero integers a and b, their **greatest common divisor**, denoted by $\mathrm{GCD}(a,b)$, is the largest positive integer that divides both a and b.

(R) The most well-known use of GCD is the reduction of a fraction, say $\frac{a}{b}$. One can find the integers a' and b' such that $\frac{a}{b} = \frac{a'}{b'}$, where $b' \neq 0$ and $GCD(a',b') = 1$.

Definition 4.1.6 — prime number. A positive integer p greater than 1 is said to be a **prime number** if it has no other positive divisors than itself and 1.

Definition 4.1.7 — floor function. The Floor of a real number x is a function that gives the largest integer below x. It is denoted $\lfloor x \rfloor$ and satisfies

$$p = \lfloor x \rfloor \in \mathbb{Z} \text{ and } p \leq x < p+1.$$

Example 4.5

$$\lfloor 2.45 \rfloor = 2, \quad \lfloor \pi \rfloor = 3, \quad \lfloor -4,67 \rfloor \neq -4 > -4.67 \text{ but } \lfloor -4,67 \rfloor = -5.$$

■

Theorem 4.1.4 Euclidean Division Algorithm
For any two integers n and $m > 0$, there exist unique integers q, r such that

$$n = mq + r, \quad 0 \leq r < m.$$

The integer r is the remainder, and q is the quotient of n by m.

Proof.

- **existence**: Let us set $q = \left\lfloor \frac{n}{m} \right\rfloor \in \mathbb{Z}$.

$$q = \left\lfloor \frac{n}{m} \right\rfloor \implies q \leq \frac{n}{m} < q+1, \qquad m > 0,$$
$$\implies m \times q \leq n < m(q+1).$$

Now, let us take $r = n - m \times q \in \mathbb{Z}$; thus, we have $0 \leq r < m$. Indeed,

$$m \times q \leq n < m(q+1) \implies n - mq \geq 0 \implies r \geq 0 \text{ and } r \in \mathbb{Z}$$

and

$$m \times q \leq n < m(q+1) = mq + m \implies n - mq < m \implies r < m.$$

Thus there exist $(q,r) \in \mathbb{Z} \times \mathbb{Z}_+$ such that $n = mq + r, \quad 0 \le r < m$.

- **uniqueness**: Assume the existence of another couple $(q',r') \in \mathbb{Z} \times \mathbb{Z}_+$: $n = mq' + r' = mq + r$, where $0 \le r' < m$ and $0 \le r < m$. Thus we have

$$\text{(i)} \quad m(q'-q) = r - r', \qquad \text{(ii)} \quad |r - r'| < m.$$

If we assume that $q' \neq q$, then we have $|q'-q| \ge 1$, which implies that $|r'-r| \ge m$ by (i). Thus we get a contradiction with (ii). Finally, $q' = q$, and then $r = r'$, since $r - r' = m(q'-q) = 0$.

■

Corollary 4.1.5 Let $m,\ n$ be two integers such that $m \neq 0$.

$$\text{If } m \nmid n, \text{ then } \exists\, k,\ r \in \mathbb{Z}:\ n = mk + r,\ \ 1 \le r < m.$$

■ **Example 4.6**

$$\forall\, n \in \mathbb{Z},\ \exists\, k,\ r \in \mathbb{Z}:\ n = 2k + r,\ \ r \in \{0,\ 1\},$$

$$\forall\, n \in \mathbb{Z},\ \exists\, k,\ r \in \mathbb{Z}:\ n = 3k + r,\ \ r \in \{0,\ 1,\ 2\},$$

$$\forall\, n \in \mathbb{Z},\ \exists\, k,\ r \in \mathbb{Z}:\ n = 5k + r,\ \ r \in \{0,\ 1,\ 2,\ 3,\ 4\},$$

$$\forall\, n \in \mathbb{Z},\ \exists\, k,\ r \in \mathbb{Z}:\ n = 21k + r,\ \ r \in \{0,\ 1,\ 2,\ 3,\ ...,\ 20\},$$

$$\forall\, n \in \mathbb{Z},\ \exists\, k,\ r \in \mathbb{Z}:\ n = mk + r,\ \ r \in \{0,\ 1,\ 2,\ ...,\ m-1\},$$

■

■ **Example 4.7** 3 *does not divide* n implies that n cannot be expressed as $3k$ for $k \in \mathbb{Z}$. $3 \nmid n$ implies that $\exists\, k,\ r \in \mathbb{Z},\ ;\ n = 3k + r,\ \ r \neq 0,\ r \in \{1,\ 2\}$

■

■ **Example 4.8** Prove that if 2 divides n then 2 also divides n^2, where $n \in \mathbb{Z}$.

Let us assume that 2 divides n. Then $n = 2k$ for some $k \in \mathbb{Z}$. Using the direct-proof method, we will show that 2 divides n^2.

Consider,

$$\begin{aligned} n^2 &= (2k)^2, \\ &= 4k^2, \\ &= 2(2k^2), \\ &= 2k', \end{aligned}$$

where $k' = 2k^2 \in Z$.
Thus 2 divides n^2 whenever 2 divides n.

■

R Obviously, the proof presented in the preceding example remains valid to prove that for any integer n, if n is even, then n^2 is also even.

■ **Example 4.9** For any integer n, if n is odd, then n^2 is also odd.

Assume that n is odd. Then there exists $k \in \mathbb{Z}$ such that $n = 2k + 1$. Using the direct-proof method, we will show that n^2 is also odd.

Consider,

$$\begin{aligned} n^2 &= (2k+1)^2, \\ &= 4k^2+4k+1, \\ &= 2(2k^2+2k)+1, \\ &= 2k', \end{aligned}$$

where $k' = (2k^2+2k) \in Z$. Thus n^2 is odd whenever n is odd. ■

■ **Example 4.10** If m and n are odd integers, then $mk + nk$ is even for every integer k.

Let m and n be odd integers. Then for some $a, b \in \mathbb{Z}$, we have

$$m = 2a+1 \text{ and } n = 2b+1.$$

Now, by direct-proof method, we will show that $mk + nk$ is even for every integer k.

Consider,

$$\begin{aligned} mk + nk &= (2a+1)k + (2b+1)k, \\ &= 2ak+2bk+2k, \\ &= 2(a+b+1)k, \\ &= 2k', \end{aligned}$$

where $k' = (a+b+1)k \in Z$. Thus $mk + nk$ is even for every integer k if m and n are both odd integers. ■

■ **Example 4.11** If m and n are perfect squares, then their product mn is also a perfect square.

Let us assume that m and n are both perfect squares, then there exist $a,\ b \in \mathbb{Z}$ such that

$$m = a^2 \text{ and } n = b^2.$$

Now, by direct proof method, we wish to show that mn is also a perfect square.

Consider,

$$\begin{aligned} mn &= a^2 b^2, \\ &= (ab)^2, \\ &= c^2 \quad \text{where } c = ab \in \mathbb{Z}. \end{aligned}$$

Thus mn is a perfect square. ■

Theorem 4.1.6 *The product of any three consecutive integers is always divisible by* 6. *That is,*

$$\forall n \in \mathbb{Z}, \exists p \in \mathbb{Z} : n(n+1)(n+2) = 6p.$$

Proof. Since $n(n+1)(n+2)$ is a product of three consecutive integers of which two integers are consecutive evens, it is sufficient to prove that $n(n+1)(n+2)$ is divisible by 3. So we need to consider three possible cases of remainders: 0, 1, 2.

That is, there exist $k,\ r \in \mathbb{Z}$ such that $n = 3k+r$, where $r \in \{0,1,2\}$:

(i) If $r = 0$, then $n = 3k$.

$$n(n+1)(n+2) = 3k(n+1)(n+2) = 3p, \text{ where } p = k(n+1)(n+2) \in \mathbb{Z}.$$

(ii) If $r = 1$, then $n = 3k+1$.

$$n(n+1)(n+2) = n(n+1)(3k+3) = 3p, \text{ where } p = n(n+1)(k+1) \in \mathbb{Z}.$$

(iii) If $r = 2$, then $n = 3k+2$.

$$n(n+1)(n+2) = n(3k+3)(n+2) = 3p, \text{ where } p = n(k+1)(n+2) \in \mathbb{Z}.$$

Thus $\forall\, n \in \mathbb{Z},\quad n(n+1)(n+2)\;$ is a multiple of 6. ■

R The preceding result can be proved by considering all the possible 6 cases. In other words, to consider $n = 6k + r,\;\; r \in \{0,1,2,3,4,5\},\; k \in \mathbb{Z}$.

1. $r = 0 \Longrightarrow n = 6k$, then
$$n(n+1)(n+2) = 6k\,(n+1)(n+2).$$
2. $r = 1 \Longrightarrow n = 6k+1$, then
$$n(n+1)(n+2) = n(6k+2)(6k+3) = 6n(3k+1)(2k+1).$$
3. $r = 2 \Longrightarrow n = 6k+2 = 2(3k+1)$, then
$$n(n+1)(n+2) = 2(3k+1)\,3(2k+1)(n+2) = 6(3k+1)(2k+1)(n+2).$$
4. $r = 3 \Longrightarrow n = 6k+3 = 3(2k+1)$, then
$$n(n+1)(n+2) = 3(2k+1)\,2(3k+2)(n+2) = 6(2k+1)(3k+2)(n+2).$$
5. $r = 4 \Longrightarrow n = 6k+4$, then
$$n(n+1)(n+2) = 6n(n+1)(k+1).$$
6. $r = 5 \Longrightarrow n = 6k+5$, then
$$n(n+1)(n+2) = 6n(k+1)(n+2).$$

Thus $\forall\, n \in \mathbb{Z},\quad n(n+1)(n+2)\;$ is a multiple of 6.

■ **Example 4.12** Prove that
$$\forall\, n \in \mathbb{Z},\quad 6 \,\big|\, n(n+1)(2n+1).$$

We present here an elegant idea based on the following simple fact

$$2n+1 \;=\; (n+2)\;+\;(n-1) \;\Longrightarrow n(n+1)(2n+1) = \underbrace{n(n+1)(n+2)}_{6k} + \underbrace{(n-1)n(n+1)}_{6k'},\;\; k,k' \in \mathbb{Z}.$$

Thus our product is the sum of two multiples of 6 that is also a multiple of 6. Note that, one may apply the same method as in the proof of Theorem 4.1.6. ■

Theorem 4.1.7 Any prime number $p > 3$ can be expressed as $6k \pm 1$ for some integer k.

Proof. Any integer can be expressed as

$$6k,\quad 6k+1,\quad 6k+2,\quad 6k+3,\quad 6k+4,\text{ and } 6k+5,$$

but four cases are not prime:

$$6k = 3(2k) = 2(3k),\;\; 6k+1,\;\; 6k+2 = 2(3k+1),\;\; 6k+3 = 3(2k+1),\;\; 6k+4 = 2(3k+2),\text{ and } 6k+5.$$

Thus any prime can be expressed as $6k+1$ and $6k+5 = 6k' - 1,\;\; (k' = k+1)$. ■

■ **Example 4.13** Prove that

$$p \neq 3, \ \ p \text{ prime} \quad \Longrightarrow \quad 3 \mid p^2 + 2.$$

If $p = 2$, we have $p^2 + 2 = 6$, a multiple of 3. For $p > 3$, we use Theorem 4.1.7 to get

$$\exists\, k \geq 1 : \ p = 6k \pm 1 \implies 2 + p^2 = 2 + 36k^2 \pm 12k + 1 = 3(1 + 12k^2 \pm 4k) \implies 3 \mid p^2 + 2.$$

■

■ **Example 4.14** Every integer greater than 7 can be expressed as $3p + 5q$, where p and q are non-negative integers. Note that any integer can be represented in one of the following forms:

$$3m - 1, \ 3m, \ 3m + 1, \ \ m \in \mathbb{Z}.$$

Thus we use these forms to express integers greater than 7 in the desired notation. Since the represented integer is greater than 7, $m \geq 3$, and we have

$$\begin{array}{lcllcl} 3m - 1 & = & 3(m-2) & + & 5(1), \\ 3m & = & 3m & + & 5(0), \\ 3m + 1 & = & 3(m-3) & + & 5(2). \end{array}$$

Also, one may consider the other possibilities with $m \geq 2$:

$$\begin{array}{lcllcl} 5m - 2 & = & 5(m-1) & + & 3(1), \\ 5m - 1 & = & 5(m-2) & + & 3(3), \\ 5m & = & 5m & + & 3(0), \\ 5m + 1 & = & 5(m-1) & + & 3(2), \\ 5m + 2 & = & 5(m-2) & + & 3(4). \end{array}$$

■

4.1.3 Congruence Modulo

Definition 4.1.8 — congruent modulo. Two integers a and b are **congruent modulo** n, written as $a \equiv b\,(mod\ n)$ if and only if n is a **divisor of their difference**. That is, a and b are **congruent modulo** n if and only if $n \mid a - b$. The given integer n is greater than 1 and is called the **modulus**. We read $a \equiv b\,(mod\ n)$ as *a is congruent to b modulo (or mod) n*.

■ **Example 4.15** $13 \equiv 1\,(mod\ 2)$, $\ 34 \equiv 2\,(mod\ 8)$, and $2021 \equiv 21\,(mod\ 50)$. Indeed,

$$2 \big| \ 13 - 1 = 12 = 2(6), \quad 8 \big| \ 34 - 2 = 32 = 8(4), \quad 50 \big| \ 2021 - 21 = 2000 = 50(40).$$

■

It is worth observing that

$$a \equiv b\,(mod\ n) \iff n \big| (a-b) \iff n \big| (a-b) - 0 \iff a - b \ \equiv \ 0\ (mod\ n).$$

■ **Example 4.16** Show that for any integer n, we have

$$n^2 \equiv n\,(mod\ 2), \ \text{ and } \ n^3 \equiv n\,(mod\ 6).$$

■

- Since $n^2 - n = n(n-1)$ is a product of two consecutive integers, it is even, and then $2 \big| (n^2 - n)$, which is equivalent to $n^2 \equiv n\,(mod\ 2)$.
- Since $n^3 - n = n(n^2-1) = n(n-1)(n+1) = (n-1)\,n\,(n+1)$ is the product of three consecutive integers, it is a multiple of 3, and the presence of $n(n-1)$ implies that $n^3 - n$ is also a multiple of 2. Therefore, $n^3 - n$ is a multiple of 6, and we get $n^3 \equiv n\,(mod\ 6)$.

Theorem 4.1.8 Let $m,\ n$ be two integers such that $n \neq 0$; then there exists and integer r such that

$$m \equiv r\ (mod\ n), \quad \text{and } 0 \leq r < n.$$

Proof. Using Theorem 4.1.4, there exist $r,\ q\ \in \mathbb{Z},\quad 0 \leq r < n$:

$$\begin{aligned} m = n\,q + r \quad &\Longrightarrow \quad m - r = n\,q \\ &\Longrightarrow \quad n \big| (m-r) \\ &\Longrightarrow \quad m \equiv r\ (mod\ n), \quad 0 \leq r < n. \end{aligned}$$

■

Corollary 4.1.9 Let $m,\ n$ be two integers such that $n > 0$; then

$$n \text{ divides } m \iff m \equiv 0\ (mod\ n).$$

Proof.

$$n \text{ divides } m \iff n \text{ divides } (m-0) \iff m \equiv 0\,(mod\ n).$$

■

Corollary 4.1.10 Let $m,\ n$ be two integers such that $n > 0$, then

$$n \text{ does not divide } m \iff m \equiv r\ (mod\ n), \quad \text{for some integer } r \in \{1, 2, ..., m-1\}.$$

Proof. Directly using Theorem 4.1.8. ■

■ **Example 4.17**

$$\forall\, n \in \mathbb{Z}, \quad \text{either } n \equiv 0\ (mod\ 2) \text{ or } n \equiv 1\ (mod\ 2),$$
$$\forall\, n \in \mathbb{Z}, \quad \text{either } n \equiv 0\ (mod\ 3) \text{ or } n \equiv 1\ (mod\ 3) \text{ or } n \equiv 2\ (mod\ 3),$$
$$\forall\, n \in \mathbb{Z},\ n \equiv r\ (mod\ 5), \quad r \in \{0,\ 1,\ 2,\ 3,\ 4\},$$
$$\forall\, n \in \mathbb{Z},\ n \equiv r\ (mod\ 21), \quad r \in \{0,\ 1,\ 2,\ 3,\ ...,\ 20\},$$

■

■ **Example 4.18** 3 *does not divide* n implies that n cannot be expressed as $3k$ for $k \in \mathbb{Z}$.

$$3 \nmid n \iff n \equiv 1\ (mod\ 3) \text{ or } n \equiv 2\ (mod\ 3).$$

■

Theorem 4.1.11 Given the integers $a, b,\ c,\ d$ and $n \neq 0$, we have the following

1. $a \equiv a\ (mod\ n)$.
2. $a \equiv b\ (mod\ n) \iff b \equiv a\ (mod\ n)$.

3. $\begin{cases} a \equiv b\,(mod\ n) \\ c \equiv d\,(mod\ n) \end{cases} \Longrightarrow \begin{cases} a + c \equiv b + d\,(mod\ n) \\ a - c \equiv b - d\,(mod\ n). \end{cases}$

4. $\begin{cases} a \equiv b\,(mod\ n) \\ c \equiv d\,(mod\ n) \end{cases} \Longrightarrow ac \equiv bd\,(mod\ n).$

5. $a \equiv b\,(mod\ n) \Longleftrightarrow ac \equiv bc\,(mod\ nc),\ c \neq 0.$

6. $a \equiv b\,(mod\ n) \Longrightarrow ac \equiv bc\,(mod\ n),\ c \neq 0.$

7. $a \equiv b\,(mod\ n) \Longrightarrow a^k \equiv b^k\,(mod\ n),$ for all positive integer, k.

Proof. We proceed by direct proofs. Using the definition of congruence, we get:

1. $a = a+0 = a+k0,\ k \in \mathbb{Z} \Longrightarrow a \equiv a\,(mod\ n)$. Also, since 0 is a multiple of any integer, $n \mid (a-a)$, and therefore, $a \equiv a\,(mod\ n)$.

2. Let $a \equiv b\,(mod\ n)$, then

$$\begin{aligned} a \equiv b\,(mod\ n) &\Longrightarrow \exists\, k \in \mathbb{Z}: a = b + nk, \\ &\Longrightarrow \exists\, k \in \mathbb{Z}: b = a - nk, \\ &\Longrightarrow \exists\, k \in \mathbb{Z}: b = a + n(-k), \\ &\Longrightarrow \exists\, k' \in \mathbb{Z}: b = a + nk',\quad k' = -k \in \mathbb{Z}, \\ &\Longrightarrow b \equiv a\,(mod\ n). \end{aligned}$$

The second implication $b \equiv a\,(mod\ n) \Longrightarrow a \equiv b\,(mod\ n)$ can be done similarly by switching a and b in the preceding proof.

3. Consider

$$\begin{aligned} \begin{cases} a \equiv b\,(mod\ n), \\ c \equiv d\,(mod\ n), \end{cases} &\Longrightarrow \begin{cases} \exists\, k \in \mathbb{Z}: a = b + kn, \\ \exists\, k' \in \mathbb{Z}: c = d + k'n, \end{cases} \\ &\Longrightarrow \begin{cases} a + c = b + kn + d + nk', \\ a - c = b + kn - d - nk', \end{cases} \\ &\Longrightarrow \begin{cases} a + c = b + d + n(k + k'), \\ a - c = b - d + n(k - k'), \end{cases} \\ &\Longrightarrow \begin{cases} \exists\, p \in \mathbb{Z}: a + c = b + d + np, \quad p = k + k', \\ \exists\, q \in \mathbb{Z}: a - c = b - d + nq, \quad q = k - k', \end{cases} \\ &\Longrightarrow \begin{cases} a + c \equiv b + d\,(mod\ n), \\ a - c \equiv b - d\,(mod\ n). \end{cases} \end{aligned}$$

4. Consider

$$\begin{aligned} \begin{cases} a \equiv b\,(mod\ n) \\ c \equiv d\,(mod\ n) \end{cases} &\Longrightarrow \begin{cases} \exists\, k \in \mathbb{Z}: a = b + nk, \\ \exists\, k' \in \mathbb{Z}: c = d + nk', \end{cases} \\ &\Longrightarrow \exists\, k, k' \in \mathbb{Z}: ac = (b + nk)(d + nk'), \\ &\Longrightarrow \exists\, k, k' \in \mathbb{Z}: ac = bd + bnk' + nkd + n^2kk', \\ &\Longrightarrow \exists\, k, k' \in \mathbb{Z}: ac = bd + (bk' + kd + nkk')n, \\ &\Longrightarrow \exists\, k'' \in \mathbb{Z}: ac = bd + nk'', \quad k'' = bk' + kd + nkk' \in \mathbb{Z}, \\ &\Longrightarrow ac \equiv bd\,(mod\ n). \end{aligned}$$

5. (i) For $c \neq 0,\ a \equiv b\,(mod\ n) \Longrightarrow ac \equiv bc\,(mod\ nc)$.

$$\begin{aligned} a \equiv b\,(mod\ n) &\Longrightarrow \exists k \in \mathbb{Z}: a = b + nk, \\ &\Longrightarrow \exists k \in \mathbb{Z}: ac = bc + nkc,\ c \neq 0, \\ &\Longrightarrow \exists k \in \mathbb{Z}: ac = bc + (nc)k, \\ a \equiv b\,(mod\ n) &\Longrightarrow ac \equiv bc\,(mod\ nc),\ c \neq 0. \end{aligned}$$

(ii) If $c \neq 0$, $ac \equiv bc\,(mod\,nc) \Longrightarrow a \equiv b\,(mod\,n)$.

$$\begin{aligned} ac \equiv bc\,(mod\,nc) &\Longrightarrow \exists k \in \mathbb{Z}:\ ac = bc + (nc)k,\\ &\Longrightarrow \exists k \in \mathbb{Z}:\ ac = c(b+nk),\\ &\Longrightarrow \exists k \in \mathbb{Z}:\ a = b+nk,\\ &\Longrightarrow a \equiv b\,(mod\,n). \end{aligned}$$

6. For $c \neq 0$, $a \equiv b\,(mod\,n) \Longrightarrow ac \equiv bc\,(mod\,n)$.

$$\begin{aligned} a \equiv b\,(mod\,n) &\Longrightarrow \exists k \in \mathbb{Z}:\ a = b + nk,\\ &\Longrightarrow \exists k \in \mathbb{Z}:\ ac = bc + nkc,\ c \neq 0,\\ &\Longrightarrow \exists k \in \mathbb{Z}:\ ac = bc + (n)kc,\\ a \equiv b\,(mod\,n) &\Longrightarrow ac \equiv bc\,(mod\,n),\ c \neq 0. \end{aligned}$$

7. See Question (48).

■

(R) Students should keep in mind and use the following results:

1. $18 \equiv 10\,(mod\ 8) \equiv 10 - 8\,(mod\ 8) \equiv 2\,(mod\ 8)$.
2. $8n^2 \equiv [8n^2 - 3(3n^2)]\,(mod\ 3) \equiv -n^2\,(mod\ 3)$, and $-5n^4 \equiv -5n^4 + 3(2n^4)\,(mod\ 3) \equiv n^4\,(mod\ 3)$.
3. Find $x > 0$ such that $63 \equiv x\,(mod\ 12)$. It suffices to subtract a multiple of 12 from 63. In other words, $63 \equiv 63 - 12(2)\,(mod\ 12) \equiv 39\,(mod\ 12) \equiv 39 - 3(12) \equiv 3\,(mod\ 12)$, or, directly, $63 \equiv 63 - 12(5)\,(mod\ 12) \equiv 63 - 60\,(mod\ 12) \equiv 3\,(mod\ 12)$. Of course, students know that x is the remainder of the Euclidean division of 63 by 12. Thus $x = 3$.

4.1.4 More Examples on Congruence Modulo

■ **Example 4.19** Prove that for all integer n, we have $n^3 - n \equiv 0\,(mod\,3)$.
For $n \in Z$, we have $n \equiv r\,(mod\,3)$, where $r \in \{0, 1, 2\}$.
We consider the following three cases:

Case 1. Let $n \equiv 0\,(mod\,3) \Longrightarrow \{\ n^3 \equiv 0\,(mod\,3) - n \equiv 0\,(mod\,3) \Longrightarrow n^3 - n \equiv 0\,(mod\,3)$.

Case 2. Let $n \equiv 1\,(mod\,3) \Longrightarrow \begin{cases} n^3 \equiv 1\,(mod\,3) \\ -n \equiv -1\,(mod\,3) \end{cases} \Longrightarrow n^3 - n \equiv 0\,(mod\,3)$.

Case 3. Let $n \equiv 2\,(mod\,3) \Longrightarrow \{\ n^3 \equiv 8\,(mod\,3) - n \equiv -2\,(mod\,3) \Longrightarrow n^3 - n \equiv 6\,(mod\,3) \equiv 0\,(mod\,3)$.

■

■ **Example 4.20** Show that $-5n^4 + 8n^2 \equiv 0\,(mod\ 3)$. ■

Let us denote $A(n) = -5n^4 + 8n^2$: *Idea 1.* Using congruence modulo 3

- $n \equiv 0\,(mod\ 3)$, then

$$n^2 \equiv 0\,(mod\ 3),\quad 8n^2 \equiv 0\,(mod\ 3),\quad n^4 \equiv 0\,(mod\ 3),\quad -5n^4 \equiv 0\,(mod\ 3),$$

which imply that $A(n) \equiv 0\,(mod\ 3)$.

- $n \equiv 1\,(mod\ 3)$ then

$$n^2 \equiv 1\,(mod\ 3),\quad 8n^2 = 8\,(mod\ 3),\quad n^4 \equiv 1\,(mod\ 3),\quad -5n^4 \equiv -5\,(mod\ 3),$$

which imply that $A(n) \equiv -5 + 8 \equiv 3\,(mod\ 3) \equiv 0\,(mod\ 3)$.

- $n \equiv 2\,(mod\ 3)$, then

$$n^2 \equiv 4\,(mod\ 3) \equiv 1\,(mod\ 3),\ \ 8n^2 \equiv 8\,(mod\ 3) \equiv 8-2(3) = 2\,(mod\ 3),$$

  and

$$n^4 \equiv 16\,(mod\ 3) \equiv 16-3(5) = 1\,(mod\ 3),\ \ -5n^4 \equiv -5\,(mod\ 3),$$

  which imply that $A(n) \equiv -5+2 \equiv -3\,(mod\ 3) \equiv 0\,(mod\ 3)$.

Idea 2. Let us rewrite $A(n)$, and then use the congruence modulo 3. Clearly

$$A(n) = -5n^4+8n^2 \ = \ -3n^4-2n^4+2n^2+3(2n^2) \ = \ 2n^2-2n^4+3(2n^2-n^4),$$

which implies that it is sufficient to prove that $2n^2-2n^4 = 0\,(mod\ 3)$.

- $n \equiv 0\,(mod\ 3)$, then

$$n^2 \equiv 0\,(mod\ 3),\ \ 2n^2 \equiv 0\,(mod\ 3),\ \ n^4 \equiv 0\,(mod\ 3),\ \ -2n^4 \equiv 0\,(mod\ 3),$$

  which imply that $2n^2-2n^4\,(mod\ 3) \equiv 0\,(mod\ 3)$.
- $n \equiv 1\,(mod\ 3)$, then

$$n^2 \equiv 1\,(mod\ 3),\ \ 2n^2 = 2\,(mod\ 3),\ \ n^4 \equiv 1\,(mod\ 3),\ \ -2n^4 \equiv -2\,(mod\ 3),$$

  which imply that $2n^2-2n^4 \equiv 2-2 \equiv 0\,(mod\ 3)$.
- $n \equiv 2\,(mod\ 3)$, then

$$n^2 \equiv 4\,(mod\ 3) \equiv 1\,(mod\ 3),\ \ 2n^2 \equiv 2\,(mod\ 3),$$

  and

$$n^4 \equiv 16\,(mod\ 3) \equiv 16-3(5) = 1\,(mod\ 3),\ \ -2n^4 \equiv -2\,(mod\ 3),$$

  which imply that $2n^2-2n^4 \equiv 2-2 \equiv 0\,(mod\ 3)$.

Idea 3. As in Idea 2,

$$\begin{aligned} A(n) &= -5n^4+8n^2 = -3n^4-2n^4+2n^2+3(2n^2) = 2n^2-2n^4+3(2n^2-n^4), \\ &= 2n^2(1-n^2)+3(2n^2-n^4) = 2n^2(1-n)(1+n)+3(2n^2-n^4), \\ &= -2n\underbrace{(n-1)n(n+1)}_{3k}+3(2n^2-n^4), \\ &= 3\left(-2nk+2n^2-n^4\right), \\ &= 3k',\ k' \in \mathbb{Z}, \end{aligned}$$

and, finally, $A(n) \equiv 0\,(mod\ 3)$.

■ **Example 4.21** Show that $n+3n^2+2n^3 \equiv 0\,(mod\ 6)$.

We will give many ideas to solve this question. Students are requested to know that

$$n(n\pm 1) \equiv 0\,(mod\ 2) \text{ and } n(n+1)(n+2) \equiv 0\,(mod\ 6).$$

In other words, the product of two consecutive integers is a multiple of 2 and the product of any three consecutive integers is a multiple of 6.

1. By rewriting $n = -2n+3n$, we get

$$n+3n^2+2n^3 \ = \ 2n^3-2n+3n^2+3n \ = \ 2n(n^2-1)+3(n^2+n) = 2\underbrace{(n-1)n(n+1)}_{\equiv 0(mod\ 6)}+\underbrace{3n(n+1)}_{\equiv 0(mod\ 6)}.$$

   Thus $n+3n^2+2n^3 \ \equiv \ 0\,(mod\ 6)$.

2. After factoring $n+3n^2+2n^3 = n(2n^2+3n+1) = n(2n+1)(n+1)$, we rewrite our expression as follows:

$$n(n+1)(2n+1) = n(n+1)(2n+4-3) = n(n+1)(2n+4) - 3n(n+1),$$

and then $n(n+1)(2n+1) = \underbrace{2n(n+1)(n+2)}_{\equiv 0 (mod\ 6)} - \underbrace{3n(n+1)}_{\equiv 0 (mod\ 6)}$

i.e.,

$$n+3n^2+2n^3 \equiv 0\,(mod\ 6).$$

3. It is based on the well-known sum of the first n squares (see Example 5.43 or Example 4.78).

$$\mathbb{N} \ni 1+2^2+3^2+\ldots+n^2 = \frac{n(n+1)(2n+1)}{6} \Longrightarrow n(n+1)(2n+1) = 6m, \quad m \in \mathbb{N}$$

Therefore, $6\,\big|\,n(n+1)(2n+1)$, and we have $n+3n^2+2n^3 = n(n+1)(2n+1) \equiv 0\,(mod\ 6)$.

■

■ **Example 4.22** Find the number of integers between 11 and 2022 for which $1+2^n$ is a multiple of 3.
Let us use the Congruence modulo 3. Since $2^1 \equiv -1(mod\ 3) \Longrightarrow 2^n \equiv (-1)^n(mod\ 3) \Longrightarrow 2^n+1 \equiv (-1)^n+1(mod\ 3)$. Thus 2^n+1 is a multiple of 3 only for odd n between 11 and 2022. ■

■ **Example 4.23** If today is Friday, which day of the week will it be 80 days later? ■

Since there are 7 days in a week, the answer only depends on the remainder of 80 when divided by 7. In fact, $7\times k$ days later, it will still be a Friday. Now $80 \equiv 3 \pmod 7$, so 80 days later, it will be Monday.

■ **Example 4.24** Find $x \le 0$: $19^{537} \equiv x\,(mod\ 7)$.
The question is about finding the remainder when dividing 19^{537} by 7. In other words, determine x such that $19^{537} \equiv x(mod\ 7)$. To this end, let us deal first by some small powers of 19

$$19 \equiv -2\,(mod\ 7) \Longrightarrow 19^3 \equiv (-2)^3 \equiv -8 \equiv -1\,(mod\ 7);$$

Thus we obtain: $19^6 \equiv (-1)^2 \equiv 1\,(mod\ 7)$. On the other hand, $537 = 6(90)-3 = 6(89)+3$, which gives us $19^{537} = 19^{6(89)+3} = 19^3 \times \left(19^6\right)^{89}$. Now, we need to evaluate the following:

$$\begin{cases} 19^3 \equiv (-2)^3\,(mod\ 7) \equiv -8 \equiv -1\,(mod\ 7), \\ \left(19^6\right)^{89} \equiv 1^{89}\,(mod\ 7) \equiv 1\,(mod\ 7). \end{cases}$$

Thus we obtain: $19^{537} \equiv -1\times 1 \equiv -1\,(mod\ 7)$ and, therefore, $x=-1$. ■

■ **Example 4.25** Find $x \ge 0$ such that $13^{1970} \equiv x\,(mod\ 9)$.
The question is about finding the remainder when dividing 13^{1970} by 9. To this end, let us, explore some small powers of 13

$$13 \equiv 4\,(mod\ 9) \Longrightarrow 13^2 \equiv 4^2\,(mod\ 9) \equiv 7\,(mod\ 9) \equiv -2\,(mod\ 9);$$

Thus we obtain $13^6 = (13^2)^3 \equiv (-2)^3\,(mod\ 9) \equiv 1\,(mod\ 9)$. On the other hand, $1970 = 6(328)+2$, which gives us $13^{1970} = 13^{6(328)+2} = 13^2 \times \left(13^6\right)^{328}$. Now, since $13^2 \equiv -2\,(mod\ 9)$, we obtain

$$\left(13^6\right)^{328} \equiv 1^{328}\,(mod\ 9) \equiv 1\,(mod\ 9) \Longrightarrow 13^{1970} \equiv (-2)(1)\,(mod\ 9) = -2\,(mod\ 9).$$

Thus we obtain $13^{1970} \equiv 7\,(mod\ 9)$ and, therefore, $x = 7$. ■

■ **Example 4.26** Show that if $a \equiv b\,(mod\ 2)$, then a and b share the same parity. ■

By definition, $2|(a-b) \Longrightarrow (a-b)$ is even, and then a and b should be of the same parity (both odd or both even).

■ **Example 4.27** Find $x \geq 0{:}17^{342} \equiv x\,(mod\ 5)$. ■

Since $17 \equiv 2\,(mod\ 5) \Longrightarrow 17^2 \equiv 4\,(mod\ 5) \equiv -1(mod\ 5) \Longrightarrow 17^4 \equiv 1(mod\ 5)$. Thus $17^{4k} \equiv 1(mod\ 5)$. Now, since $342 = 4(85)+2$, we get $17^{342} = 17^2 \times 17^{4(85)} \equiv 17^2 \times 1\,(mod\ 5) \equiv 4\,(mod\ 5)$ directly. Therefore, $x = 4$.

■ **Example 4.28** Find $x \geq 0{:}5^{903} \equiv x\,(mod\ 60)$. ■

We start with small exponents of 5:

$$5^3 = 125 = 120+5 = 2(60)+5 \Longrightarrow 5^3 \equiv 5\,(mod\ 60) \Longrightarrow 5^{3k} \equiv 5^k\,(mod\ 60).$$

Thus

$$\begin{aligned} 5^{903} &= 5^{3(301)} \equiv 5^{301} = 5 \times 5^{3(100)} \equiv 5 \times 5^{100} = 5^2 \times 5^{99} \equiv 5^2 \times 5^{3(33)}\,(mod\ 60), \\ &\equiv 5^2 \times 5^{33} = 5^2 \times 5^{3(11)} \equiv 5^2 \times 5^{11} = 5 \times 5^{3(4)} \equiv 5 \times 5^4\,(mod\ 60), \\ &= 5^2 \times 5^3 = 5^2 \times 5^1 \equiv 5\,(mod\ 60). \end{aligned}$$

Therefore, $x = 5$.

4.1.5 Additional Examples of Direct Proofs

■ **Example 4.29** Let x, y, and z be nonzero real numbers such that xy, yz, and $zx \in \mathbb{Q}$.

(i) Prove that $x^2+y^2+z^2 \in \mathbb{Q}$.

(ii) Furthermore we assume that $x^3+y^3+z^3 \in \mathbb{Q}\setminus\{0\}$, prove that x, y, and z are also rational numbers.

Assume that xy, yz, $zx \in \mathbb{Q}$, and $xyz \neq 0$.

(i) Let us use this simple idea:

$$x^2 = \frac{(xy)(zx)}{(yz)} \in \mathbb{Q},\ y^2 = \frac{(yz)(xy)}{(xz)} \in \mathbb{Q},\ z^2 = \frac{(zx)(yz)}{(xy)} \in \mathbb{Q},$$

Then $x^2+y^2+z^2 \in \mathbb{Q}$.

(ii) Now, we assume the following:

$$xy,\ yz,\ zx,\ x^2,\ y^2,\ z^2,\ (x^3+y^3+z^3) \in \mathbb{Q}.$$

Thus we can affirm that

$$x\,(x^3+y^3+z^3) = x^4+xy^3+xz^3 = (x^2)^2+(xy)(y^2)+(xz)(z^2) = RHS \in \mathbb{Q}$$

because x^2, (xy), y^2, (xz), and z^2 are all rational numbers and then

$$x = \frac{RHS}{x^3+y^3+z^3} \in \mathbb{Q}.$$

It is similarly for y and z.

■

■ **Example 4.30** Prove that for any real number x, we have

$$\boxed{x \leq |x|}$$

Using a direct proof, we consider two cases:

- $x \geq 0$, then $x = |x| \leq |x|$, which is always true.
- $x < 0$, then $x < 0 < |x|$, which is always true.

Therefore, $x \leq |x|$ for all $x \in \mathbb{R}$. ■

■ **Example 4.31** For any real numbers x and y, we have

$$\boxed{|x+y| \leq |x|+|y|.}$$

Using a direct proof, consider

$$\begin{aligned} |x+y|^2 &= |(x+y)^2|, \\ &= (x+y)^2 = x^2+y^2+2xy, \\ &= |x^2|+|y^2|+2xy, \\ &\overbrace{\leq}^{xy\leq|xy|} |x^2|+|y^2|+2|xy|, \quad (\textit{Example } 4.30) \\ &\leq |x|^2+|y|^2+2|x||y|, \\ &\leq (|x|+|y|)^2, \\ |x+y| &\leq |x|+|y|, \end{aligned}$$

■

■ **Example 4.32** For any real numbers x and y, we have

$$\boxed{xy > 0 \Longrightarrow |x+y| = |x|+|y|.}$$

Sometimes, we may start with brief exploration of the result to be proved. For example $|(-2)+(-3)| = |-2|+|-3|$, and here, $(-2)(-3) > 0$. Now, to prove the equality, it is sufficient to discuss only two cases instead of four, since x and y are symmetric and play the same role in the equation:

1. $x > 0,\ y > 0 \Longrightarrow x+y > 0 \Longrightarrow |x+y| = x+y = |x|+|y|$.
2. $x < 0,\ y < 0 \Longrightarrow x+y < 0 \Longrightarrow |x+y| = -(x+y) = (-x)+(-y) = |x|+|y|$.

■

(R) By replacing y with $(-y)$ in the triangle inequality, we obtain

$$\boxed{|x-y| \leq |x|+|y|.}$$

■ **Example 4.33** For any real number $x > 0$, we have

$$\boxed{x + \frac{1}{x} \geq 2.}$$

Let $E(x) = x + \frac{1}{x} - 2$, and we would like to show that $E(x) \geq 0$ for any positive x. It is easy to see that we have for any $x \neq 0$:

$$xE(x) = x^2+1-2x = x^2-2x+1 = (x-1)^2 \geq 0 \Longrightarrow xE(x) \geq 0,$$

$$x > 0 \Longrightarrow E(x) \geq 0.$$

■

■ **Example 4.34** Let $a,\ b$, and c be the side lengths of a triangle, then

$$\begin{aligned} a &> |b-c|, \\ b &> |c-a|, \\ c &> |a-b|, \end{aligned}$$

To prove these inequalities, let us use the triangle inequalities: $\begin{cases} a+b>c, \\ b+c>a, \\ c+a>b, \end{cases}$.

Thus

$$\begin{cases} a+b>c \Longrightarrow b-c > -a, \\ c+a>b \Longrightarrow b-c < a \end{cases} \Longrightarrow -a<b-c<a \Longrightarrow |b-c|<a.$$

Similarly,

$$\begin{cases} a+b>c \Longrightarrow a-c > -b \\ b+c>a \Longrightarrow a-c < a \end{cases} \Longrightarrow -b<a-c<b \Longrightarrow |a-c|<b \Longrightarrow |c-a|<b$$

and

$$\begin{cases} c+a>b \Longrightarrow a-b > -c, \\ b+c>a \Longrightarrow a-b < c \end{cases} \Longrightarrow -c<a-b<c \Longrightarrow |a-b|<c.$$

■

■ **Example 4.35** Let $a,\ b$, and c be the side lengths of a triangle; then we have

$$\boxed{a^2+b^2+c^2 < 2\,(ab+bc+ca).}$$

Using the inequalities proved in Example 4.34, we obtain:

$$\begin{cases} a>|b-c|, \\ b>|c-a|, \\ c>|a-b| \end{cases} \Longrightarrow \begin{cases} a^2>(b-c)^2, \\ b^2>(c-a)^2, \\ c^2>(a-b)^2 \end{cases} \Longrightarrow \begin{cases} a^2>b^2-2bc+c^2, \\ b^2>c^2-2ca+a^2, \\ c^2>a^2-2ab+b^2, \end{cases}$$

and by adding all three together, we get $a^2+b^2+c^2 > 2(a^2+b^2+c^2)-2(bc+ca+ab)$; Thus we have $a^2+b^2+c^2 < 2\,(ab+bc+ca)$. ■

■ **Example 4.36** Let P be a polynomial of degree n in x with integer coefficients.
For $a,b\in\mathbb{Z}$, we have

$$(a-b)\,\Big|\,P(a)-P(b).$$

We divide $P(x)$ by $(x-b)$ through long division, and there exists a polynomial $Q(x)$ with integer coefficients and a constant[1] r satisfying $P(x) = (x-b)\,Q(x) + r$. Thus by evaluating $P(b)$, we get $P(b)=(b-b)\,Q(b)+r = r$ and then $P(x)=(x-b)Q(x)+P(b)$. Finally, since for $x=a$, we have $Q(a)\in\mathbb{Z}$, we obtain

$$P(a)=(a-b)Q(a)+P(b) \Longrightarrow P(a)-P(b) = (a-b)\,Q(a) \Longrightarrow (a-b)\,\Big|\,P(a)-P(b).$$

■

■ **Example 4.37** Show that there is no nonnegative integer n such that $\sqrt{n^2+13n+42}\in\mathbb{N}$. Since we have for any nonnegative integer n

$$n^2+12n+36 < n^2+13n+42 < n^2+14n+49 \Longrightarrow (n+6)^2 < n^2+13n+42 < (n+7)^2,$$

Thus we obtain $n+6<\sqrt{n^2+13n+42}<n+7$, which means $\sqrt{n^2+13n+42}$ is between two consecutive integers, it is not an integer. ■

[1] r is a polynomial of degree zero.

4.2 Proof by Contraposition

Definition 4.2.1 — proof by contraposition. In mathematics, **proof by contraposition** is an extremely powerful method for proving implications. Basically, the method is based on the concept of **contrapositive** of a logical statement, which is introduced in Chapter 2:

$$P \implies Q \equiv \neg Q \implies \neg P.$$

In other words, proving the implication $P \implies Q$ is the same as proving the implication $\neg Q \implies \neg P$. Using a proof by contraposition, we assume that the conclusion Q of the conditional statement $P \implies Q$ is false. Then using a direct proof, we show that the hypothesis P must also be false. Consequently, we have proved the implication $P \implies Q$.

We will now illustrate various examples of the contrapositive proof method.

■ **Example 4.38** Prove that if n is even, then n^2 is even. We wish to prove the following implication:

$$n \text{ is even} \implies n^2 \text{ is even.}$$

The contrapositive of the above statement is

$$n^2 \text{ is not even} \implies n \text{ is not even.}$$

Since

$$\begin{aligned} n^2 \text{ is odd} &\implies \exists\, k \in \mathbb{Z} : n^2 = 2k+1, \\ &\implies \exists\, k \in \mathbb{Z} : n^2 - 1 = 2k, \\ &\implies \exists\, k \in \mathbb{Z} : (n-1)(n+1) \text{ is even}, \\ &\implies (n-1) \text{ is even or } (n+1) \text{ is even}, \\ &\implies n \text{ is odd.} \end{aligned}$$

Thus equivalently we proved that

$$\text{If } n^2 \text{ is odd, then } n \text{ is also odd.}$$

This completes the proof. ■

■ **Example 4.39** Prove that if n^2 is divisible by 2, then n is also divisible by 2.
Let $n \in Z$, we wish to prove the following implication

$$n^2 \text{ is divisible by } 2 \implies n \text{ is divisible by } 2.$$

We proceed by Contraposition. Assume that

$$\begin{aligned} n \text{ is not divisible by } 2 &\implies \exists\, k \in \mathbb{Z}, : n = 2k+1, \\ &\implies \exists\, k \in \mathbb{Z}, : n^2 = (2k+1)^2, \\ &\implies \exists\, k \in \mathbb{Z}, : n^2 = 4k^2 + 4k + 1, \\ &\implies \exists\, k' \in \mathbb{Z}, : n^2 = 2(2k^2 + 2k) + 1 = 2k' + 1, \\ &\implies n^2 \text{ is not divisible by } 2. \end{aligned}$$

Thus the contraposition proves the given implication. ■

■ **Example 4.40** Prove that if n^2 is odd, then n is also odd. Let $n \in Z$, we wish to prove the following implication n^2 is odd $\implies n$ is odd. We proceed by contraposition. Assume that

$$\begin{aligned} n \text{ is not odd} &\implies \exists\, k \in \mathbb{Z}, : n = 2k, \\ &\implies \exists\, k \in \mathbb{Z}, : n^2 = 4k^2 = 2(2k^2), \\ &\implies \exists\, k' \in \mathbb{Z}, : n^2 = 2k', \\ &\implies n^2 \text{ is not odd.} \end{aligned}$$

Thus the contraposition proves the given implication. ■

■ **Example 4.41** Prove that if 4 does not divide x^2-1, then 2 does not divide x^2-1.

$$4 \nmid (x^2-1) \implies 2 \nmid (x^2-1)$$

We proceed by contraposition.

$$\begin{aligned} 2 \mid (x^2-1) &\implies \exists k \in \mathbb{Z}:\ x^2-1=2k, \\ &\implies \exists k \in \mathbb{Z}:\ x^2=2k+1, \qquad \text{(using the result in Example ??} \\ &\implies x \text{ is odd}, \\ &\implies \exists m \in \mathbb{Z}:\ x=2m+1, \\ &\implies \exists m \in \mathbb{Z}:\ x^2-1=(2m+1)^2-1=4m^2+4m+1-1, \\ &\implies \exists m \in \mathbb{Z}:\ x^2-1=4(m^2+m), \\ &\implies \exists p \in \mathbb{Z}:\ x^2-1=4p,\ \ p=m^2+m, \\ &\implies 4 \mid (x^2-1). \end{aligned}$$

■

■ **Example 4.42** Prove that if 3 divides n^2, then 3 divides n.
We will proceed by contraposition. 3 does not divide $n \implies 3$ does not divide n^2. Let $n \in \mathbb{Z}$; assume that

$$3 \text{ does not divide } n \implies \exists\, k \in \mathbb{Z} : n=3k+1 \text{ or } n=3k+2.$$

Thus let us consider both cases:

$$\begin{aligned} \exists\, k \in \mathbb{Z}, n=3k+1 \qquad &\implies n^2=(3k+1)^2, \\ &\implies n^2=9k^2+6k+1, \\ &\implies n^2=3(3k^2+2k)+1, \\ &\implies \exists\, k' \in \mathbb{Z}, : n^2=3k'+1, \\ &\implies 3 \text{ does not divide } n^2. \end{aligned}$$

Similarly,

$$\begin{aligned} \exists\, k \in \mathbb{Z}, n=3k+2 \qquad &\implies n^2=(3k+2)^2, \\ &\implies n^2=9k^2+12k+4, \\ &\implies n^2=3(3k^2+4k+1)+1, \\ &\implies \exists\, k' \in \mathbb{Z}, : n^2=3k'+1, \\ &\implies 3 \text{ does not divide } n^2. \end{aligned}$$

Thus the contraposition proves the given implication. ■

■ **Example 4.43** Prove that if $11n+8$ is an odd integer, then n is odd. We wish to prove that

$$11n+8 \text{ is odd} \implies n \text{ is odd}.$$

Assume that

$$\begin{aligned} n \text{ is not odd} &\implies \exists\, k \in \mathbb{Z}, :\ n=2k, \\ &\implies \exists\, k \in \mathbb{Z}, :\ 11n+8 = 11(2k)+8 = 2(11k+4), \\ &\implies \exists\, k' \in \mathbb{Z},\ 11n+8 = 2k', \\ &\implies 11n+8 \text{ is not odd}. \end{aligned}$$

This completes the proof by contraposition. ■

(R) Alternatively, one can solve the Example 4.43 using direct proof method. Assume that

$$\begin{aligned} N = 11n+8 \text{ is odd} &\implies N = n + 10n + 8 \text{ is odd},\\ &\implies N = n + \overbrace{2(5n+4)}^{even} \text{ is odd},\\ &\implies n = \overbrace{N}^{odd} - \overbrace{2(5n+4)}^{even},\\ &\implies n \text{ is odd.} \end{aligned}$$

■ **Example 4.44** Prove that if $x > 0$ is an irrational number, then $\sqrt{x}$ is also an irrational number. We will proceed by contraposition. Assume that

$$\begin{aligned} \sqrt{x} \text{ is not irrational} &\implies \sqrt{x} \in \mathbb{Q},\\ &\implies \exists\, a, b \in \mathbb{N},\ b \neq 0,:\ \sqrt{x} = \frac{a}{b},\\ &\implies \exists\, a, b \in \mathbb{N},\ b \neq 0,:\ x = (\sqrt{x})^2 = \frac{a^2}{b^2},\\ &\implies x \in \mathbb{Q},\\ &\implies x \text{ is not irrational.} \end{aligned}$$

Thus the contraposition proves the given implication. ■

■ **Example 4.45** Prove that $21n-9$ is even if and only if n is odd.

Proof.

- If $21n-9$ is even then n is odd. Assume that

$$\begin{aligned} n \text{ is even} &\implies \exists\, k \in \mathbb{Z}:\ n = 2k,\\ &\implies 21n - 9 = 21\,(2k) - 9,\\ &\implies 21n - 9 = 42k - 9 = 2(21k-5)+1,\ \ 21k-5 \in \mathbb{Z}\\ &\implies 21n - 9 \text{ is odd.} \end{aligned}$$

- If n is odd, then $21n-9$ is even. Assume that

$$\begin{aligned} 21n-9 \text{ is odd} &\implies \exists\, k \in \mathbb{Z}:\ 21n - 9 = 2k+1,\\ &\implies n + 20n = 2k + 10\\ &\implies n = 2(k+5-10n),\ \ k+5-10n \in \mathbb{Z}\\ &\implies n \text{ is even.} \end{aligned}$$

Thus the given implication is proved by the contrapositions.

■

(R) Alternatively, the second implication in preceding example can be done with direct proof. Let us assume that n is odd.

$$\begin{aligned} n \text{ is odd} &\implies \exists\, h \in \mathbb{Z}: n = 2h+1,\\ &\implies \exists\, h \in \mathbb{Z}: 21n = 21(2h+1),\\ &\implies \exists\, h \in \mathbb{Z}: 21n - 9 = 2(21h) + 21 - 9,\\ &\implies \exists\, m \in \mathbb{Z}: 21n - 9 = 2m,\ \ m = 21h + 6 \in \mathbb{Z},\\ &\implies 21n - 9 \text{ is even.} \end{aligned}$$

■ **Example 4.46** Let m, n be integers. Prove that mn is even if and only if m is even or n is even.

The given biconditional has the same equivalence to the conjunction as the following two implications. That is,

- mn is even $\implies$ m is even or n is even. We will proceed by contrapositive. The negation of m is even or n is even is

$$\begin{aligned} m \text{ is odd } \textbf{and } n \text{ is odd} &\implies \exists\, k,\ k' \in \mathbb{Z}:\ m = 2k+1 \text{ and } n = 2k'+1 \\ &\implies \exists\, k,\ k' \in \mathbb{Z}: mn\ =\ (2k+1)(2k'+1), \\ &\implies \exists\, k,\ k' \in \mathbb{Z}: mn\ =\ 4kk'+2k+2k'+1, \\ &\implies \exists\, k,\ k' \in \mathbb{Z}: mn\ =\ 2\,(2kk'+k+k')+1, \\ &\implies \exists\, h = 2kk'+k+k' \in \mathbb{Z}: mn\ =\ 2h+1, \\ &\implies mn \text{ is odd.} \end{aligned}$$

- m is even or n is even $\implies$ mn is even. We will proceed by direct proof considering the two possible cases:
  (i) m is even, so $\exists\, k \in \mathbb{Z}$: $m = 2k$, then $mn = (2k)n\ =\ 2\,(km)$ and mn is even.
  (ii) n is even, so $\exists\, k \in \mathbb{Z}$: $n = 2k$, then $mn = m(2k)\ =\ 2\,(mk)$ and mn is even.
  Thus we have proven that mn is even if and only if m is even or n is even.

■

(R) After this example, we may extract the following results:

$$\boxed{mn \text{ even} \implies m \text{ is even or } n \text{ is even.}}$$

$$\boxed{mn \text{ odd} \implies m \text{ and } n \text{ cannot be even.}}$$

■ **Example 4.47** Let $x \in \mathbb{R}$ show that

$$\forall\, \varepsilon > 0,\ |x| < \varepsilon \implies x = 0.$$

Let us proceed by contrapositive, where we need to prove that

$$x \neq 0 \implies \exists\, \varepsilon > 0,\ |x| \geq \varepsilon.$$

Let us consider

$$x \neq 0 \implies |x| > 0 \implies \exists\, \varepsilon = \frac{|x|}{2} > 0:\ |x| \geq \frac{|x|}{2} = \varepsilon.$$

■

4.3 Proof by Contradiction

We now introduce a technique called **proof by contradiction**. The technique is not limited to proving a conditional statement; it can be used to prove other statements as well. Suppose we want to show that a certain mathematical statement R is true. We have already seen a few techniques: direct and contrapositive proof methods that can be used to establish the truthfulness of R. Here we will use a third method that can be used to prove that R is true, *regardless of whether R is expressed in terms of an implication*. Suppose we assume R is a false statement and, from this assumption, we are able to deduce a statement that contradicts some assumptions we made in the proof or some well-known fact.

Definition 4.3.1 — proof by contradiction. In a proof by contradiction, we start with the assumption that the given statement is false. By using theorems and rules of logic, or some well-known facts, we reach a conclusion that contradicts the assumption, and hence, what we assumed is wrong, so the given statement is true.

■ **Example 4.48** For example, we wish to prove that there is no smallest positive real number.

When using proof by contradiction, we begin by assuming that the statement is false, and we attempt to show that this leads us to a contradiction. So we begin by assuming there is a smallest positive real number – say r. We aim to produce a contradiction. To that end, let us consider $r/2$. Since $r > 0$, we have $0 < r/2 < r$. This leads us to a contradiction to the assumption that r is the smallest positive real number. Therefore, the statement *there is no smallest positive real number* is true.

The remaining part of this section illustrates contradiction proof technique with a variety of examples. ■

■ **Example 4.49** Prove that $\forall\, x \in \mathbb{R} : \; x^2 - x + 1 > 0$.

We can prove this inequality directly by completing the square or through the quadratic formula, but here, we invite students to proceed with a contradiction proof. To this goal, we assume that $\exists\, x \in \mathbb{R} : \; x^2 - x + 1 \;\leq\; 0$. Then

$$0 \geq x^2 - x + 1 = \left(x - \frac{1}{2}\right)^2 + 1 - \frac{1}{4} = \left(x - \frac{1}{2}\right)^2 + \frac{3}{4} > 0,$$

which is impossible. Thus our assumption is wrong, which proves that $\forall\, x \in \mathbb{R} : \; x^2 - x + 1 > 0$. ■

■ **Example 4.50** Prove that if n is even, then n^2 is even.

By the contradiction approach, we consider the negation of the conclusion, n^2 is not even, is true, assuming that n is even. Since

$$\begin{aligned}
n^2 \text{ is odd} \quad &\Longrightarrow \quad \exists\, k \in \mathbb{Z} : n^2 = 2k + 1, \\
&\Longrightarrow \quad \exists\, k \in \mathbb{Z} : n^2 - 1 = 2k, \\
&\Longrightarrow \quad \exists\, k \in \mathbb{Z} : (n-1)(n+1) \text{ is even}, \\
&\Longrightarrow \quad (n-1) \text{ is even or } (n+1) \text{ is even}, \\
&\Longrightarrow \quad n \text{ is odd}.
\end{aligned}$$

which contradicts the hypothesis that n is even. Thus what we assumed is wrong, so n^2 is even for any even integer n. ■

(R) For the preceding example, a direct proof is more natural choice.

■ **Example 4.51** Let $n \in \mathbb{Z}$ prove that

$$n^2 \text{ is divisible by } 2 \Longrightarrow n \text{ is divisible by } 2.$$

We proceed by contradiction. Let us assume that n^2 is divisible by 2 and

$$\begin{aligned}
n \text{ is not divisible by } 2 \quad &\Longrightarrow \quad \exists\, k \in \mathbb{Z}, : n = 2k + 1, \\
&\Longrightarrow \quad \exists\, k \in \mathbb{Z}, : n^2 = (2k+1)^2, \\
&\Longrightarrow \quad \exists\, k \in \mathbb{Z}, : n^2 = 4k^2 + 4k + 1, \\
&\Longrightarrow \quad \exists\, k' \in \mathbb{Z}, : n^2 = 2(2k^2 + 2k) + 1 = 2k' + 1, \\
&\Longrightarrow \quad n^2 \text{ is not divisible by } 2.
\end{aligned}$$

which contradicts with the fact that n^2 is divisible by 2. Thus what we assumed is wrong, so n is divisible by 2. ■

■ **Example 4.52** Let $n \in \mathbb{Z}$ prove that

$$3 \text{ divides } n^2 \implies 3 \text{ divides } n.$$

We proceed by contradiction. Let us assume that 3 divides n^2 and

$$\begin{aligned} n \text{ is not divisible by } 3 &\implies \exists\, k \in \mathbb{Z},\ r \in \{1,\ 2\}: n = 3k + r, \\ &\implies n^2 = 9k^2 + 6kr + r^2, \\ &\implies n^2 = 3(3k^2 + 2kr) + r^2, \quad r^2 \in \{1,\ 4\}, \\ &\implies \exists\, k' \in \mathbb{Z},: n^2 = 3k' + 1, \\ &\implies n^2 \text{ is not divisible by } 3. \end{aligned}$$

which contradicts the hypothesis. Thus what we assumed is wrong, so 3 divides n. ■

■ **Example 4.53** Let $n \in \mathbb{Z}$ prove that

$$11n + 8 \text{ is odd} \implies n \text{ is odd.}$$

We proceed by contradiction. Let us assume that $11n + 8$ is odd and

$$\begin{aligned} n \text{ is not odd} &\implies \exists\, k \in \mathbb{Z},:\ n = 2k, \\ &\implies \exists\, k \in \mathbb{Z},:\ 11n + 8 = 11(2k) + 8 = 2(11k + 4), \\ &\implies \exists\, k' \in \mathbb{Z},\ 11n + 8 = 2k', \\ &\implies 11n + 8 \text{ is not odd.} \end{aligned}$$

which contradicts the hypothesis. Thus what we assumed is wrong, and n is odd is proved. ■

■ **Example 4.54** Prove that if $x > 0$ is an irrational number, then $\sqrt{x}$ is also an irrational number.

We proceed by contradiction. Let us assume that $x > 0$ is an irrational number and

$$\begin{aligned} \sqrt{x} \text{ is not irrational} &\implies \sqrt{x} \in \mathbb{Q}, \\ &\implies \exists\, a, b \in \mathbb{N},\ b \neq 0,:\ \sqrt{x} = \frac{a}{b}, \\ &\implies \exists\, a, b \in \mathbb{N},\ b \neq 0,:\ x = (\sqrt{x})^2 = \frac{a^2}{b^2}, \\ &\implies x \in \mathbb{Q}, \\ &\implies x \text{ is not irrational.} \end{aligned}$$

which contradicts the hypothesis. Thus what we assumed is wrong, so $\sqrt{x}$ is an irrational number. ■

■ **Example 4.55** Prove that $\sqrt{2}$ is not a rational number.

We proceed by contradiction. Let us assume that $\sqrt{2}$ is a rational number, then

$$\begin{aligned} \exists\, a,\ b \in \mathbb{Z},\ b \neq 0,\ GCD(a,b) = 1 : \sqrt{2} = \frac{a}{b} &\implies 2 = \frac{a^2}{b^2}, \\ &\implies a^2 = 2b^2, \\ &\implies a^2 \text{ is even} \quad \text{(using Example 4.51)}, \\ &\implies a \text{ is even}, \\ &\implies \exists\, k \in \mathbb{Z} :\ a = 2k, \\ &\implies (2k)^2 = a^2 = 2b^2, \\ &\implies 4k^2 = 2b^2, \\ &\implies 2k^2 = b^2, \\ &\implies b^2 \text{ is even}, \quad \text{(using Example 4.51)} \\ &\implies b \text{ is even}, \\ &\implies GCD(a,b) \geq 2. \end{aligned}$$

which contradicts the hypothesis that $GCD(a,b) = 1$. Thus $\sqrt{2}$ is not rational. ∎

■ Example 4.56 Prove that $\sqrt{3}$ is not a rational number.

We proceed by contradiction. Let us assume that $\sqrt{3}$ is a rational number. Then,

$$\begin{aligned}
\exists\, a,\ b \in \mathbb{Z},\ b \neq 0,\ GCD(a,b) = 1 : \sqrt{3} = \frac{a}{b} &\implies 3 = \frac{a^2}{b^2}, \\
&\implies a^2 = 3b^2, \\
&\implies a^2 \text{ is a multiple of } 3, \qquad \text{(using Example 4.52)} \\
&\implies a \text{ is is a multiple of } 3, \\
&\implies \exists\, k \in \mathbb{Z} :\ a = 3k, \\
&\implies (3k)^2 = a^2 = 3b^2, \\
&\implies 9k^2 = 3b^2, \\
&\implies 3k^2 = b^2, \\
&\implies b^2 \text{ is a multiple of } 3, \qquad \text{(using Example 4.52)} \\
&\implies b \text{ is a multiple of } 3, \\
&\implies GCD(a,b) \geq 3.
\end{aligned}$$

which contradicts the hypothesis that $GCD(a,b) = 1$. Thus $\sqrt{3}$ is not rational. ∎

(R) Addition (or subtraction) of two irrational numbers may not be an irrational number. For instance,

$$\sqrt{2}+3 \text{ and } 1-\sqrt{2} \notin \mathbb{Q} \quad \text{but} \quad (\sqrt{2}+3)+(1-\sqrt{2}) = 4 \in \mathbb{Q}.$$

■ Example 4.57 Prove that $\sqrt{2}+\sqrt{3}+\sqrt{5}$ is not rational.
Let $t = \sqrt{2}+\sqrt{3}+\sqrt{5}$. We proceed by contradiction, assuming that $t \in \mathbb{Q}$. Consider the following binomial expansion of $(t-\sqrt{2})^2$:

$$\begin{aligned}
(t-\sqrt{2})^2 = t^2 - 2t\sqrt{2} + 2 &= (\sqrt{3}+\sqrt{5})^2, \\
&= 3+5+2\sqrt{3\times 5}, \\
&= 8+2\sqrt{15}.
\end{aligned}$$

Thus [2]

$$\begin{aligned}
t^2 - 2t\sqrt{2} - 6 &= 2\sqrt{15}, \\
(t^2 - 2t\sqrt{2} - 6)^2 &= 60, \\
t^4 + 8t^2 + 36 - 4t^3\sqrt{2} + 24t\sqrt{2} - 12t^2 &= 60, \\
t^4 - 4t^2 - 4t\sqrt{2}(t^2-6) &= 24.
\end{aligned}$$

Using the fact that[3] $(t^2-6) > 2+3+5-6 = 4$, we can express $\sqrt{2}$ as follows:

$$t^4 - 4t^2 - 24 = 4t\sqrt{2}(t^2-6) \implies \sqrt{2} = \frac{t^4 - 4t^2 - 24}{4t\,(t^2-6)} \in \mathbb{Q},$$

which contradicts the fact that $\sqrt{2}$ is irrational. Thus $\sqrt{2}+\sqrt{3}+\sqrt{5}$ is not rational. ∎

[2] The known identity $(x+y+z)^2 = x^2+y^2+z^2+2xy+2yz+2zx$.
[3] Here, $t^2 = (\sqrt{2}+\sqrt{3}+\sqrt{5})^2 > 2+3+5$.

■ **Example 4.58** [2]
Let a,b,c be real numbers. Prove that at least one of the three equations has real roots.

$$\begin{cases} x^2+(a-b)x+(b-c)=0, \\ x^2+(b-c)x+(c-a)=0, \\ x^2+(c-a)x+(a-b)=0. \end{cases}$$

As a contradiction, let us presume that the three quadratic equations have negative discriminant, that is,

$$\begin{cases} (a-b)^2 \; - \; 4(b-c) \; < \; 0, \\ (b-c)^2 \; - \; 4(c-a) \; < \; 0, \\ \underline{(c-a)^2 \; - \; 4(a-b) \; < \; 0,} \\ (a-b)^2+(b-c)^2+(c-a)^2 \; < \; 4[(b-c)+(c-a)+(a-b)]=0. \end{cases}$$

which is impossible. Thus our assumption was wrong, and at least one of the three quadratic equation has real roots.

■

■ **Example 4.59** Let (L) and (L') be two distinct parallel lines in a plane. Prove that if a third line (L'') crosses (L), then (L'') will also cross (L').
Let us assume

$$p: \; L//L' , \quad q: L''\cap L \neq \emptyset, \text{ and } r: L''\cap L' \; \neq \; \emptyset.$$

Our problem reduces to prove: $(p \wedge q) \implies r$ is true. We will use the proof by contradiction method and work on the statement $(p\wedge \; q) \; \wedge \neg r$. Since the lines are in the plane, $\neg r: \; L''//L'$, we get

$$(p\wedge \; q) \; \wedge \neg r \equiv \underbrace{(p\wedge \; \neg r)}_{False} \; \wedge \; q \quad \text{will give a contradiction.}$$

Thus what we assumed is wrong. ■

■ **Example 4.60** Let A and B be two square matrices. Without using determinants, prove that

$$AB \text{ is invertible} \implies A \text{ and } B \text{ are invertible.}$$

Since the negation of (A and B are invertible) is (A is not invertible or B is not invertible), we may assume:

$$\begin{aligned} B \text{ is not invertible} &\implies \exists x \neq 0: \; Bx \; = \; 0, \\ &\implies \exists x \neq 0: \; A\,Bx \; = A\,0 \; = 0, \\ &\implies \exists x \neq 0: \; (AB)x \; = \; 0, \\ &\implies AB \text{ is not invertible.} \end{aligned}$$

This completes the proof by contraposition.
Here we have used the equivalence between the following

$$M \text{ is not invertible} \iff \exists x \neq 0: \; Mx \; = \; 0.$$

■

R If we have considered in the previous proof that A was not invertible, we get

$$A \text{ is not invertible} \implies \exists x \neq 0 : \; Ax = 0.$$

Thus we cannot get ABx, which is why we may proceed differently. We will use the following result:

$$M \text{ invertible} \iff M^\top \text{ is invertible},$$

$$\begin{aligned}
A \text{ is not invertible} &\implies A^\top \text{ is not invertible},\\
&\implies \exists x \neq 0 : \; A^\top x = 0,\\
&\implies \exists x \neq 0 : \; B^\top A^\top x = B^\top 0 = 0,\\
&\implies \exists x \neq 0 : \; (AB)^\top x = 0,\\
&\implies (AB)^\top \text{ is not invertible},\\
&\implies (AB) \text{ is not invertible}.
\end{aligned}$$

■ **Example 4.61** Let M_n be the integer defined as follows:

$$\begin{cases} M_1 = 1, \; M_2 = 11, \\ M_n = 1111\cdots 11, \quad n \text{ ones}. \end{cases}$$

Show that if M_n is prime, then n is prime.

To solve this question, we apply the formula of sum of geometric progression as follows [4]:

$$\sum_{k=0}^{m} r^k = r^0 \times \left(\frac{r^{m+1} - 1}{r - 1} \right), \quad r \neq 1.$$

Moreover, it is worth observing that

$$\begin{cases} M_1 = 1 = 10^0 = 10^{1-1}, \\ M_2 = 11 = 10 + 1 = 1 + 10 = 10^0 + 10^1 = 10^0 + 10^{2-1}, \\ M_3 = 111 = 100 + 10 + 1 = 1 + 10 + 100 = 10^0 + 10^1 + 10^2 = 10^0 + 10^1 + 10^{3-1}, \end{cases}$$

and for any $k \geq 4$, we have

$$M_k = 10^0 + 10^1 + 10^2 + \ldots + 10^{k-1} = \sum_{i=0}^{k-1} 10^i = 10^0 \times \frac{1 - 10^k}{1 - 10} = \frac{10^k - 1}{10 - 1}.$$

Let us proceed by contradiction. Assume that n is not prime, and there exist p and q positive integers greater than 1 such that $n = pq$. Thus we have

$$M_n = M_{pq} = \frac{10^{pq} - 1}{10 - 1} \hookrightarrow \text{We will use a elegant and simple idea.}$$

$$\begin{aligned}
M_n = M_{pq} &= \frac{10^{pq} - 1}{10 - 1} = \frac{[10^p]^q - 1}{10 - 1},\\
&= \left(\frac{[10^p]^q - 1}{10^p - 1} \right) \left(\frac{10^p - 1}{10 - 1} \right).
\end{aligned}$$

For more details, since $r \neq 1$ and if we set $r = 10^p$, we have

$$\frac{[10^p]^q - 1}{10^p - 1} = \frac{r^q - 1}{r - 1} = \sum_{k=0}^{q-1} r^k = \sum_{k=0}^{q-1} (10^p)^k$$

[4] See the geometric progression sum formula in section 6.6.2.

and the last factor $\dfrac{10^p-1}{10-1} = \sum_{k=0}^{p-1} 10^k$.
Thus we conclude that if n is not prime, then

$$M_n = \underbrace{\sum_{k=0}^{q-1} (10^p)^k}_{s} \times \underbrace{\sum_{k=0}^{p-1} 10^k}_{t} = s \times t, \qquad \text{where } s,\, t \in \mathbb{N}, \quad s,t > 1,$$

which proves that M_n is also not prime, which contradicts the fact that M_n is assumed to be prime. Thus what we assumed is wrong. Finally, M_n is prime, then n is prime. ■

■ **Example 4.62** Prove that there doesn't exist a polynomial P in x with integer coefficients such that $P(1) = 12$ and $P(8) = 2022$.

On the contrary, we assume that there exists a polynomial P satisfying

$$P(1) = 12 \text{ and } P(8) = 2022.$$

Then according to Example 4.36, we have $(8-1) \mid P(8) - P(1) \iff 7 \mid 2010$, which is not true $(2010 = 287 \times 7 + 1)$. Thus what we assumed is wrong, and the polynomial P does not exist. ■

■ **Example 4.63** Let P be a polynomial in x with integer coefficients. If $P(0)$ and $P(1)$ are odd numbers, prove that $P(x)$ cannot be zero for any integer x.

On the contrary, suppose there exists $m \in \mathbb{Z}$ such that $P(m) = 0$. Using long division, we can write the polynomial P as follows:

$$P(x) = (x-m)\, Q(x), \text{ where } Q \text{ has integer coefficients} \implies \begin{cases} P(0) = -m\, Q(0), & \text{(odd)} \\ P(1) = (1-m)\, Q(1). & \text{(odd)} \end{cases}$$

Here are the possible cases:
- $-m$ is even $\implies P(0)$ is even.
- $-m$ is odd $\implies 1-m$ is even $\implies P(1)$ is even.

In both cases, we conclude that either $P(0)$ or $P(1)$ is even, which contradict the hypotheses. Thus what we assumed is wrong. ■

■ **Example 4.64** If A is invertible, then A^{-1} is unique.
Let us proceed by contradiction and assume that A has two distinct inverses B and C. That is,

$$\exists B \neq C: \begin{cases} AB = BA = I, \\ AC = CA = I. \end{cases}$$

So, we can write

$$B = B \times I = B \times (AC) = (BA) \times C = I \times C = C \text{ which contradicts } B \neq C.$$

Thus what we assumed is wrong, so the inverse of A is unique. ■

Summary: Three Proof Techniques

If $P \Longrightarrow Q$ is being proved,

direct proof	contrapositive proof	contradiction proof
begins by assuming P is true.	begins by assuming $\neg Q$ is true.	begins by assuming $P \wedge \neg Q$ is true.
...	...	...
until Q is concluded.	until $\neg P$ is concluded.	until a contradiction is obtained.

4.4 Additional Proof Techniques

4.4.1 Proof by Cases

In some situations, establishing a statement's truth value can be accomplished by examining a number of cases. The **proof by cases** method consists of breaking the claim down into subcases or smaller understandable units before proving that the overall claim remains true.

■ **Example 4.65 — triangle inequality.** Prove that for all real numbers a and b,

$$|a+b| \leq |a| + |b|.$$

We proceed with the proof by cases according to the signs of a and b:

- *Case 1.* $a \geq 0$ and $b \geq 0$,
  Then $|a| = a$, $|b| = b$, and $a+b \geq 0$.
  $$|a+b| = a+b = |a|+|b| \Longrightarrow |a+b| \leq |a| + |b| \text{ is true.}$$
- *Case 2.* $a < 0$ and $b < 0$,
  Then $|a| = -a$, $|b| = -b$, and $a+b < 0$.
  $$|a+b| = -(a+b) = (-a) + (-b) = |a|+|b| \Longrightarrow |a+b| \leq |a| + |b|.$$
- *Case 3.* Without loss of generality, we assume that $a \geq 0$ and $b < 0$. Then $|a| = a$ and $|b| = -b$.
  If $|a| \geq |b| \Longrightarrow |a+b| = a+b = |a| - |b| \leq |a| + |b|$.
  If $|a| \leq |b| \Longrightarrow |a+b| = -(a+b) = (-a)+(-b) = -|a|+|b| \leq |a|+|b|$.

■

Let us consider *Case 3* to provide more clarity for the reader, assuming $a \geq 0$ and $b < 0$:

- $|a| \geq |b| \Longrightarrow a \geq -b \Longrightarrow a+b \geq 0 \Longrightarrow |a+b| = a+b$.
- $|a| \leq |b| \Longrightarrow a \leq -b \Longrightarrow a+b \leq 0 \Longrightarrow |a+b| = -(a+b)$.

■ **Example 4.66** Prove that 4 does not divide $2+m^2$ for any integer m.

Here we consider the following two cases:

- *Case 1.* m is even $\Longrightarrow \exists k \in \mathbb{Z}$: $m = 2k$, and then $m^2+2 = 4k^2+2 = 2(2k^2+1)$. Thus we have $m^2+2 = 2 \times$ (odd number), which implies that m^2+2 is not a multiple of 4.
- *Case 2.* m is odd $\Longrightarrow \exists k \in \mathbb{Z}$: $m = 2k+1$, and then $m^2+2 = 4k^2+4k+3 = 4(k^2+k)+3$, which implies that m^2+2 is not a multiple of 4.

■

4.4.2 Proof by Equivalence

The **proof by equivalence** has to be carried out in using many equivalent statements until we reach a statement that is always true (tautology). Let us present a few simple yet interesting examples using this method.

■ **Example 4.67** For $a, b, c \in \mathbb{R}$,

$$a^2+b^2+c^2 \geq ab+bc+ca.$$

The given inequality has the following equivalences:

$$\begin{aligned} a^2+b^2+c^2 \geq ab+bc+ca &\iff 2(a^2+b^2+c^2) \geq 2(ab+bc+ca), \\ &\iff 2(a^2+b^2+c^2)-2(ab+bc+ca) \geq 0, \\ &\iff (a^2+b^2-2ab)+(b^2+c^2-2bc)+(c^2+a^2-2ca) \geq 0, \\ &\iff (a-b)^2+(b-c)^2+(c-a)^2 \geq 0. \end{aligned}$$

which is always true. ■

■ **Example 4.68** Prove that $\sqrt{11} \geq 1+\sqrt{5}$.

Since the function $x \to x^2$ is increasing on $[0, \infty)$, the given inequality has the following equivalences:

$$\begin{aligned} \sqrt{11} \geq 1+\sqrt{5} &\iff (\sqrt{11})^2 \geq (1+\sqrt{5})^2, \\ &\iff 11 \geq 1+5+2\sqrt{5}, \\ &\iff 11-6 \geq 2\sqrt{5}, \\ &\iff 5 \geq 2\sqrt{5}, \\ &\iff 5^2 \geq \left(2\sqrt{5}\right)^2, \\ &\iff 25 \geq 20. \end{aligned}$$

which is always true. ■

■ **Example 4.69** For $a, b \in [0, \infty)$, we have $\dfrac{a+b}{2} \geq \sqrt{ab}$.
The given inequality has the following equivalences:

$$\begin{aligned} \frac{a+b}{2} \geq \sqrt{ab} &\iff \left(\frac{a+b}{2}\right)^2 \geq (\sqrt{ab})^2, \\ &\iff \frac{(a+b)^2}{4} \geq ab, \\ &\iff (a+b)^2 \geq 4ab, \\ &\iff (a+b)^2-4ab \geq 0, \\ &\iff (a-b)^2 \geq 0. \end{aligned}$$

which is always true.

It is clear that the equality case holds when $a=b$. This inequality is well known as the arithmetic mean - geometric mean (AM - GM). ■

It is sometimes necessary to prove that several statements are equivalent. Now, we show that the equivalences can be accomplished in a number of efficient ways.

■ **Example 4.70** Prove that the following statements are equivalent:

(a) Statement S_1
(b) Statement S_2
(c) Statement S_3
(d) Statement S_4

We proceed through the following ways:

- *Method 1.* One may prove

$$S_1 \Longleftrightarrow S_2, \quad S_2 \Longleftrightarrow S_3, \quad S_3 \Longleftrightarrow S_4.$$

- *Method 2.* One may prove

$$\begin{array}{ccc} S_1 & \Longrightarrow & S_2 \\ \Uparrow & & \Downarrow \\ S_4 & \Longleftarrow & S_3 \end{array}$$

- *Method 3.* One may prove

$$\begin{cases} S_1 \Longleftrightarrow S_2, \\ S_2 \Longrightarrow S_3 \Longrightarrow S_4 \Longrightarrow S_2. \end{cases}$$

- *Method 4.*One may prove

$$\begin{cases} S_2 \Longleftrightarrow S_3, \\ S_1 \Longrightarrow S_3 \Longrightarrow S_4 \Longrightarrow S_1. \end{cases}$$

■

■ **Example 4.71** Let A be a square matrix. The following statements are equivalent:

(a) A is invertible.

(b) $det(A) \neq 0$.

(c) $\lambda = 0$ is not an eigenvalue for A.

For a brief review of Linear Algebra results, we refer to Section 5.6. We will use Method 2: $(a) \Longrightarrow (b) \Longrightarrow (c) \Longrightarrow (a)$

1. $(a) \Longrightarrow (b)$ We have

$$\begin{aligned} A \text{ is invertible} &\Longrightarrow A \times A^{-1} = I_n, \\ &\Longrightarrow det\left(A \times A^{-1}\right) = det(I_n), \\ &\Longrightarrow det(A) \times det(A^{-1}) = 1, \\ &\Longrightarrow det(A) \neq 0. \end{aligned}$$

2. $(b) \Longrightarrow (c)$ Since

$$det(A) \neq 0 \Longrightarrow det(A - 0I_n) \neq 0 \Longrightarrow \lambda = 0 \text{ is not an eigenvalue.}$$

3. $(c) \Longrightarrow (a)$ Let us proceed by contradiction, assuming that

$$\begin{aligned} A \text{ is not invertible} &\Longrightarrow \exists\, \bar{x} \neq 0 \colon A\bar{x} = 0, \\ &\Longrightarrow \exists\, \bar{x} \neq 0 \colon A\bar{x} = 0\bar{x}, \\ &\Longrightarrow 0 \text{ is an eigenvalue for } A, \end{aligned}$$

which is a contradiction to our hypothesis that 0 is not an eigenvalue for A. Thus the given statements are equivalent.

■

R All the proofs we have discussed in this chapter prove universal statements about forms:

$$\forall x,\ P(x);\quad P(x) \Longrightarrow Q(x);\quad P(x) \iff Q(x)$$

In the following section, we will discuss how to prove statements of the type $\exists x,\ P(x)$. There are many interesting and important theorems in the form $\exists x,\ P(x)$– that is, there exists a number x satisfying some statement $P(x)$. If you are proving the existence of something, be as specific as you can. Sometimes it is impossible, or extremely difficult, to think of a specific example. **Existence proofs** that are useful and satisfying usually provide a concrete example or explicitly describe how to find x that makes $P(x)$ a valid statement. Sometimes an existence proof can be composed indirectly, or by combining evidence from other existence results.

■ **Example 4.72** Prove that there exists a continuous function is not differentiable.

It is sufficient to consider $f(x) = |x|$, which is continuous on $\mathbb{R}$ and not differentiable at $x = 0$. For completeness, let us show the nondifferentiability of f at the origin. To achieve this end, we proceed by contradiction, assuming that the following limit exists:

$$\begin{aligned} f'(0) &= \lim_{h\to 0} \frac{f(0+h)-f(0)}{h} \\ &= \lim_{h\to 0} \frac{|h|-|0|}{h} \\ &= \lim_{h\to 0} \frac{|h|}{h}. \end{aligned}$$

Now, let's consider left and right limits.

$$\lim_{h\to 0^-} \frac{|h|}{h} = \lim_{h\to 0^-} \frac{-h}{h} = -1,$$

and

$$\lim_{h\to 0^+} \frac{|h|}{h} = \lim_{h\to 0^+} \frac{h}{h} = +1.$$

Since both limits are different, we conclude that $\lim_{h\to 0} \frac{|h|}{h}$ does not exist. Thus f is not differentiable at $x = 0$. It is important to remember that providing an example does not prove a conditional statement contrary to an existence proof. ■

4.4.3 Proofs by Counterexample

We prove a property or statement false by using a **proof by counterexample** providing an example of when the statement does not hold. In general, we say that the counterexample **disproves the statement**.

Definition 4.4.1 — counterexample. The statement $\forall x \in A\ P(x)$ is false if

$$\exists\, \tilde{x} \in A :\ P(\tilde{x}) \text{ is false.}$$

Such an element $\tilde{x}$ is called a **counterexample** of the statement.

■ **Example 4.73** Show that $\forall\, x \in \mathbb{R}, P(x) :\ x^6(2x^2+1) > 0$ is false.

Since $x^6 \geq 0$ and $2x^2+1 > 0$ are always true, it may appear that the statement is true for all $x \neq 0$. Thus the counterexample is $x = 0$, which makes the statement false. ■

■ **Example 4.74** Prove or disprove

$$\forall\, x \in \mathbb{R} :\ 1 + \cot^2(x) = \csc^2(x).$$

We tend to find a counterexample, and since the above statement contains denominators, it is natural to pay attention to *division by zero* scenarios. That is, if we take $x = \pi$, for instance, we obviously get $\cot(\pi)$ undefined because $\sin(\pi) = 0$. Therefore, $x = k\pi,\ k \in \mathbb{Z}$ is one counterexample that makes the statement false. ■

(R) Statistically speaking, most students find this problem complex and, therefore, get a wrong answer, even though it is really an easy problem. The following statement is true:

$$\forall x \in \mathbb{R} \setminus \{k\pi,\ k \in \mathbb{Z}\} :\ 1 + \cot^2(x) = \csc^2(x).$$

■ **Example 4.75** Prove or disprove that for every odd positive integer n, we have $7|(n^3 - 1)$.

We tend to find a counterexample. Let $n = 3$, then $3^3 - 1 = 26 = 3 + 7 \times 3$, which is not a multiple of 7. Thus $7|(n^3 - 1)$ is false. ■

■ **Example 4.76** ([**1, 6**])
Disprove that for every $n \in \mathbb{Z}$, the integer $f(n) = n^2 - n + 11$ is prime.

Let $n = 11 \in \mathbb{Z}$, then $f(11) = 11^2 - 11 + 11 = 11 \times 11$. Thus $f(11)$ is not prime.

(R) We may want to evaluate $f(n)$ for a few values of n. Take the following, for instance

n	-1	0	1	2	3	4	5	6	7	8	9	10
$f(n)$	13	11	11	13	17	23	31	41	53	67	83	101

Here, one can jump to conclude that $f(n)$ is prime; however, if we evaluate $f(11)$ it would be obvious that $f(n)$ is not a prime.

■

(R)

- To *disprove* $\forall x \in A,\ P(x)$, it is sufficient to provide a counterexample by finding an $\tilde{x} \in A$ that makes $P(\tilde{x})$ false.
- To *disprove* $\forall x,\ P(x) \implies Q(x)$, it is sufficient to find an $\tilde{x}$ that makes $P(\tilde{x})$ true, and $Q(\tilde{x})$ false.

4.5 Mathematical Induction Proof

The inductive proof is a powerful and elegant way to demonstrate certain mathematical propositions, offering a structured and systematic approach for establishing the validity of statements and theorems in mathematics. To help you understand how *induction* works, the following sequence serves as an introduction:

$$\begin{cases} u_0 = -1, \\ u_{n+1} = 3u_n + 4,\ n > 0. \end{cases}$$

We wish to express the sequence (u_n) in terms of n values as the following:

k	0	1	2	3	4	5	6	7	8
u_k	-1	1	7	25	79	241	727	2185	6559

after observing the pattern we conjecture that $u_k = 3^k - 2$.
Let us find the value of u_9, knowing that $u_8 = 6559$.

$$u_9 = 3u_8 + 4 = 3 \times 6559 + 4 = 19681 = 3^9 - 2,$$

which turns out to be true, but still this doesn't not prove that our conjecture is true. We note that

k	0	1	2	3	4	5	6	7	8
u_k	-1	1	7	25	79	241	727	2185	6559
u_k	3^0-2	3^1-2	3^2-2	3^3-2	3^4-2	3^5-2	3^6-2	3^7-2	3^8-2

$$u_9 = 3^9 - 2 \implies u_{10} = 3(3^9 - 2) + 4 = 3^{10} - 2.$$

Thus by using the iterative values of u_n, we arrive at the following conclusion:

$$\text{If } u_n = 3^n - 2, \text{ then } u_{n+1} = 3(3^n - 2) + 4 = 3^{n+1} - 2 \text{ is a true statement.}$$

Thus this observation shows that for $n = 0$, u_0 implies that u_1 is true, which further implies that u_2 is true, and so on .

So, inductively our conjecture $u_n = 3^n - 2$ is true for all $n \geq 0$. This proof technique is known as the **principle mathematical induction**.

4.6 The Principle of Mathematical Induction

An inductive proof requires each step to be justified, just as an ordinary proof does. There is, however, a unique algorithm that allows you to prove a statement about a number n by first showing it is true when $n = 1$ and then assuming it is true for $n = k$ and proving also true for $n = k+1$. For example, if you want to show that someone is capable of climbing to any of the floors in a fire escape, it's enough to show that one can climb the ladder up to first floor, and then prove that one knows how to climb the stairs from kth fire escape floor to the $(k+1)$th floor.

Theorem 4.6.1 For each positive integer n, let $P(n)$ be a statement. If
Base step. $P(1)$ is true,
Inductive step. For all $k \in \mathbb{N}$, $P(k) \implies P(k+1)$ is true,
then, $P(n)$ is true $\forall n \in \mathbb{N}$.

A proof using the principle of mathematical induction (PMI) is also called *induction proof.*

(R) Mathematical induction can be easily related to the *Domino's* effect phenomena. In other words, when first domino falls down, it pushes the second, which pushes the third, and the process is repeated until the last one is down.

To explore further results and properties related to sequences, we encourage students to refer to Chapter 6. Within this chapter, readers will encounter numerous illustrative examples that delve into the application of induction proofs within the context of sequences.

■ **Example 4.77** Let $\forall n \in \mathbb{N}$:

$$P(n): \ 1 + 2 + \ldots + (n-1) + n = \sum_{k=1}^{n} k = \frac{n(n+1)}{2}.$$

Prove that $P(n)$ is true for every positive integer n.

We use the PMI to establish the given implication:
Base step. For $n=1$, since $1=\frac{1\times 2}{2}=1$, the statement $P(1)$ is true.
Inductive step. Assume now that $P(k)$ is true for an arbitrary integer $k\geq 1$; that is,

$$P(k)\text{: } 1+2+\ldots+(k-1)+k=\frac{k(k+1)}{2}.$$

We wish to show that $P(k+1)$ remains true; that is,

$$P(k+1):\ 1+2+\ldots+(k-1)+k+(k+1)=\frac{(k+1)(k+2)}{2}.$$

Then,

$$\begin{aligned} 1+2+\ldots+k+(k+1) &= (1+2+\ldots+k)+(k+1), \\ &= \underbrace{\frac{k(k+1)}{2}}_{P(k)}+(k+1), \\ &= \frac{(k+1)(k+2)}{2}. \end{aligned}$$

Thus $P(k+1)$ is true, as desired. Consequently, $P(n)$ is true for all positive integers n. ∎

Example 4.78 Prove that

$$P(n):\ \sum_{j=1}^{n} j^2 = \frac{n(n+1)(2n+1)}{6}, \qquad \forall n\in\mathbb{N}.$$

We employ PMI.
Base step. For $n=1$, since $1^2=\frac{1\times 2\times(2(1)+1)}{6}=1$, $P(1)$ is true.
Inductive step. Assume that $P(k)$ is true for an arbitrary positive integer k; that is,

$$1^2+2^2+\ldots+k^2=\frac{k(k+1)(2k+1)}{6}.$$

We wish to show that $P(k+1)$ remains true.
Then,

$$\begin{aligned} 1^2+2^2+\ldots+k^2+(k+1)^2 &= (1^2+2^2+\ldots+k^2)+(k+1)^2, \\ &= \underbrace{\frac{k(k+1)(2k+1)}{6}}_{P(k)}+(k+1)^2, \\ &= \frac{(k+1)}{6}\left[k(2k+1)+6(k+1)\right], \\ &= \frac{(k+1)(2k^2+k+6k+6)}{6}, \\ &= \frac{(k+1)(2k^2+7k+6)}{6}, \\ &= \frac{(k+1)(k+2)(2k+3)}{6}. \end{aligned}$$

Thus $P(k+1)$ is true, as desired. Consequently, $P(n)$ is true for all positive integers n.

That is, $\forall n \in \mathbb{N}$,

$$P(n)\colon 1^2+2^2+\ldots+n^2+(n+1)^2 = \frac{(n+1)(2n^2+7n+6)}{6} = \frac{(n+1)(n+2)(2n+3)}{6}.$$

■

■ **Example 4.79** Prove that 5 divides $n^5-n, \quad \forall n \geq 0$.

Let $P(n) : n^5 - n = 5t$, where $t \in \mathbb{Z}$. We employ induction on $P(n)$:

Base step. For $n=0$, since $0=5(0)$, $P(0)$ is true.

For $n=1$, $1-1=0=5(0)$, $P(1)$ is true.

Inductive step. Assume that $P(k)$ is true for an arbitrary positive integer k. We wish to show that $P(k+1)$ remains true; that is, $\exists m \in \mathbb{Z}\colon (n+1)^5-(n+1) = 5m$.

Then

$$\begin{aligned}
(k+1)^5-(k+1) &= k^5+5k^4+10k^3+10k^2+5k+1-k-1,\\
&= \underbrace{(k^5-k)}_{P(k)}+5(k^4+2k^3+2k^2+k),\\
&= 5t+5(k^4+2k^3+2k^2+k),\\
&= 5(t+k^4+2k^3+2k^2+k),\\
(k+1)^5-(k+1) &= 5m,
\end{aligned}$$

where $m=t+k^4+2k^3+k \in \mathbb{Z}$, thus $P(k+1)$ is true as desired. Hence we conclude that for any integer, 5 divides n^5-n for all $n \geq 0$. ■

■ **Example 4.80** Let $P(n)\colon 2^n > n+3, \quad n \geq 3$.

We employ induction on n.

Base Step: For $n=3$, $2^3 = 8 > 3+3=6$ implies that $P(3)$ is true.

Inductive Step: Assume that $P(k)\colon 2^k > k+3$ is true for an arbitrary positive integer $k \geq 3$. We want to show that $P(k+1) : 2^{k+1} > k+4$ remains true. Now consider

$$2^{k+1} = 2\times 2^k \underbrace{>}_{P(k)} 2\times(k+3) = 2k+6 = (k+4)+k+2 > k+4.$$

Thus $P(k+1)$ is true. Hence $2^n > n+3$ for any $n \geq 3$. ■

■ **Example 4.81** Let (u_n) be a sequence defined as $u_0=2, \ u_{n+1}=2u_n-3$.

Let's check whether (u_n) is bounded and monotone:

1. (u_n) is bounded.

   Let us consider some of the initial terms in the sequence (u_n):

   $$u_1=2u_0-3=1,\ u_2=2u_1-3=2(1)-3=-1,\ u_3=2u_2-3=2(-1)-3=-5.$$

   These suggest that (u_n) is decreasing. Since $u_0=2$, we infer that (u_n) is bounded from above by 2. This can be proved by induction.

   Let $P(n)\colon u_n \leq 2, \ n \geq 0$:

   Base step. For $n=0$, $u_0=2 \leq 2$ then $P(0)$ is true.

   Inductive step. Assume $P(k)$ is true for an arbitrary integer $k \geq 0$. We wish to show that $P(k+1) : u_{k+1} \leq 2$ is also true. Then $u_{k+1}=2u_k-3 \leq 2\times 2-3=1 < 2$; thus $P(k+1)$ is true, and hence, $u_n \leq 2, \ \forall n \geq 0$.

2. (u_n) is monotone.

   Note that

   $$u_{n+1}-u_n = 2u_n-3-u_n = u_n-3 \leq 2-3=-1<0.$$

Thus (u_n) is strictly decreasing and, hence, monotone.

■

■ **Example 4.82** Prove that for any $n \geq 0$

$$7 \text{ divides } 3^{2n+1} + 2^{n+2}.$$

Let $P(n)$: $7|\ 3^{2n+1} + 2^{n+2}$, for $n \geq 0$.
We employ induction on $P(n)$.

Base step. For $n = 0$, $3^1 + 2^2 = 7(1)$ implies that $P(0)$ is true.

Inductive step. Assume that for some arbitrary positive integer k, there exists $t \in \mathbb{Z}$ such that $P(k)$: $3^{2k+1} + 2^{k+2} = 7t$ is true.

We want to show that $P(k+1)$ also remains true. Then

Idea 1.

$$\begin{aligned} P(k+1)\colon 3^{2(k+1)+1} + 2^{(k+1)+2} &= 3^{2k+3} + 2^{k+3}, \\ &= 3^2 \times 3^{2k+1} + 2 \times 2^{k+2}, \\ &= (7+2) \times 3^{2k+1} + 2 \times 2^{k+2}, \\ &= 7 \times 3^{2k+1} + 2\left(\underbrace{3^{2k+1} + 2^{k+2}}_{7t}\right), \\ &= 7 \times 3^{2k+1} + 2 \times 7t, \\ &= 7\left(3^{2k+1} + 2t\right). \end{aligned}$$

Thus $P(k+1)$ is true, so $7|3^{2n+1} + 2^{n+2}$ for any $n \geq 0$.

Idea 2. From $P(k)$ we get $3^{2k+1} = 7t - 2^{k+2}$, and then $P(k+1)$ becomes

$$\begin{aligned} P(k+1)\colon 3^{2(k+1)+1} + 2^{(k+1)+2} &= 3^{2k+3} + 2^{k+3}, \\ &= 3^2 \times 3^{2k+1} + 2 \times 2^{k+2}, \\ &= 9 \times (7t - 2^{k+2}) + 2 \times 2^{k+2}, \\ &= 9 \times (7t) - 7 \times 2^{k+2}, \\ &= 7\left(9k - 2^{k+2}\right). \end{aligned}$$

Thus $P(k+1)$ is true, so $7|3^{2n+1} + 2^{n+2}$ for any $n \geq 0$.

■

■ **Example 4.83** Prove that for any $n \geq 0$, 133 divides $11^{n+2} + 12^{2n+1}$.
Let $P(n)$: $133|11^{n+2} + 12^{2n+1}$, $n \geq 0$.

Base step. For $n = 0$, $11^2 + 12^1 = 133(1)$ implies that $P(0)$ is true.

Inductive step. Assume that $P(k)$ is true for some arbitrary positive integer k, and we wish to show that $P(k+1)$ is also true. Then

$$11^{k+3} + 12^{2k+3} = 11^{k+2} \times 11 \ + \ 12^{2k+1} \times 12^2.$$

Since $P(k)$ is true, $\exists\, t \in \mathbb{Z}$ such that $12^{2k+1} = 133t - 11^{k+2}$.

Then

$$\begin{aligned} 11^{k+3} + 12^{2k+3} &= 11^{k+2} \times 11 \ + \ (133t - 11^{k+2}) \times 144, \\ &= 11^{k+2}(11 - 144) + 133(144t), \\ &= 11^{k+2}(-133) + 133(144t), \\ &= 133\left(-11^{k+2} + 144t\right), \\ &= 133m, \quad \text{where } m = 144k - 11^{k+2} \in \mathbb{Z}, \\ 11^{n+3} + 12^{2n+3} &= 133\, m. \end{aligned}$$

Thus $P(k+1)$ is true by induction. Hence 133 divides $11^{n+1} + 12^{2n+1}$ for any $n \geq 0$. ■

■ **Example 4.84** Prove that for any natural number $n \geq 3$, $2n+1 < 2^n$.
Let $P(n)$: $2n+1 < 2^n$. We employ induction on $P(n)$:

Base step. For $n=3$, $2(3)+1 = 7 < 2^3 = 8$ implies that $P(3)$ is true.

Inductive step. Assume that $P(k)$ is true for some integer $k \geq 3$ and we wish show that $P(k+1):\ 2(k+1)+1 < 2^{k+1}$ is also true. Then

$$2k+3 = (2k+1)+2 \underbrace{<}_{P(k)} 2^k+2 < 2^k+2^k = 2^{k+1}.$$

Thus $P(k+1)$ is true therefore, $2n+1 < 2^n$ for any $n \geq 3$. ■

■ **Example 4.85** Prove that for any natural number $n \geq 5$, $n^2 < 2^n$. Let $P(n):\ n^2 < 2^n$. We employ induction on n.

Base step. For $n=5$, $5^2 = 25 < 2^5 = 32$ then $P(5)$ is true.

Inductive step. Assume that $P(k)$ is true for some integer $k \geq 5$. We wish to show that $P(k+1):\ (k+1)^2 < 2^{k+1}$ is also true.

Idea 1 Note that

$$5 \leq k \Longrightarrow 1 < k \quad \text{and} \quad 5k \leq k^2.$$

Then

$$(k+1)^2 = k^2+2k+1 < k^2+2k+k = k^2+3k < k^2+5k \leq k^2+k^2 = 2k^2 \underbrace{<}_{P(k)} 2\times 2^k.$$

Thus $(k+1)^2 < 2^{k+1}$, implies that $P(k+1)$ is true. Hence $n^2 < 2^n$ for any $n \geq 5$.

Idea 2 As shown in Example 4.84, $2k+1 < 2^k$ for any $k \geq 3$.
Then for $k \geq 5$, we have

$$(k+1)^2 = k^2+(2k+1) \underbrace{<}_{P(k)} 2^k + 2^k = 2\times 2^k = 2^{k+1}.$$

Thus $(k+1)^2 < 2^{k+1}$, which implies that $P(k+1)$ is true. Hence, $n^2 < 2^n$ for any $n \geq 5$.

■

■ **Example 4.86** If A is a square matrix, show that for any positive integer n, we have

$$\left(A^n\right)^\top = \left(A^\top\right)^n.$$

Let us proceed by induction and let $P(n):\ (A^n)^\top = \left(A^\top\right)^n$.

Base step. For $n=1$, $\left(A^1\right)^\top = A^\top = \left(A^\top\right)^1$, then $P(1)$ is true.

Inductive step. Assume now that $P(k)$ is true for some positive integer k, and we wish to show that
$P(k+1):\ \left(A^{k+1}\right)^\top = \left(A^\top\right)^{k+1}$ remains true. Then

$$\begin{aligned}\left(A^{k+1}\right)^\top &= \left(A^kA\right)^\top \\ &= A^\top \left(A^k\right)^\top \\ &= A^\top \left(A^\top\right)^k, \quad \text{by } P(k) \\ &= \left(A^\top\right)^{k+1}.\end{aligned}$$

Thus $P(k+1)$ is true, and by induction we conclude that $(A^n)^\top = \left(A^\top\right)^n$ for all $n \geq 1$. ■

■ **Example 4.87** Prove that

- $(1+\sqrt{2})^{2n}+(1-\sqrt{2})^{2n}$ is an even integer.
- $(1+\sqrt{2})^{2n}-(1-\sqrt{2})^{2n}=b\sqrt{2}$ for some $b>0$ and $\forall n\geq 1$.

We will use only one proof by induction for both expressions. To this end, for $n\geq 1$, let

$$P(n):\ \begin{cases} \exists\, m\in\mathbb{Z}, & a_n=(1+\sqrt{2})^{2n}+(1-\sqrt{2})^{2n}=2m,\\ \exists\, b\in\mathbb{Z}, & b_n=(1+\sqrt{2})^{2n}-(1-\sqrt{2})^{2n}=b\sqrt{2}.\end{cases}$$

We employ induction on n:

Base step. For $n=1$, $\begin{cases} a_1=(1+\sqrt{2})^2+(1-\sqrt{2})^2=6=3(2), & m=2\\ b_1=(1+\sqrt{2})^2-(1-\sqrt{2})^2=4\sqrt{2}, & b=4\end{cases}$

thus $P(1)$ is true.

Inductive step. Assume that $P(k)$ is true for some arbitrary positive integer k and $\exists\, m,\ b\in\mathbb{Z}$ such that

$$P(k):\ \begin{cases} a_k=(1+\sqrt{2})^{2k}+(1-\sqrt{2})^{2k} = 2m,\\ \\ b_k=(1+\sqrt{2})^{2k}-(1-\sqrt{2})^{2k} = b\sqrt{2}.\end{cases}$$

We want to show that $P(k+1)$ is also true.

$$\begin{aligned} a_{k+1}=(1+\sqrt{2})^{2k+2}+(1-\sqrt{2})^{2k+2} &= (1+\sqrt{2})^{2k}(1+\sqrt{2})^2+(1-\sqrt{2})^{2k}(1-\sqrt{2})^2,\\ &= (1+\sqrt{2})^{2k}(3+2\sqrt{2})+(1-\sqrt{2})^{2k}(3-2\sqrt{2}),\\ &= 3\,a_k+2\sqrt{2}\,b_k,\\ &= 3(2m)\ +\ 2\sqrt{2}b\sqrt{2},\\ &= 2\,(3m\ +\ 2b),\\ &= 2c, \quad \text{where } c=(3m\ +\ 2b)\in\mathbb{Z}.\end{aligned}$$

Moreover,

$$\begin{aligned} b_{k+1}=(1+\sqrt{2})^{2k+2}-(1-\sqrt{2})^{2k+2} &= (1+\sqrt{2})^{2k}(1+\sqrt{2})^2-(1-\sqrt{2})^{2k}(1-\sqrt{2})^2,\\ &= (1+\sqrt{2})^{2k}(3+2\sqrt{2})-(1-\sqrt{2})^{2k}(3-2\sqrt{2}),\\ &= 3\,b_k+2\sqrt{2}\,a_k,\\ &\underbrace{\overset{P(k)}{=}}\ 3(b\sqrt{2})\ +\ 2\sqrt{2}(2m),\\ &= (3b\ +\ 4m)\sqrt{2},\\ &= d\sqrt{2}, \quad \text{where } d=(3b\ +\ 4m)\in\mathbb{Z}.\end{aligned}$$

Thus $P(k+1)$ is true, as desired. Hence, $P(n)$ is true for all $n\in\mathbb{N}$. ■

■ **Example 4.88** Prove that for all $n\geq 0$, we have: $12\mid n^4-n^2$.
Let $P(n)$: $\exists\, m\in\mathbb{N}$: $n^4-n^2\ =\ 12m$:

Base step. For $n=0$, $0-0=0=12(0)$, $m=0$ so $P(0)$ is true.

Inductive step. Assume now that $P(k)$ is true, and let us show that $P(k+1)$: $\exists\, m'\in\mathbb{Z}$: $(k+1)^4-(k+1)^2\ =\ 12m'$ is also true:

$$\begin{aligned} (k+1)^4-(k+1)^2 &= (k^4+4k^3+6k^2+4k+1)-(k^2+2k+1),\\ &= (k^4-k^2)+4k^3+6k^2+2k,\\ &= 12m+\left(4k^3+6k^2+2k\right),\\ &= 12m+2k\left(2k^2+3k+1\right),\\ &= 12m+2k\,(k+1)\,(2k+1).\end{aligned}$$

Here, we need to prove that $6\mid k\,(k+1)\,(2k+1)$. Since we know that

$$\sum_{i=1}^{k} i^2\ =\ \frac{k(k+1)(2kn+1)}{6}\ \in\mathbb{N},$$

we conclude that 6 divides
$k(k+1)(2k+1)$, and $\exists k' \in \mathbb{Z}:\ 2k(k+1)(2k+1) = 12k'$.
Finally, we obtain

$$(k+1)^4-(k+1)^2 = 12m+12k' = 12(m+k') = 12m',\ m' = m+k' \in \mathbb{Z}.$$

Therefore, $P(k+1)$ is true. Hence, $P(n)$ is true for all $n \geq 0$. ■

> R One may use induction to prove that 12 divides $2n(2n+1)(n+1)$. In other words to prove that $6|n(n+1)(2n+1)$. But since $n(n+1)$ is always even, it is sufficient to prove that 3 divides $n(n+1)(2n+1)$.

■ **Example 4.89** Prove that

$$\underbrace{\sqrt{2+\sqrt{2+\sqrt{2+\ldots+\sqrt{2}}}}}_{n \text{ radicals}} = 2\cos\left(\frac{\pi}{2^{n+1}}\right) \quad \text{for } n \in \mathbb{N}.$$

Let $P(n)$: $\sqrt{2+\sqrt{2+\sqrt{2+\ldots+\sqrt{2}}}} = 2\cos\left(\frac{\pi}{2^{n+1}}\right)$, with n radicals.

Base step. For $n=1$, $\sqrt{2} = 2\cos\left(\frac{\pi}{2^{1+1}}\right) = 2\cos\left(\frac{\pi}{4}\right) = \sqrt{2}$ so $P(1)$ is true.

Inductive step. Assume now that $P(k)$ is true, and let us show that

$P(k+1)$: $\sqrt{2+\sqrt{2+\sqrt{2+\ldots+\sqrt{2}}}} = 2\cos\left(\frac{\pi}{2^{k+2}}\right)$, with $(k+1)$ radicals also true:

$$\begin{aligned}
\underbrace{\sqrt{2+\sqrt{2+\sqrt{2+\ldots+\sqrt{2}}}}}_{k+1 \text{ radicals}} &= \sqrt{2+2\cos\left(\frac{\pi}{2^{k+1}}\right)}, \\
&= \sqrt{4\left[\frac{1}{2}+\frac{\cos\left(\frac{\pi}{2^{k+1}}\right)}{2}\right]}, \\
&= 2\sqrt{\frac{1+\cos\left(\frac{\pi}{2^{k+1}}\right)}{2}}, \qquad \frac{\pi}{2^{k+1}} = 2\left(\frac{\pi}{2^{k+2}}\right), \\
&= 2\sqrt{\cos^2\left(\frac{\pi}{2^{k+2}}\right)}, \\
&= 2\left|\cos\left(\frac{\pi}{2^{k+2}}\right)\right|, \qquad \frac{\pi}{2^{k+2}} \text{ in first quadrant}, \\
&= 2\cos\left(\frac{\pi}{2^{k+2}}\right).
\end{aligned}$$

Thus $P(k+1)$ is also true, and hence $P(n)$ is true for all $n \in \mathbb{N}$. ■

We close this section with a discussion about a fascinating property of the Fibonacci sequence.

4.7 Strong Induction

An advanced method of induction called the **strong induction** or **complete induction** provides the same advantages as mathematical induction. The main advantage of strong induction is that it is suitable for proving propositions, where the truth of the statement for an element depends on the truth of the statement for multiple smaller elements. This approach allows us to establish the statement's validity for each element in a well-ordered set in a systematic manner.

Theorem 4.7.1 strong induction
Let $m \in \mathbb{N}$, and let $P(n)$ be a proposition about n. If we suppose that
Base step. $P(m)$ is true, and
Inductive step. for all $n \geq m$, if $P(k)$ is true for all $m \leq k \leq n \implies P(n+1)$ is true. Then $P(n)$ is true, $\forall n \geq m$.

■ **Example 4.90** Let (λ_n) be a sequence defined by

$$\begin{cases} \lambda_1 = 1, \\ \lambda_2 = 8, \\ \lambda_n = \lambda_{n-1} + 2\lambda_{n-2},\ n \geq 3. \end{cases}$$

Prove that $\lambda_n = 3\left(2^{n-1}\right) + 2(-1)^n$ for all $n \in \mathbb{N}$.

Let $P(n):\ \lambda_n = 3\left(2^{n-1}\right) + 2(-1)^n$.
Base step For $n = 1,\ \lambda_1 = 1 = 3 \times 2^0 - 2 = 1$, so $P(1)$ is true.
For $n = 2,\ \lambda_2 = 8 = 3 \times 2^1 + 2 = 8$, so $P(2)$ is true.
Inductive step. Assume that $P(k)$ and $P(k-1)$ are true, and we wish to show that $P(k+1)$: $\lambda_{k+1} = 3\left(2^k\right) + 2(-1)^{k+1}$ is also true. Then

$$\begin{aligned} \lambda_{k+1} &= \lambda_k + 2\lambda_{k-1}, \\ &= \underbrace{3\left(2^{k-1}\right) + 2(-1)^k}_{P(k)} + 2\left(\underbrace{3(2^{k-2}) + 2(-1)^{k-1}}_{P(k-1)}\right), \\ &= 3\left(2^{k-1}\right) + 2 \times 3\left(2^{k-2}\right) + 2(-1)^k + 2 \times 2(-1)^{k-1}, \\ &= 3\left(2^{k-1}\right) + 3\left(2^{k-1}\right) + 2(-1)^{k-1}[(-1)+2], \\ &= 3\left(2 \times 2^{k-1}\right) + 2(-1)^{k-1} \times 1, \\ &= 3\left(2^k\right) + 2(-1)^{k-1} \times (-1)^2, \\ \lambda_{k+1} &= 3\left(2^k\right) + 2(-1)^{k+1}. \end{aligned}$$

Thus $P(k+1)$ is also true, and then $\forall\, n \in \mathbb{N},\ \lambda_n = 3\left(2^{n-1}\right) + 2(-1)^n$. ■

We invite students to check the base step at least for two cases. Above, we checked for $n = 1$ and $n = 2$.

■ **Example 4.91** Let $p, n \in \mathbb{N}$. Prove that if $p + \frac{1}{p} \in \mathbb{N}$, then $p^n + \frac{1}{p^n} \in \mathbb{N}$ for any $n > 0$.

We note that

$$\left(p^n + \frac{1}{p^n}\right)\left(p + \frac{1}{p}\right) = p^{n+1} + \frac{1}{p^{n+1}} + p^{n-1} + \frac{1}{p^{n-1}}$$

then

$$p^{n+1} + \frac{1}{p^{n+1}} = \left(p^n + \frac{1}{p^n}\right)\left(p + \frac{1}{p}\right) - \left(p^{n-1} + \frac{1}{p^{n-1}}\right), \tag{4.1}$$

This suggests using a strong induction to prove the following implication, assuming that $p + \frac{1}{p} \in \mathbb{N}$.

$$P(n):\ \exists\, t_n \in \mathbb{N},\ p^n + \frac{1}{p^n} = t_n.$$

Base step. For $n = 1,\ p + \frac{1}{p} = t_1 \in \mathbb{N}$, then $P(1)$ is true.

For $n = 2,\ \ p^2 + \frac{1}{p^2} = \left(p + \frac{1}{p}\right)^2 - 2 = (t_1)^2 - 2 \in \mathbb{N}$, then $P(2)$ is true.

Inductive step. Assume now that $P(k)$ and $P(k-1)$ are true for some $1 \leq k \leq n$. We wish to show that $P(k+1):\ p^{k+1} + \frac{1}{p^{k+1}} \in \mathbb{N}$ is also true. From Equation (4.1), we have

$$\begin{aligned} p^{k+1} + \frac{1}{p^{k+1}} &= \underbrace{\left(p^k + \frac{1}{p^k}\right)}_{P(k)} \left(p + \frac{1}{p}\right) - \underbrace{\left(p^{k-1} + \frac{1}{p^{k-1}}\right)}_{P(k-1)} \\ &= t_k\, t_1 - t_{k-1} \in \mathbb{N}. \end{aligned}$$

Thus $P(k+1)$ is true. Hence, $P(n)$ is true for any natural number n. ■

■ **Example 4.92** Show that $F_n = \dfrac{\varphi^n - \varphi'^n}{\sqrt{5}}, \quad \forall n \geq 0$, where

$$\begin{cases} F_0 = 0,\ F_1 = 1, \\ F_{n+1} = F_n + F_{n-1},\ \ n \geq 1, \end{cases}$$

and φ, φ' are the roots of $x^2 - x - 1 = 0$.

Since the roots of the equation $x^2 - x - 1 = 0$ are

$$\varphi = \frac{1+\sqrt{5}}{2},\ \ \varphi' = \frac{1-\sqrt{5}}{2}.$$

Then,

$$\varphi^2 = \varphi + 1,\ \ \varphi - \varphi' = \sqrt{5}.$$

Let $P(n)$: $F_n = \dfrac{\varphi^n - \varphi'^n}{\sqrt{5}}$.

Using strong induction, $P(n)$ is derived as follows:

Base step. For $n = 0$

$$F_0 = \frac{\varphi^0 - \varphi'^0}{\sqrt{5}} = 0,$$

which implies $P(0)$ is true.

For $n = 1$,

$$F_1 = 1 = \frac{\varphi^1 - \varphi'^1}{\sqrt{5}} = \frac{\sqrt{5}}{\sqrt{5}} = 1.$$

Thus $P(1)$ is also true.

Inductive step. Assume that $P(k)$ and $P(k-1)$ are true for all $0 \leq k \leq n$. We will show that

$$P(k+1):\ F_{k+1} = \frac{\varphi^{k+1} - \varphi'^{k+1}}{\sqrt{5}}$$

is also true.

Since

$$\begin{aligned} F_{k+1} &= F_k + F_{k-1} \\ &= \frac{\varphi^k - \varphi'^k}{\sqrt{5}} + \frac{\varphi^{k-1} - \varphi'^{k-1}}{\sqrt{5}}, \\ &= \frac{1}{\sqrt{5}}\varphi^{k-1}(\varphi+1) - \frac{1}{\sqrt{5}}\varphi'^{k-1}(\varphi'+1), \\ &= \frac{1}{\sqrt{5}}\varphi^{k-1}(\varphi^2) - \frac{1}{\sqrt{5}}\varphi'^{k-1}(\varphi'^2), \\ &= \frac{1}{\sqrt{5}}\varphi^{k+1} - \frac{1}{\sqrt{5}}\varphi'^{k+1}. \end{aligned}$$

Thus $P(k+1)$ is true. Hence, $P(n)$ is true for any $n \geq 0$. ■

R Using the preceding result, one can write

$$\frac{F_{n+1}}{F_n} = \frac{\dfrac{\varphi^{n+1} - \varphi'^{n+1}}{\sqrt{5}}}{\dfrac{\varphi^n - \varphi'^n}{\sqrt{5}}} = \frac{\varphi^{n+1} - \varphi'^{n+1}}{\varphi^n - \varphi'^n} = \frac{\varphi^{n+1}\left(1 - \dfrac{\varphi'^{n+1}}{\varphi^{n+1}}\right)}{\varphi^n\left(1 - \dfrac{\varphi'^n}{\varphi^n}\right)} = \frac{\varphi\left(1 - \dfrac{\varphi'^{n+1}}{\varphi^{n+1}}\right)}{\left(1 - \dfrac{\varphi'^n}{\varphi^n}\right)}.$$

Using the fact that

$$|r| = \left|\frac{\varphi'}{\varphi}\right| = 0.38196601 < 1 \implies \lim_{n\to\infty} r^n = 0,$$

we obtain

$$\lim_{n\to\infty} \frac{F_{n+1}}{F_n} = \lim_{n\to\infty} \frac{\varphi\left(1 - r^{n+1}\right)}{1 - r^n} = \varphi \,, \lim_{n\to\infty} \frac{1 - r^{n+1}}{1 - r^n} = \varphi.$$

The limit, $\varphi = \dfrac{1+\sqrt{5}}{2}$ is well known as the **Golden ratio** or the **beauty number**.

■ **Example 4.93** Show that $F_n \geq \left(\dfrac{3}{2}\right)^{n-2}, \quad \forall n \geq 1$, where

$$\begin{cases} F_0 = F_1 = 1, \\ F_{n+1} = F_n + F_{n-1}, \quad n \geq 1. \end{cases}$$

Let

$$P(n):\ F_n \geq \left(\frac{3}{2}\right)^{n-2}.$$

Using strong induction, $P(n)$ is derived as follows:

Base step. For $n = 1$, $F_1 = 1 \geq \dfrac{2}{3} = \left(\dfrac{3}{2}\right)^{-1}$ then $P(1)$ is true.

For $n = 2$, $F_2 = 1 + 1 = 2 \geq \left(\dfrac{3}{2}\right)^{0}$ then $P(2)$ is true.

Inductive step. Assume now that

$$P(k):\ F_k \geq \left(\frac{3}{2}\right)^{k-2}$$

and

$$P(k-1):\ F_{k-1} \geq \left(\frac{3}{2}\right)^{k-3}$$

are both true for all $1 \le k \le n$. We wish to show that $P(k+1)$: $F_{k+1} \ge \left(\frac{3}{2}\right)^{k-1}$ is also true.

Since

$$\begin{aligned}
F_{k+1} &= \underbrace{F_k}_{P(k)} + \underbrace{F_{k-1}}_{P(k-1)}, \\
&\ge \left(\frac{3}{2}\right)^{k-2} + \left(\frac{3}{2}\right)^{k-3} \ge \left(\frac{3}{2}\right)^{k-3}\left[\frac{3}{2}+1\right] \ge \frac{5}{2}\times\left(\frac{3}{2}\right)^{k-3}, \\
&\ge \frac{10}{4}\times\left(\frac{3}{2}\right)^{k-3} \ge \frac{9}{4}\times\left(\frac{3}{2}\right)^{k-3} \ge \left(\frac{3}{2}\right)^{2}\times\left(\frac{3}{2}\right)^{k-3}, \\
F_{k+1} &\ge \left(\frac{3}{2}\right)^{k-1}.
\end{aligned}$$

Thus $P(k+1)$ is true, and then $P(n)$ is true for any $n \ge 1$. ■

The next example presents an other strong result that can be used to solve some complex situations.

■ **Example 4.94** Let a and b be any integer numbers, then

$$a+b \,\Big|\, a^{2n+1}+b^{2n+1}, \quad \forall\, n \ge 0.$$

Let $P(n): \exists\, m \in \mathbb{Z}: \; a^{2n+1}+b^{2n+1} = (a+b)m$. Using strong induction, $P(n)$ is derived as follows:

Base step. For $n=0$, $a+b=(a+b)(1)$, $m=1$ then $P(0)$ is true.

For $n=1$, $a^3+b^3=(a+b)(a^2-ab+b^2) = (a+b)m$,

Inductive step. Assume that $P(k)$ and $P(k-1)$ are both true for any $0 \le k \le n$.

$$P(k): \exists\, q \in \mathbb{Z}: \; a^{2k+1}+b^{2k+1} = (a+b)\,q,$$

$$P(k-1): \exists\, q' \in \mathbb{Z}: \; a^{2k-1}+b^{2k-1} = (a+b)\,q'.$$

We wish to show that

$$P(k+1): \exists\, m' \in \mathbb{Z}: \; a^{2k+3}+b^{2k+3} = (a+b)\,m'$$

is also true. Then

$$\begin{aligned}
(a^{2k+1}+b^{2k+1})(a^2+b^2) &= a^{2k+3}+b^{2k+3}+a^2b^{2k+1}+a^{2k+1}b^2, \\
&= a^{2k+3}+b^{2k+3}+a^2b^2\left(b^{2k-1}+a^{2k-1}\right).
\end{aligned}$$

That is,

$$a^{2k+3}+b^{2k+3} = (a^{2k+1}+b^{2k+1})(a^2+b^2) - a^2b^2\left(b^{2k-1}+a^{2k-1}\right).$$

Then

$$\begin{aligned}
a^{2k+3}+b^{2k+3} &= \underbrace{(a^{2k+1}+b^{2k+1})}_{P(k)}(a^2+b^2) - a^2b^2\underbrace{\left(a^{2k-1}+b^{2k-1}\right)}_{P(k-1)}, \\
&= (a+b)\,q\,(a^2+b^2) - a^2b^2\,(a+b)q', \\
&= (a+b)\left[q\,(a^2+b^2) - a^2b^2\,q'\right] \\
&= (a+b)\,m', \qquad \text{where } m' = q\,(a^2+b^2) - a^2b^2\,q' \in \mathbb{Z}.
\end{aligned}$$

Thus $P(k+1)$ is true. Therefore, $a+b \,\Big|\, a^{2n+1}+b^{2n+1}, \quad \forall\, n \geq 0,\ a,\ b \in \mathbb{Z}.$ ■

R Using classical identities, we may prove the above result directly from

$$x^n + y^n = (x+y)[\,x^{n-1} - x^{n-2}y + x^{n-3}y^2 - \ldots - xy^{n-2} + y^{n-1}\,],$$

where n is odd, by replacing $n, x,$ and y in the identity with $2n+1,\ a,$ and, b respectively.

The following three examples present the direct application of the above result.

■ **Example 4.95** Prove that

$$8^n + 1 \text{ is a multiple of } 9, \quad \forall \text{ odd integer } n \geq 1.$$

The proof can be done by induction. We will use the result proved in Example 4.94 as an alternate method here.

Since $8^n+1 = 8^n + 1^n,\ 8+1 \,\Big|\, 8^n + 1^n$, where n is odd. Thus $9 \big| 8^n+1$ for any odd integer $n \geq 1$. ■

■ **Example 4.96** Prove that

$$85 \,\Big|\, 9^{70} + 2^{70}.$$

A first observation gives $9+2 = 11 \neq 85$. Also the power 70 is even; thus we cannot directly apply the result from Example 4.94. But a second observation suggests we use $70 = 2 \times 35$, and then

$$9^{70} + 2^{70} = 9^{2\times 35} + 2^{2\times 35} = 81^{35} + 4^{35} \implies 81+4 = 85 \,\Big|\, 9^{70} + 2^{70}.$$

■

■ **Example 4.97** Prove that

$$2^{2n} - 1 \text{ is a multiple of } 3, \quad \forall \text{ integer } n \geq 1.$$

Here we solve this example directly by using the result in Question 57. In other words,

$$2^{2n} - 1 = 4^n - 1^n \implies 4-1 = 3 \,\Big|\, 4^n - 1^n = 2^{2n} - 1;$$

thus $2^{2n} - 1$ is a multiple of 3, $\forall$ integer $n \geq 1$. ■

■ **Example 4.98** Show that $\cos(1°)$ is irrational. On the contrary, let us assume that $\cos(1°)$ is a rational number and let $P(n)$: $\cos(n°) \in \mathbb{Q}$.

We apply strong induction on $P(n)$:

Base step. For $n = 1$, $\cos(1°)$ is rational, so $P(1)$ is true.

For $n = 2$, $\cos(2°) = 2\cos^2(1°) - 1 \in \mathbb{Q}$, thus $P(2)$ is true.

Inductive step. Assume $P(k)$ and $P(k-1)$ are true. We wish to show that $P(k+1)$: $\cos[(k+1)°] \in \mathbb{Q}$ is also true.

Since

$$\cos[(k+1)°] + \cos[(k-1)°] = 2\cos(1°)\,\cos(k°).$$

$$\cos[(k+1)^\circ] = 2\underbrace{\cos(1^\circ)}_{\in\mathbb{Q}}\underbrace{\cos(k^\circ)}_{\in\mathbb{Q}} - \underbrace{\cos[(k-1)^\circ]}_{\in\mathbb{Q}} \in \mathbb{Q}.$$

Thus $P(k+1)$ is also true. Hence, $\cos(k^\circ)$ is rational for any positive integer k.

Now, if we take $k = 30^\circ$, we have

$$\cos(30^\circ) = \frac{\sqrt{3}}{2} \in \mathbb{Q} \implies \sqrt{3} \in \mathbb{Q},$$

which is not true. Thus our assumption was wrong, and we conclude that $\cos(1^\circ)$ is irrational. ■

(R) In the preceding example, students were confronted with a difficult question, and the suggested idea was quiet new. It can be summarized as follows:

1. Consider the first step in the contradiction proof: ($\cos(1^\circ)$ is rational).
2. Use this assumption in the initial (or base) step in a strong induction proof of a general case: $\cos(n^\circ)$ is rational.
3. Take a particular angle value that we know in advance is cosine is irrational. In Example4.98 we took $n = 30^\circ$ and one may take, for instance, $n = 45^\circ$, where

$$\cos(45^\circ) = \frac{1}{\sqrt{2}} \in \mathbb{Q} \implies \sqrt{2} \in \mathbb{Q},$$

which is impossible.

(R) *Cauchy's principle of induction.* It is not widely mentioned in many mathematical analysis textbooks. This induction principle can be summarized as follows: to prove that $P(n)$ is true for all $n \geq 1$, we need to prove that:

- $P(1)$ is true.
- $P(n) \implies P(n-1)$ is true.
- $P(n) \implies P(2n)$ is true.

We close Chapter 4 with this interesting problem, where we will face a serious issue when applying an induction directly.

■ **Example 4.99** Let us try to prove by induction that

$$\frac{1}{1^2} + \frac{1}{2^2} + \frac{1}{3^2} + \ldots + \frac{1}{n^2} < 2.$$

Let

$$P(n)\colon \sum_{m=1}^{n} \frac{1}{m^2} < 2:$$

Base step. For $n = 1$, $\sum_{m=1}^{1} \frac{1}{m^2} = 1 < 2$, then $P(1)$ is true.

Inductive step. Assume now that $P(k)$ is true, and let us show that

$$P(k+1)\colon \sum_{m=1}^{k+1} \frac{1}{m^2} < 2$$

is also true.

Since

$$\sum_{m=1}^{k+1} \frac{1}{m^2} = \sum_{m=1}^{k} \frac{1}{m^2} + \frac{1}{(k+1)^2} < \underbrace{2 + \frac{1}{(k+1)^2}}_{>2},$$

thus we cannot use the preceding technique to prove it. Now, let us prove another inequality that will easily lead to our main question. To this end let us consider the following result to prove by induction:

$$\frac{1}{1^2}+\frac{1}{2^2}+\frac{1}{3^2}+\ldots+\frac{1}{n^2} \le 2-\frac{1}{n},\ \forall\, n\ge 1.$$

Let

$$Q(n):\sum_{m=1}^{n}\frac{1}{m^2}\le 2-\frac{1}{n}.$$

Base step. For $n=1$, $\sum_{m=1}^{1}\frac{1}{m^2}=1\le 2-1=1$, so $Q(1)$ is true.

Inductive step. Assume now that $Q(k)$ is true, and let us show that

$$Q(k+1):\sum_{m=1}^{k+1}\frac{1}{m^2}<2-\frac{1}{k+1}$$

is also true.

Since

$$\sum_{m=1}^{k+1}\frac{1}{m^2}=\sum_{m=1}^{k}\frac{1}{m^2}+\frac{1}{(k+1)^2}\underbrace{\le}_{Q(k)}2-\frac{1}{k}+\frac{1}{(k+1)^2},$$

it suffices to prove that

$$\begin{aligned}2-\frac{1}{k}+\frac{1}{(k+1)^2}\le 2-\frac{1}{k+1}\quad &\iff\quad -\frac{1}{k}+\frac{1}{(k+1)^2}\le-\frac{1}{k+1},\\ &\iff\quad \frac{1}{k}-\frac{1}{(k+1)^2}\ge\frac{1}{k+1},\\ &\iff\quad \frac{1}{k}-\frac{1}{(k+1)^2}-\frac{1}{k+1}\ge 0,\\ &\iff\quad \frac{(k+1)^2-k-k(k+1)}{k(k+1)^2}\ge 0,\\ &\iff\quad \frac{1}{k(k+1)^2}\ge 0.\end{aligned}$$

Then $Q(k+1)$ is true, and consequently $Q(n)$ is true for all $n\in\mathbb{N}$. Finally, we have

$$\sum_{m=1}^{n}\frac{1}{m^2}\le 2-\frac{1}{n}<2,$$

which completes the proof. ■

4.8 Chapter 4 Practice Problems

Question 1 Let $x,y\in\mathbb{Z}$. Prove that $x-y$ is even if and only if x and y are of the same parity.

Question 2 Let $a,\,b\in\mathbb{Z}$. Prove that if ab is odd, then a^2+b^2 is even.

Question 3 For $a,b\in\mathbb{Z}$, prove that if $a^2(b^2-2b)$ is odd, then a and b are odd.

Question 4 Let $x\in\mathbb{Z}$. Prove that if 2^{2x} is odd, then 2^{-2x} is also odd.

Question 5 Prove in $\mathbb{Z}$ that if $xy = 9$, then $(x-y)^3 = 64(x-y)$.

Question 6 Use contraposition to prove that

$$x^3 + x^2 - 2x < 0 \Longrightarrow x < 1.$$

Question 7 Show that

$$\forall\, x,\, y \in \mathbb{R}: \ \big||x| - |y|\big| \leq |x - y|.$$

Question 8 Prove that

$$\forall\, k \in \mathbb{Z}, \ 5 \mid k^2(k^4 - 1).$$

Question 9 Show that $(n^2 + 7n + 12)(7 + 2n)$ is divisible by 6.

Question 10 Show that $(n+2)(n+3)(n+4)(n+5) + 1$ is a perfect square.

Question 11 Prove that

$$\forall\, p \in \mathbb{Z}, \ 12 \mid (p^4 - p^2).$$

Question 12 Show that $n^3 + (n-1)^3 + (n-2)^3$ is a multiple of 9 for any integer n.

Question 13 Let $n \in \mathbb{Z}$. If $4 \nmid (n^2 + 3)$, then $4 \nmid (n-1)$ and $4 \nmid (n+1)$.

Question 14 If today is Monday, which day of the week will it be 211 days later, including today?

Question 15 Prove that $(2 + 3n)(n^3 - n)$ is a multiple of 24 for any integer n.

Question 16 Prove that for any prime number p $(p > 3)$, there exists an integer m such that $p = 24m + 1$.

Question 17 Prove that for every even integer m such that 4 does not divide m, there exist no integers x and y for which $x^2 + 3y^2 = m$.

Question 18 Prove that if a and b are real numbers, where $b \neq 0$, then $\left|\dfrac{a}{b}\right| = \dfrac{|a|}{|b|}$.

Question 19 Show that if for $x \in \mathbb{R}$ there exists a prime number m such that $x^2 = m$, then x is irrational.

Question 20 Prove that for any integers a, b, we have: $(a+b)^3 \equiv (a^3 + b^3) \ (mod\ 3)$.

Question 21 Use modular (congruence) to prove that

$$\forall\, k \in \mathbb{Z}, \ 5 \mid k^2(k^4 - 1).$$

Question 22 Use modular (congruence) to prove that

$$\forall\, k \in \mathbb{Z}, \ 5 \mid 5(k^6 - k^3).$$

Question 23 Assume that $a \equiv b\,(mod\ 1)$. What does this imply for a and b?

Question 24 Let $a, b \in \mathbb{Z}$, prove that

$$a = b(mod\,3) \Longleftrightarrow 2a + b = 0(mod\,3).$$

Question 25 Use modular (congruence) to prove that

$$p \neq 3, \ p \text{ prime} \quad \Longrightarrow \quad 3 \,\Big|\, (p^2 + 2).$$

Question 26 Complete the proof of the following statement by filling in the blanks, and answer the questions after the proof.
If two integers have opposite parity, then their sum is odd.

Proof. Suppose m and n are two integers with opposite parity. We wish to show that $----$ $----$ is odd. Without loss of generality, suppose $---$ is even and $-----$ is odd. Thus $------ = --------$ and $-------- = --------$, for some integers k and h. Therefore, $------------- = 2(----)+1$ which is odd.

Questions.

1. Which proof technique is being used?

2. What is the conclusion part of the proof?

3. What should be mentioned in a complete proof?

Question 27 Prove that $(n+1)^3 - 2(n+2)^2 + (n+3)$ is even for all $n \in \mathbb{Z}$.

Question 28 Prove that for $n \in \mathbb{N}$, if $4|n$, then $5|\ 3^n - 1$.

Question 29 Prove that if $a^4 - 4a^3b + 6a^2b^2 - 4ab^3 + b^4$ is odd, then a or b is odd.

Question 30 Prove that for $a \in \mathbb{Z}$, if $5 \nmid (a^4 - 81)$, then $5 \nmid (a-3)$ and $5 \nmid (a+3)$.

Question 31 For $n \in \mathbb{Z}$, prove that if $1 - n^2 > 0$, then $5n^5 - 4n^4 + 3n^3 - 2n^2 + n - 1$ is odd.

Question 32 Show that $2^{6n} + 3^{2n-2}$ is divisible by 5 for all $n \geq 1$.

Question 33 Show that $5^{2n} + 7$ is divisible by 8 for any $n \geq 0$.

Question 34 Show that $11^{n+2} + 12^{2n+1}$ is divisible by 133 for any $n \geq 0$.

Question 35 Prove Theorem 4.1.4.

Question 36 Prove by mathematical induction, the formula for the sum of the first n terms of an *arithmetic progression*:

$$x_0 + (x_0+r) + (x_0+2r) + (x_0+3r) + \ldots + (x_0+(n-1)r) = \frac{n}{2}(2x_0+(n-1)r).$$

Question 37 Prove by mathematical induction, the formula for the sum of the first n terms of an *geometric progression*:

$$x_0 + x_0 r + x_0 r^2 + x_0 r^3 + \ldots + x_0 r^{n-1} = \frac{x_0(1-r^n)}{1-r},\ r \neq 1.$$

Question 38 Prove that

$$\sum_{i=1}^{m} (-1)^{i-1} i^2 = \frac{(-1)^{m-1} m(m+1)}{2}.$$

Question 39 Prove that $2^{n+2} \geq 2n+5$ for any $n \geq 1$.

Question 40 Show that

$$12 \mid 1 + 2^{2021} + 3^{2021} + 4^{2021} + 5^{2021} + 6^{2021} + 7^{2021} + 8^{2021} + 9^{2021} + 10^{2021} + 11^{2021}.$$

Question 41 Prove by Induction

$$\left(1-\frac{1}{2}\right)\left(1-\frac{1}{3}\right)\left(1-\frac{1}{4}\right)\ldots\left(1-\frac{1}{n}\right) = \frac{1}{n},\ \forall n \geq 2.$$

Question 42 Show in two ways that $n^3 - n$ is divisible by 3 for any non-negative integer. What about if n is a negative integer?

Question 43 Show in two ways that $n^5 - n$ is divisible by 5 for any non-negative integer. What about if n is a negative integer?

Question 44 Show that $n^5 - n$ is divisible by 30 for any non-negative integer.

Question 45 Show in two ways that $n^{11} - n$ is divisible by 11 for any non-negative integer. What about if n is a negative integer?

Question 46 Show in two ways that $n^3 - 7n + 3$ is divisible by 3 for any non-negative integer.

Question 47 Show that $3^{2n} - 1$ is divisible by 8 for any nonnegative integer.

Question 48 Show that $7^n - 2^n$ is divisible by 5 for any nonnegative integer.

Question 49 Prove in two ways that $8 \mid 3^{2n} + 7$.

Question 50 Prove in two ways that $6 \mid n(n-1)(2n-1)$.

Question 51 Prove in two ways that $10 \mid 1 + 8^n - 6^n - 3^n$.

Question 52 Prove that $4 \mid n - 2n^2 + n^3$ for any odd integer $n > 1$.

Question 53 Prove that $5^{2n+1} \times 2^{n+2} + 3^{n+2} \times 2^{2n+1}$ is divisible by 19 for any $n \geq 0$.

Question 54 Show that for all $n \geq 1$,

$$\forall\, n \geq 1, \quad n! > n^n e^{-n}.$$

Question 55 Prove that $n^2 \leq 2^n, \ \forall\, n \geq 4$.

Question 56 Prove that $n^{n+1} > (n+1)^n, \ \forall\, n \geq 3$.

Question 57 Let x and $y \in \mathbb{Z}$; then

$$x - y \,\big|\, x^n - y^n, \quad \forall\, n \geq 1.$$

Question 58 How many pairs (m,n) of integers satisfy the equation

$$|mn| + |m-n| = 1?$$

Question 59 Show that $\sin(1^\circ)$ is irrational.

Question 60 Prove that for $n \in \mathbb{N}$, $1^3 + 2^3 + 3^3 + \ldots + n^3 = \left[\frac{n(n+1)}{2}\right]^2$.
Complete the following proof. (Hint: use PMI).
Proof. Let

$$P(n) : 1^3 + 2^3 + 3^3 + \ldots + n^3 = \left[\frac{n(n+1)}{2}\right]^2$$

for some $n \in \mathbb{N}$.

We will prove that $P(n)$ is true by using the principle of mathematical induction (PMI).

Base step.
Observe that for $n = 1$, we have

$$P(1)\colon 1^3 = \left[\frac{1(1+1)}{2}\right]^2 = 1,$$

which is obviously true.

Step 2 (Inductive):

We must prove that if $P(k)$, then $P(k+1)$ for any $k \in \mathbb{N}$. That is, we want to show that $P(k+1)$:_____________, whenever $P(k)$:_____________.

Assume that $P(k): 1^3+2^3+3^3+\ldots+k^3 = \left[\frac{k(k+1)}{2}\right]^2$ holds for some $k \in \mathbb{Z}$. Consider,

$$\begin{aligned} 1^3+2^3+3^3+\ldots+k^3+(k+1)^3 &= \left[\frac{k(k+1)}{2}\right]^2 + \underline{\qquad\quad}, \\ &= \underline{\qquad\qquad\qquad}, \\ &= \underline{\qquad\qquad\qquad}, \\ &= \underline{\qquad\qquad\qquad}. \end{aligned}$$

Thus we have shown that

$P(k)$: _______________ = _______________, which implies that

$P(k+1)$: _______________ = _______________.

This completes the proof that $P(k) \rightarrow P(k+1)$ for any $k \in \mathbb{N}$.

Hence, by using (PMI), we have shown that

$$P(n):\ 1^3+2^3+3^3+\ldots+n^3 = \left[\frac{n(n+1)}{2}\right]^2$$

holds for all natural numbers n. ■

4.9 Chapter 4 Additional Practice Problems with Solutions

Question 1 Prove that for every integer x, the integers $3x+1$ and $5x+2$ are of opposite parity.

Solution. Let us consider the two possible cases:

(i) If $3x+1$ is odd, then $3x$ is even, which implies that x is even. Now, if we consider $5x+2$, we conclude that $5x$ is even and $5x+2$ is also even. Thus in this case, $3x+1$ and $5x+2$ are of opposite parity.

(ii) If $3x+1$ is even, then $3x$ is odd, which implies that x is odd. Now, if we consider $5x+2$, we conclude that $5x$ is odd and $5x+2$ is also odd. Thus in this case, $3x+1$ and $5x+2$ are of opposite parity.

We may apply another approach by cases:

(i) If x is odd, then $\exists k \in \mathbb{Z}:\ x = 2k+1$, and then

$$\begin{cases} 3x+1 = 3(2k+1)+1 = 6k+4 = 2(3k+2) \ \text{ is even.} \\ 5x+2 = 5(2k+1)+2 = 10k+7 = 2(5k+3)+1 \ \text{ is odd.} \end{cases}$$

(ii) If x is even, then $\exists k \in \mathbb{Z}:\ x = 2k$, and then

$$\begin{cases} 3x+1 = 3(2k)+1 = 6k+1 = 2(3k)+1 \ \text{ is odd.} \\ 5x+2 = 5(2k)+2 = 10k+2 = 2(5k+1) \ \text{ is even.} \end{cases}$$

Therefore, for both cases $3x+1$ and $5x+2$ have opposite parity.

Question 2 Prove that p^2-1 is divisible by 8 for all odd integers p.

Solution. We proceed by direct proof. If p is odd, then there exists $k \in \mathbb{Z}:\ p = 2k-1$, and then

$$p^2-1 = (2k-1)^2-1 = (2k-1+1)(2k-1-1) = 2k(2)(k-1) = 4k(k-1).$$

Since the product of two consecutive integers is an even number, then $\exists k' \in \mathbb{Z}:\ k(k-1) = 2k'$, and then $p^2-1 = 4(2k') = 8k'$ for some integer k'. Finally, p^2-1 is divisible by 8.

Question 3 Prove that the sum of the first n natural numbers ($n \geq 3$) is not a prime number.

Solution. We proceed by cases. To this end, let us discuss in terms of the parity of n.

(i) If $\exists\, k \in \mathbb{N}:\ n = 2k-1,\ k \geq 2$, then

$$1+2+3+\ldots+n \;=\; n \times \frac{(2k-1+1)}{2} = n \times \frac{2k}{2} = (2k-1)\times k.$$

(ii) If $\exists\, k \in \mathbb{N}:\ n = 2k,\ k \geq 2$, then

$$1+2+3+\ldots+n = \frac{(2k)(2k+1)}{2} = k \times (2k+1).$$

Thus since in both cases $k \geq 2$, we conclude that the sum is a composite integer, and it cannot be prime.

Question 4 Let $n \in \mathbb{Z}$. Prove that $(n+1)^2 - 1$ is even if and only if n is even.

Solution. Since we have to prove an equivalence, let us consider the two implications:

(i) n is even, so $(n+1)^2 - 1$ is even.

$$\exists\, k \in \mathbb{Z}:\ n = 2k \implies (n+1)^2 - 1 = (2k+1)^2 - 1 = 2(2k^2+2k) \implies (n+1)^2 - 1 \text{ is even.}$$

(ii) $(n+1)^2 - 1$ is even, so n is even.
Since $(n+1)^2 - 1 = n^2 + 2n$ is even, we conclude that n^2 is also even implying n is even according to the result in Example 4.39.

R Another solution for the converse may be given:
$(n+1)^2 - 1 = n^2 + n + n = \underbrace{n(n+1)}_{even} + n$ is even, so n must be even.

Question 5 Prove that for any natural numbers m and n, we have:

$$mn\,(m^2 - n^2) \text{ is a multiple of 3.}$$

Solution. We proceed by cases.

1. $m \equiv 0\,(mod\ 3)$ or $n \equiv 0\,(mod\ 3)$, then $mn \equiv 0\,(mod\ 3)$ and then $mn\,(m^2 - n^2) \equiv 0\,(mod\ 3)$.
2. $m = n \equiv r\,(mod\ 3)$, then $m^2 - n^2 \equiv 0\,(mod\ 3)$ and then $mn\,(m^2 - n^2) \equiv 0\,(mod\ 3)$.
3. $m \equiv 1\,(mod\ 3)$ and $n \equiv 2\,(mod\ 3)$, then $m^2 - n^2 \equiv 1 - 4 = -3 = 0\,(mod\ 3)$ and then $mn\,(m^2 - n^2) \equiv 0\,(mod\ 3)$.
4. $m \equiv 2\,(mod\ 3)$ and $n \equiv 1\,(mod\ 3)$, then $m^2 - n^2 \equiv 4 - 1 = 3 = 0\,(mod\ 3)$ and then $mn\,(m^2 - n^2) \equiv 0\,(mod\ 3)$.

Question 6 Show that $n^3 + (n+1)^3 + (n+2)^3$ is a multiple of 9.

Solution.

Idea 1. We have

$$n^3 + (n+1)^3 + (n+2)^3 = 3n^3 + 9n^2 + 9 + 15n = 3n^3 + 15n + 9(n^2+1).$$

To complete the proof it is enough to show that $3n^3 + 15n$ is a multiple of 9. To do this, we note that $3n^3 + 15n = 3n(n^2+5)$. We have following possible cases:

- $n = 3k,\ k \in \mathbb{Z}$. In this case, obviously, $3n^3 + 15n$ is a multiple of 9.
- $n = 3k+1,\ k \in \mathbb{Z}$. In this case, $n^2+5 = 9k^2+6k+6$, which is a multiple of 3, and thus $3n^3 + 15n$ is a multiple of 9.
- $n = 3k+2,\ k \in \mathbb{Z}$. In this case, $n^2+5 = 9k^2+12k+9$, which is a multiple of 3, and thus $3n^3 + 15n$ is a multiple of 9.

Idea 2. To complete the proof, from $n^3 + (n+1)^3 + (n+2)^3 = 3(n^3+2n) + 9(n^2+n+1)$, it is enough to show that n^3+2n is a multiple of 3. We have following possible cases:

- $n \equiv 0 (mod\ 3) \Longrightarrow n^3+2n \equiv 0\ (mod\ 3)$.
- $n \equiv 1\ (mod\ 3) \Longrightarrow n^3+2n \equiv 3 (mod\ 3) \equiv 0 (mod\ 3)$.
- $n \equiv 2 (mod\ 3) \Longrightarrow n^3+2n \equiv 12 (mod\ 3) \equiv 0 (mod\ 3)$.

Idea 3. To complete the proof it is enough to show that $n(n^2+2)$ in $3n^3+6n+9n$ is a multiple of 3. To do this, we note that $n(n^2+2) = n(n^2-1\ +\ 3)\ =\ n(n-1)(n+1)\ + 3n$, which is the sum of a multiple of 3 term with a product of three consecutive integers. Thus $3 \mid n(n^2+2)$.

Question 7 Prove that for any integers a, b, we have: $(a+b)^3 \equiv (a^3+b^3)\ (mod\ 3)$.

Solution. We know $(a+b)^3 = a^3+b^3+3ab(a+b) \equiv a^3+b^3+0 (mod 3) \equiv (a^3+b^3)\ (mod\ 3)$.

Question 8 Prove that

$$\forall\, k \in \mathbb{Z},\ \ 5 \mid k^5 - 5k^3 + 4k.$$

Solution.

Idea 1. Solve by cases using congruence modulo n:

1. $k \equiv 0\ (mod\ 5) \Longrightarrow k^5-5k^3+4k \equiv 0\ (mod\ 5)$.
2. $k \equiv 1\ (mod\ 5) \Longrightarrow k^5-5k^3+4k \equiv 1-0+4 \equiv 0\ (mod\ 5)$.
3. $k \equiv 2\ (mod\ 5) \Longrightarrow k^5-5k^3+4k \equiv 2^5-0+4(2)\ (mod\ 5)$.
   Since $2^4 \equiv 1\ (mod\ 5)$, $2^5+8 \equiv\ 2+8 \equiv 0\ (mod\ 5)$.
4. $k \equiv 3\ (mod\ 5) \Longrightarrow k^5-5k^3+4k \equiv 3^5-0+4(3)\ (mod\ 5)$.
   Since $3^2 \equiv -1\ (mod\ 5) \Longrightarrow 3^4 \equiv 1\ (mod\ 5)$, $3^5+12 \equiv\ 3+12 \equiv 0\ (mod\ 5)$.
5. $k \equiv 4\ (mod\ 5) \Longrightarrow k^5-5k^3+4k \equiv 4^5-0+4(4) \equiv 0\ (mod\ 5)$.
   Since $4^2 \equiv 1\ (mod\ 5) \Longrightarrow 4^4 \equiv 1\ (mod\ 5)$, $4^5+16 \equiv\ 4+16 \equiv 0\ (mod\ 5)$.

Idea 2. Solve by cases using factorization and congruence modulo n:

$$P(k)\ =\ k^5-5k^3+4k = k(k^4-5k^2+4) = k(k^2-4)(k^2-1) =\ k(k-2)(k+2)(k-1)(k+1).$$

1. $k \equiv 0\ (mod\ 5) \Longrightarrow P(k) \equiv 0\ (mod\ 5)$.
2. $k \equiv 1\ (mod\ 5) \Longrightarrow k-1\ \equiv 0 (mod\ 5) \Longrightarrow P(k)\ \equiv 0\ (mod\ 5)$.
3. $k \equiv 2\ (mod\ 5) \Longrightarrow k-2\ \equiv 0 (mod\ 5) \Longrightarrow P(k)\ \equiv 0\ (mod\ 5)$.
4. $k \equiv 3\ (mod\ 5) \Longrightarrow k+2\ \equiv 0 (mod\ 5) \Longrightarrow P(k)\ \equiv 0\ (mod\ 5)$.
5. $k \equiv 4\ (mod\ 5) \Longrightarrow k+1\ \equiv 0 (mod\ 5) \Longrightarrow P(k)\ \equiv 0\ (mod\ 5)$.

Question 9 Prove that if 7 divides m^2+n^2, then 7 divides m and 7 divides n.

Solution. Let us proceed by contrapositive: If(7 does not divide m or 7 does not divide n) $\Longrightarrow$ 7 does not divide m^2+n^2. Let us consider $m^2 = p\ (mod\ 7)$ and $n^2 = q\ (mod\ 7)$ when $m =$ $1,2,3,4,5,6\ (mod 7)$,
$n = 0,1,2,3,4,5,6\ (mod 7)$:

m	1	2	3	4	5	6
m^2	1	4	9	16	25	36
p	1	4	2	2	4	1

and

n	0	1	2	3	4	5	6
n^2	0	1	4	9	16	25	36
q	0	1	4	2	2	4	1

p/q	1	2	4
0	1	2	4
1	2	3	5
2	3	4	6
4	5	6	1

One can easily see that $m^2+n^2 = p+q \neq 0\,(mod\ 7)$.

Question 10 Let $X = \{n \in \mathbb{Z} : n > 2 \text{ and } n \text{ is odd}\}$ and $Y = \{n \in \mathbb{Z} : n < 11\}$. Prove that if $n \in X \cap Y$, then $n^2 - 2$ is a prime number.

Solution. By direct observation, we get: $n \in \{3,5,7,9\} \Longrightarrow n^2 - 2 \in \{7,23,47,79\}$, which are prime numbers.

Question 11 Show that if

$$x+y \geq 100 \implies x \geq 50 \text{ or } y \geq 50.$$

Solution.

(i) Solve by the direct method. The fact that for any numbers $x,\ y \in \mathbb{R} :\ x \leq\ y$, we have $x \leq \frac{x+y}{2} \leq y$. Indeed,

$$x - \frac{x+y}{2} = \frac{2x-x-y}{2} = \frac{x-y}{2} \leq 0 \Longrightarrow x \leq \frac{x+y}{2}.$$

Similarly, we have $\frac{x+y}{2} - y = \frac{x-y}{2} \leq 0$, which implies that $\frac{x+y}{2} \leq y$.
Now, assuming that $x+y \geq 100$, we have

$$50 = \frac{100}{2} \leq \frac{x+y}{2} \leq y, \text{ then } y \geq 50.$$

which is sufficient to make $x \geq 50$ or $y \geq 50$ true, and the proof is complete.

(ii) Solve by contraposition. Let us assume that $x \geq 50$ or $y \geq 50$ is wrong, then its negation is true:

$$[(x < 50) \text{ and } (y < 50)] \implies x+y < 100 \text{ which is the negation of } (x+y \geq 100).$$

Question 12 Prove that if x and y are positive real numbers, then $\sqrt{x+y} \neq \sqrt{x} + \sqrt{y}$.

Solution. Let us proceed by contradiction and assume that $\sqrt{x+y} = \sqrt{x} + \sqrt{y}$. Squaring both sides and simplifying, we get $2\sqrt{xy} = 0$, which implies either $x = 0$ or $y = 0$, which is a contradiction with the fact that x and y are positive real numbers.

Question 13 Use proof by contradiction to show

1. $\sqrt{11} \geq 1 + \sqrt{5}$.
2. $\sqrt{10} \geq 1 + \sqrt{3}$.

Solution. We proceed by contradiction.

1. $\sqrt{11} \geq 1+\sqrt{5}$.
   Suppose by contradiction that $\sqrt{11} < 1+\sqrt{5}$. Then $11 < (1+\sqrt{5})^2 = 1+5+2\sqrt{5}$. This further gives $5 < 2\sqrt{5}$, implying that $25 < 20$, which is a contradiction. Thus what we assumed is wrong.

2. $\sqrt{10} \geq 1+\sqrt{3}$.
   Suppose by contradiction that $\sqrt{10} < 1+\sqrt{3}$. Then $10 < (1+\sqrt{3})^2 = 1+3+2\sqrt{3}$. This further gives $6 < 2\sqrt{3}$, implying that $36 < 12$, which is a contradiction. Thus what we assumed is wrong.

Question 14 Prove that $2n^2+n$ is odd if and only if $\cos(n\pi/2)$ is even.

Solution. We proceed by direct proof.

1. $2n^2+n$ is odd, so $\cos(n\pi/2)$ is even.
   Suppose that $2n^2+n = n(2n+1)$ is odd, then n is odd, and thus $\cos(\frac{n\pi}{2}) = 0$[5].
2. $\cos(n\pi/2)$ is even, then $2n^2+n$ is odd.
   Suppose that $\cos\left(\frac{n\pi}{2}\right)$ is even and since it is between -1 and 1, $cos(\frac{n\pi}{2}) = 0$, and then n must be odd.

Question 15 Prove that if two of the three integers $x,\ y,\ z$ are even, then $3x+5y+7z$ is odd.

Solution. We proceed by direct proof. If two of the three integers x, y, z are even, then exactly two of the three terms in $3x+5y+7z$ are even, and since the third one is odd, $3x+5y+7z$ is odd.

Question 16 Show that between any two rational numbers, there exists always another rational number.

Solution. We proceed by direct proof. Let x and y be two rational numbers. Then $z = \dfrac{x+y}{2}$ is a rational number as half the sum of x and y. It is obvious that the average between x and y is always between x and y (see the solution of Question 11).

Question 17 Show that $7n^3-3n^2-4n$ is divisible by 6.

Solution. We have

$$\begin{array}{rcl} 7n^3-3n^2-4n &=& n(7n^2-3n-4), \\ &=& n\left(7n^2-7n+4n-4\right), \\ &=& n[\,7n(n-1)+4(n-1)\,], \\ &=& n(n-1)(7n+4). \end{array}, \quad \text{or use discriminant to factor.}$$

Since $n(n-1)$ is even as the product of two consecutive integers, we need to prove that $n(n-1)(7n+4)$ is a multiple of 3. To reach this objective, we may use a proof by cases, considering $n = 3k+r,\ r \in \{0,1,2\}$, or we may use the following nice simple idea based on observing that

$$\begin{array}{rcl} n(n-1)(7n+4) &=& n(n-1)\,[(n+1)\ +\ (6n+3)], \\ &=& \underbrace{n(n-1)(n+1)}_{3p}\ +\ \underbrace{3n(n-1)(2n+1)}_{3q};\ , \end{array}$$

therefore, $6\,|\,7n^3-3n^2-4n$.

Question 18 Prove that

$$\forall\, k \in \mathbb{Z},\ \ 5\ |\ k^5-5k^3+4k.$$

[5]Students need to know that the $\cos\left(\frac{(2k+1)\pi}{2}\right) = 0,\ \ \forall k \in \mathbb{Z}$.

Solution. Since we have $k^5 - 5k^3 + 4k = k(k^4 - 5k^2 + 4)$,

$$k\left[(k^2)^2 - 5(k^2) + 4\right] = k(k^2-1)(k^2-4) = k(k-1)(k+1)(k-2)(k+2).$$

Consider the following possible cases ($k = 5m + r,\ r \in \{0,1,2,3,4\}$) :

- $k = 5m+0,\ m \in \mathbb{Z}$. In this case, $5|k$, and then $5|k^5 - 5k^3 + 4k$.
- $k = 5m+1,\ m \in \mathbb{Z}$. In this case, $5|(k-1)$, and then $5|k^5 - 5k^3 + 4k$.
- $k = 5m+2,\ m \in \mathbb{Z}$. In this case, $5|(k-2)$, and then $5|k^5 - 5k^3 + 4k$.
- $k = 5m+3,\ m \in \mathbb{Z}$. In this case, $5|(k+2)$, and then $5|k^5 - 5k^3 + 4k$.
- $k = 5m+4,\ m \in \mathbb{Z}$. In this case, $5|(k+1)$, and then $5|k^5 - 5k^3 + 4k$.

Question 19 Show that $1 + (-1)^n(2n-1)$ is a multiple of 4 for all natural integer n.

Solution. We proceed by cases:

- If n is even, then $1 + (-1)^n(2n-1) = 2n$. Since n is even, $2n$ is a multiple of 4.
- If n is odd, then $1 + (-1)^n(2n-1) = -2(n-1)$. Since n is odd, $n-1$ is even, and thus $-2(n-1)$ is a multiple of 4.

Question 20 Show that $n^4 - n^2$ is a multiple of 12.

Solution. We have $n^4 - n^2 = n^2(n^2-1) \implies n^4 - n^2$ is even as a product of two consecutive integers. To complete the proof it is sufficient to prove that $n^4 - n^2$ is a multiple of 6. To do this, we note that $n^4 - n^2 = n^2(n^2-1) = n\Big(n(n-1)(n+1)\Big)$ is always a multiple of 6 because of the presence of the product of three consecutive integers: $(n-1)n(n+1)$.

Question 21 If $4 \nmid (n^2-4)$, then $4 \nmid (n-2)$ and $4 \nmid (n+2)$.

Solution. We proceed by contradiction. Let us assume that the negation of $4 \nmid (n-2)$ and $4 \nmid (n+2)$ is true, in other words

$$4 \mid (n-2) \text{ or } 4 \mid (n+2).$$

If we consider $4 \mid (n-2) \Longrightarrow 4 \big| \ (n-2)(n+2) = n^2 - 4$, which contradicts the assumption. Thus what we assumed is wrong.

Also, if we consider $4 \mid (n+2) \Longrightarrow 4 \mid (n+2)(n-2) = n^2 - 4$, which contradicts the assumption. Thus what we assumed is wrong.

Question 22 Let p and q be two integers. If $4p - q$ is divisible by 3, prove that

$$9 \,\Big|\, 2q^2 - 7pq - 4p^2 = E.$$

Solution. We proceed by direct proof. The assumption $4p - q$ is divisible by 3 implies

$$\exists\, m \in \mathbb{Z}:\ 4p - q = 3m \implies p - q = 3(m-p) = 3k \Longrightarrow p = q + 3k, \text{ for some integer } k.$$

By a direct long division (or a factorization),[6] we get

$$E = 2q^2 - 7pq - 4p^2 = (4p-q)(-p-2q).$$

Now, using $p = q + 3k$ for some integer k, the expression E becomes

$$\begin{aligned} E &= (4p-q)(-p-2q) = (3m)(-q-3k-2q) \\ &= 3m(-3)(q+k) = -9\,m(q+k), \end{aligned}$$

which completes the proof.

[6]We try to use the $(4p-q)$ term as follows:
$2q^2 - 7pq - 4p^2 = -2q(4p-q) + pq - 4p^2 = -2q(4p-q) - p(4p-q) = (4p-q)(-p-2q)$.

Question 23 Prove that for any positive integer we have $10 \mid 1+2^{3n}-3^n-6^n$.

Solution. We proceed by direct proof to show that our expression $m = 1+2^{3n}-3^n-6^n$ is divisible by 5 and 2 based on the observation that m can be expressed as: $m = (8^n-3^n) - (6^n-1) = (1-3^n) + (8^n-6^n)$.

$$6^n - 1 = (6-1)\left(6^{n-1}+\ldots+6^2+6+1\right) \Longrightarrow 5 \,\Big|\, (6^n-1).$$

On the other hand, we have

$$8^n - 3^n = (8-3)\left(8^{n-1}+8^{n-2}(3)+\ldots+8(3^{n-2})+3^{n-1}\right) \Longrightarrow 5 \,\Big|\, (8^n-3^n).$$

Therefore, $5 \mid m$. Moreover,

$$1-3^n = (1-3)(1+3+\ldots+3^{n-1}),$$

then $2|1-3^n$. In addition, noting that $2|2^{3n}$ and $2|6^n$, we get $2|m$. Finally, since $5|m$ and $2|m$, we conclude that $10|m$.

Question 24 Let a, b be odd integers. Prove that

if 4 does not divide $(a-b)$, then 4 cannot divide a^3-b^3.

Solution. We proceed by direct proof. Since a and b are odd, then $\exists\, k,\ k' \in \mathbb{Z}:\ a-b = 2k$ and $a^2+ab+b^2 = 2k'+1$. Moreover, since 4 does not divide $(a-b) = 2k$, then k must be odd. Thus we conclude

$$a^3-b^3 = (a-b)(a^2+ab+b^2) \;=\; 2k\,(2k'+1),\ \ k \text{ odd} \;\Longrightarrow 4 \text{ cannot divide } a^3-b^3.$$

Question 25 If n is a positive integer, prove that $(n^3-n)(5n+2)$ is a multiple of 24.

Solution. We proceed by direct proof.

$$\begin{aligned}(n^3-n)(5n+2) &= n(n-1)(n+1)(5n+2),\\ &= n(n-1)(n+1)(\ 5(n+2)\ -\ 8\ ),\\ &= 5n(n-1)(n+1)(n+2) - 8n(n-1)(n+1).\end{aligned}$$

Since $n(n-1)(n+1)$ is a multiple of 3, $8n(n-1)(n+1)$ is a multiple of 24. Therefore, to complete the proof, it is sufficient to show that $5n(n-1)(n+1)(n+2)$ is also a multiple of 24. This is implied by the fact that the product of 4 consecutive integers is divisible by 2, 3 and 4, which completes the proof.

(R) In the product of four consecutive integers we observe the following:

$$(n-1)\ \underbrace{n(n+1)(n+2)}_{6k} \quad \text{and} \quad \underbrace{(n-1)\,n}_{2p}.\ \underbrace{(n+1)(n+2)}_{2q}$$

Question 26 Prove that if an odd number m could be expressed as a sum of two squares, then it can be also expressed as $4n+1$ for some integer n.

Solution. We proceed by direct proof. Since m is an odd integer,[7] then we need to express it as a sum of a square of an odd number and a square of an even number as follows:

$$\exists\, p,\, q \in \mathbb{Z}:\ m \;=\; (2p)^2 \;+\; (2q+1)^2 \;=\; 4p^2+4q^2+4q+1 = 4\,(p^2+q^2+q) \;+\; 1 = 4n+1,$$

where $n = p^2+q^2+q \in \mathbb{Z}$.

Question 27 Prove that the sum of an irrational number and a rational number is an irrational number.

Solution. We proceed by contradiction. Let x be a rational number and y be an irrational number. Suppose that $z = x+y$ is rational. Then $y = z-x$ is rational, which is a contradiction what we assumed is wrong.

Question 28 Assume that $a,\ b,\ (a \neq b)$, and $\sqrt{a}-\sqrt{b}$ are three rational numbers. Prove that $\sqrt{a}$ and $\sqrt{b}$ are also rational.

Solution. We proceed by direct proof. Given $a \neq b$, we have $\sqrt{a}+\sqrt{b} = \dfrac{a-b}{\sqrt{a}-\sqrt{b}}$. Since a, b and $\sqrt{a}-\sqrt{b}$ are rational, $\sqrt{a}+\sqrt{b}$ is also rational. On the other hand, we have

$$\sqrt{a} = \frac{(\sqrt{a}-\sqrt{b})+(\sqrt{a}+\sqrt{b})}{2}, \qquad \sqrt{b} = \frac{(\sqrt{a}+\sqrt{b})-(\sqrt{a}-\sqrt{b})}{2},$$

implying that $\sqrt{a}$ and $\sqrt{b}$ are rational.

Question 29 Let x, y and z be non-zero real numbers. Prove that if

$$\begin{cases} x^2+y^2 \in \mathbb{Q}, \\ x^3+y^3 \in \mathbb{Q}, \\ x^4+y^4 \in \mathbb{Q} \end{cases} \implies xy,\ x+y \in \mathbb{Q}.$$

Solution. We proceed by direct proof. Using $(x^2+y^2)(x^4+y^4) \in \mathbb{Q}$ and $(x^3+y^3)^2 \in \mathbb{Q}$ and the fact that terms with a power of 6 can be simplified, we get:

$$(x^2+y^2)(x^4+y^4)-(x^3+y^3)^2 = x^2y^2(x^2+y^2)-2x^3y^3 = x^2y^2(x^2+y^2-2xy) \in \mathbb{Q}.$$

Now, since

$$\begin{cases} x^4+y^4 = (x^2+y^2)^2-2x^2y^2 \in \mathbb{Q}, \\ (x^2+y^2)^2 \in \mathbb{Q} \end{cases} \implies x^2y^2 \in \mathbb{Q}.$$

Thus

$$\begin{cases} x^2y^2(x^2+y^2-2xy) \in \mathbb{Q} \\ x^2y^2 \in \mathbb{Q} \end{cases} \implies x^2+y^2-2xy = \frac{x^2y^2(x^2+y^2-2xy)}{x^2y^2} \in \mathbb{Q},$$

and

$$xy = \frac{(x^2+y^2)-(x^2+y^2-2xy)}{2} \in \mathbb{Q}.$$

Finally, $x+y = \dfrac{x^3+y^3}{x^2+y^2-xy} \in \mathbb{Q}$.

Question 30 For any prime P, prove that $\sqrt{P}$ is not a rational number.

Solution. Follow the same idea of proving that $\sqrt{2}$ is irrational. You only need the following argument, if $a = mq$ and m is prime, then we conclude that m divides a.

[7] If m is odd, then m^2 is odd it cannot be a sum of two odd or two even numbers.

Question 31 Let $n \in \mathbb{N}$. Prove that

$$|n-1|+|n+1| \le 1 \implies |n^2-1| \le 4.$$

Solution. If we consider our implication as $p \implies q$, and since it is clear that the statement $p:\ |n-1|+|n+1| \le 1$ is always false, then the implication is always true, without considering the validity of statement q, which completes the proof.

Question 32 Prove that $\log_3(2)$ is irrational.

Solution. By contradiction, we assume that $\exists\, p,q \in \mathbb{N}, GCD(p,q)=1\colon \log_3(2) = \frac{p}{q}$.

Since

$$\log_3(2) = \frac{p}{q} \iff 2 = 3^{p/q} \iff \underbrace{2^q}_{\text{even}} = \underbrace{3^p}_{\text{odd}}$$

is impossible, so what we assumed is wrong.

Question 33 Prove that $\sqrt[3]{2}$ is irrational.

Solution. Solve by contradiction (Follow the same idea of proving that $\sqrt{2}$ is irrational).

Question 34 Prove in two ways that for all $n \in \mathbb{N}$

$$\frac{1}{2\sqrt{n+1}} < \sqrt{n+1}-\sqrt{n} < \frac{1}{2\sqrt{n}}.$$

Solution.

- *Method 1.* We have

$$\sqrt{n+1}-\sqrt{n} = (\sqrt{n+1}-\sqrt{n})\frac{\sqrt{n+1}+\sqrt{n}}{\sqrt{n+1}+\sqrt{n}} = \frac{1}{\sqrt{n+1}+\sqrt{n}},$$

and since

$$\sqrt{n+1}+\sqrt{n} > \sqrt{n}+\sqrt{n} = 2\sqrt{n} \implies \frac{1}{\sqrt{n+1}+\sqrt{n}} < \frac{1}{2\sqrt{n}},$$

and

$$\sqrt{n+1}+\sqrt{n} < \sqrt{n+1}+\sqrt{n+1} = 2\sqrt{n+1} \implies \frac{1}{\sqrt{n+1}+\sqrt{n}} > \frac{1}{2\sqrt{n+1}},$$

we obtain,

$$\frac{1}{2\sqrt{n+1}} < \sqrt{n+1}-\sqrt{n} < \frac{1}{2\sqrt{n}}.$$

- *Method 2.* Let us consider the difference

$$\begin{aligned}
\sqrt{n+1}-\sqrt{n}-\frac{1}{2\sqrt{n}} &= \frac{1}{2\sqrt{n}}\left(2\sqrt{n(n+1)}-2n-1\right),\\
&= \frac{-1}{2\sqrt{n}}\left(2n+1-2\sqrt{n(n+1)}\right),\\
&= \frac{-1}{2\sqrt{n}}\left(n-2\sqrt{n}\sqrt{n+1}+(n+1)\right),\\
&= \frac{-1}{2\sqrt{n}}\left(\sqrt{n}-\sqrt{n+1}\right)^2,\\
&< 0.
\end{aligned}$$

Hence, $\sqrt{n+1}-\sqrt{n} < \frac{1}{2\sqrt{n}}$. The second inequality, can be done similarly.

Question 35 Prove that for any positive integer n, we have

$$3 \mid n \iff 7 \,\Big|\, (2^n - 1).$$

Solution. Let us do the first implication by direct proof and the second implication by contraposition:

- $3 \mid n \implies 7 \,\Big|\, (2^n - 1)$.
  Since $3|n$, there exists $k \in \mathbb{Z}$ such that $n = 3k$. Therefore,
  $$2^n - 1 = 2^{3k} - 1 = 8^k - 1 = (8-1)(1+8+8^2+...+8^{n-1}),$$
  which implies that $7|2^n - 1$.
- $7 \,\Big|\, (2^n - 1) \implies 3 \mid n$.
  Assume that $3 \nmid n \Longrightarrow \exists\, k \in \mathbb{Z},\ r \in \{1,2\} :\ n = 3k + r$. Thus we have
  $$2^n - 1 = 2^{3k+r} - 1 = 2^r\,(8^k) \;-\; 1 = 2^r\left(8^k - 1 + 1\right) - 1 = 2^r(8^k - 1) + 2^r - 1.$$
  Using the fact that $(8^k - 1) = 7m$ for some $m \in \mathbb{Z}$ and as well as that $2^r - 1 \in \{1,\ 3\}$, we can conclude that for $m' = 2^r m$
  $$\left(2^n - 1 = 7m' + 1 \;\text{ or }\; 7m' + 3\right) \implies 7 \nmid (2^n - 1),$$
  which contradicts our hypothesis, so what we assumed is wrong.

Question 36 Prove that

$$p \neq 3,\ \ p \text{ prime} \quad \implies \quad 3 \mid (p^2\ +\ 2).$$

Solution. We will proceed by direct proof. The result is true for $p = 2$, since $3 \mid p^2 + 2 = 6$. Let us consider a prime number $p > 3$, so it can be expressed as $3k \pm 1$ for some positive integer $k \geq 2$. Then we have the following situation:

$$2 + p^2 = 2 + (3k \pm 1)^2 \;=\; 2 +\, 9k^2 \pm 6k + 1 \;=\; 3(3k^2 \pm 2k + 1) \;=\; 3\,k',\ \ k' > 0 \implies 3 \mid p^2\ +\ 2.$$

Question 37 Show that $n(n+1)(n+2)(n+3)+1$ is a perfect square.

Solution. If we start expanding the product, we may proceed as follows:

$$(n^2+n)(n^2+5n+6)+1 \;\text{ or }\; (n^2+2n)(n^2+4n+3)+1 \;\text{ or }\; (\mathbf{n^2}+\mathbf{3n})(\mathbf{n^2}+\mathbf{3n}+2)+1.$$

Thus the idea came directly and we obtain, if we set $m = n^2 + 3n$, thus

$$n(n+1)(n+2)(n+3)+1 = m(m+2)+1 = m^2+2m+1 = (m+1)^2 \;=\; (n^2+3n+1)^2.$$

Question 38 Prove by mathematical induction that $n^2 - n$ is even for any positive integer n.

Solution. Let $P(n)$: $\exists\, k' \in \mathbb{N} :\ n^2 - n = 2k'$.

Base step. $n = 1,\ 1^2 - 1 = 0$, then $P(1)$ is true.

Inductive step. Assume that $P(k)$ is true, and let us show that $P(k+1)$ is also true:

$$P(k+1):\ (k+1)^2 - (k+1) = k^2 + 2k + 1 - k - 1 = (k^2 - k) + 2k = 2k' + 2k = 2(k' + k).$$

Thus $P(k+1)$ is true. Hence, $P(n)$ is true for any positive integer.

Question 39 Prove that 7 divides $3^{2n+1} + 2^{4n+2}$ for any $n \geq 0$.

Solution. Let $P(n)$: $\exists\, t \in \mathbb{Z}:\ 3^{2n+1} + 2^{4n+2} = 7t$. Note that $2^{4n+2} = 2^{2(2n+1)} = 4^{2n+1}$.

Base step. $n = 0$, $3^{2(0)+1} + 2^{0+2} = 3 + 4 = 7$, so $P(0)$ is true.

Inductive step. Assume that $P(k)$ is true, and let us show that $P(k+1)$ is also true.

- *Idea 1.*

$$\begin{aligned}
P(k+1)\colon 3^{2(k+1)+1} + 2^{4(k+1)+2} &= 3^{2k+3} + 2^{4k+6},\\
&= 3^2 \times 3^{2k+1} + 4^{2k+3},\\
&= 9 \times 3^{2k+1} + 4^2 \times 4^{2k+1},\\
&= 9 \times 3^{2k+1} + 16 \times 4^{2k+1},\\
&= 9 \times 3^{2k+1} + (9+7) \times 4^{2k+1},\\
&= 7 \times 4^{2k+1} + 9\left(\underbrace{3^{2k+1} + 4^{2k+1}}_{7t}\right),\\
&= 7 \times \left(4^{2k+1} + 9 \times t\right),\\
&= 7\left(4^{2k+1} + 9t\right) \implies P(k+1) \text{ is true.}
\end{aligned}$$

Thus $7 \mid 3^{2n+1} + 2^{4n+2}$ for any $n \geq 0$.

- *Idea 2.* From $P(k)$ we get: $3^{2k+1} = 7t - 4^{2k+1}$, and then $P(k+1)$ becomes

$$\begin{aligned}
P(k+1):\ 3^{2(k+1)+1} + 4^{2(k+1)+1} &= 3^{2k+3} + 4^{2k+3},\\
&= 3^2 \times 3^{2k+1} + 4^2 \times 4^{2k+1},\\
&= 9 \times (7t - 4^{2k+1}) + 4^2 \times 4^{2k+1},\\
&= 9 \times (7t)\ +\ 7 \times 4^{2k+1},\\
&= 7\left(9t + 4^{2k+1}\right) \implies P(k+1) \text{ is true,}
\end{aligned}$$

Thus $7 \mid 3^{2n+1} + 2^{4n+2}$ for any $n \geq 0$.

Question 40 Let a, b be two real numbers. Show that
$a^n - b^n = (a-b)\left[a^{n-1} + a^{n-2}b + a^{n-3}b^2 + \ldots + ab^{n-2} + b^{n-1}\right]$ for any $n \geq 2$.

Solution. Let $P(n)$: $a^n - b^n = (a-b)\ \left[a^{n-1} + a^{n-2}b + a^{n-3}b^2 + \ldots + ab^{n-2} + b^{n-1}\right]$.

Base step. $n = 2$, $a^2 - b^2 = (a-b)(a+b) = (a-b)\left[a^{2-1} + b^{2-1}\right]$, so $P(2)$ is true.

Inductive step. Assume that $P(k)$ is true, and let us show that $P(k+1)$ is also true.

$$\begin{aligned}
P(k+1)\colon a^{k+1} - b^{k+1} &= a \times \left(a^k - b^k + b^k\right) - b^{k+1},\\
&= a \times (a-b) \sum_{i=0}^{k-1} a^{k-1-i} b^i + a \times b^k - b^{k+1},\\
&= a(a-b) \sum_{i=0}^{k-1} a^{k-1-i} b^i + b^k(a-b),\\
&= (a-b)\left[\sum_{i=0}^{k-1} a \times a^{k-1-i} b^i + b^k\right],\\
&= (a-b)\left[\sum_{i=0}^{k-1} a^{k-i} b^i + b^k\right],\\
&= (a-b)\left[\sum_{i=0}^{k} a^{k-i} b^i\right],\\
&= (a-b)\left[a^k + a^{k-1}b + a^{k-2}b^2 + \ldots + ab^{k-1} + b^k\right].
\end{aligned}$$

Thus $P(k+1)$ is true. Hence, $P(n)$ is true for any $n \geq 2$.

Question 41 Let a, b be two integers. Show that $(a-b) \mid a^n - b^n$ for any $n \geq 1$.

Solution. It is sufficient to apply the result in the previous question by pointing to the fact that

$$\sum_{i=0}^{n-1} a \times a^{n-1-i} b^i$$

is an integer; and thus, we have

$$a^n - b^n = (a-b)\left[a^{n-1} + a^{n-2}b + a^{n-3}b^2 + \ldots + ab^{n-2} + b^{n-1}\right] = (a-b)k, \; k \in \mathbb{Z}.$$

Question 42 Show by induction that $5^{2n} - 1$ is divisible by 8 for any $n \geq 0$.

Solution.

- *Idea 1.* By using (PMI): let $P(n)$: $\exists\, m \in \mathbb{N}$: $5^{2n} - 1 = 8m$.
  Base step. $n = 0$, $5^{2(0)} - 1 = 0$, so $P(0)$ is true.
  Inductive step. Assume that $P(k)$ is true; and let us show that $P(k+1)$ is also true.
  From $P(k)$, we have: $5^{2k} = 1 + 8m$, and then

$$\begin{aligned} P(k+1)\colon 5^{2(k+1)} - 1 &= 5^{2k+2} - 1 = 5^2 \times 5^{2k} - 1, \\ &= 25 \times (1+8m) - 1 = 25 + 8(25m) - 1, \\ &= 24 + 8(25k) \;=\; 8(3+25k) \;\Longrightarrow\; P(k+1) \text{ is true.} \end{aligned}$$

  Thus $8 \mid 5^{2n} - 1$ for any $n \geq 0$.
- *Idea 2.* Using the identity:

$$a^n - 1 = (a-1)\left(a^{n-1} + a^{n-2} + \ldots + a^3 + a^2 + a + 1\right).$$

  Now,

$$5^{2n} - 1 \;=\; (5^2)^n - 1 \;=\; 25^n - 1 \;=\; (25-1)\left(25^{n-1} + 25^{n-2} + \ldots + 25^3 + 25^2 + 25 + 1\right),$$

  we obtain

$$5^{2n} - 1 \;=\; 24\,k \;= 8\,k', \;\; k' = 3\sum_{i=0}^{n-1} 25^i \;\in\; \mathbb{N}.$$

  Hence, $8 \mid 5^{2n} - 1$ for any $n \geq 0$.

Question 43 Show that $10^n + 3 \times 4^{n+2} + 5$ is divisible by 9 for any $n \geq 0$.

Solution. Let $P(n)$: $\exists\, k \in \mathbb{N}:\; 10^n + 3 \times 4^{n+2} + 5 = 9k$.
Base step. $n = 0$, $10^0 + 3 \times 4^{0+2} + 5 = 1 + 3(16) + 5 = 54 = 9(6)$, so $P(0)$ is true.
Inductive step. Assume $P(n)$ is true, we will show that $P(n+1)$ is also true.
From $P(n)$, we have $10^n = 9k - 5 - 3 \times 4^{n+2}$.
Now,

$$\begin{aligned} P(n+1):\; 10^{n+1} + 3 \times 4^{n+1+2} + 5 &= 10 \times 10^n + 3 \times 4^{n+2} \times 4 + 5, \\ &= 10 \times (9k - 5 - 3 \times 4^{n+2}) + 12 \times 4^{n+2} + 5, \\ &= 90k - 50 - 30 \times 4^{n+2} + 12 \times 4^{n+2} + 5, \\ &= 90k - 45 - 18 \times 4^{n+2}, \\ &= 9\left(10k - 5 - 2 \times 4^{n+2}\right) = 9\,k', \;\; k' \in \mathbb{Z} \;\Longrightarrow\; P(k+1) \text{ is true.} \end{aligned}$$

Thus, $9 \mid 10^n + 3 \times 4^{n+2} + 5$ for any $n \geq 0$.

Question 44 Prove by induction

$$1 + 2^3 + 3^3 + \ldots + n^3 = \frac{n^2(n+1)^2}{4}, \;\; \forall n \geq 1.$$

Solution. Let $P(n)$: $1+2^3+3^3+\ldots+n^3=\dfrac{n^2(n+1)^2}{4}$.

Base step. $n=1$, $1^3=1=\dfrac{1^2(1+1)^2}{4}=1$, so $P(1)$ is true.

Inductive step. Assume $P(k)$ is true, we will show that $P(k+1)$ is also true.

$$\begin{aligned}
P(k+1):\ \sum_{k=1}^{k+1} k^3 &= \sum_{k=1}^{k} k^3 + (k+1)^3, \quad (\text{using } P(k))\\
&= \frac{k^2(k+1)^2}{4}+(k+1)^3 = (k+1)^2\left[\frac{k^2}{4}+(k+1)\right],\\
&= \frac{(k+1)^2}{4}\left[k^2+4(k+1)\right] = \frac{(k+1)^2(k^2+4k+4)}{4},\\
&= \frac{(k+1)^2(k+2)^2}{4} \quad \Longrightarrow P(k+1) \text{ is also true.}
\end{aligned}$$

Hence, $1+2^3+3^3+\ldots+n^3=\dfrac{n^2(n+1)^2}{4}$ for any $n\geq 1$.

Question 45 Prove by Induction

$1\times 2+2\times 3+3\times 4+\ldots+(n-1)\times n=\dfrac{n(n-1)(n+1)}{3}$, $\forall n\geq 2$.

Solution. Let $P(n)$: $\displaystyle\sum_{k=2}^{n}(k-1)k = \frac{n(n-1)(n+1)}{3}$.

Base step. $n=2$, $(2-1)2=2=\dfrac{2(2-1)(2+1)}{3}=2$. Then $P(2)$ is true.

Inductive step. Assume that $P(m)$ is true, and let us show that $P(m+1)$ is also true.

$$\begin{aligned}
P(m+1):\ \sum_{k=2}^{m+1}(k-1)k &= \textstyle\sum_{k=2}^{m}(k-1)k + m(m+1),\\
&= \frac{m(m-1)(m+1)}{3} + m(m+1),\\
&= \frac{m(m+1)}{3}((m-1)+3),\\
&= \frac{m(m+1)}{3}(m+2),\\
&= \frac{(m+1)m(m+2)}{3} \quad \Longrightarrow P(m+1) \text{ is also true.}
\end{aligned}$$

Thus $\displaystyle\sum_{k=2}^{n}(k-1)k = \frac{n(n-1)(n+1)}{3}$ for any $n\geq 2$.

Question 46 Prove by induction

$\dfrac{1}{1\times 3}+\dfrac{1}{3\times 5}+\dfrac{1}{5\times 7}+\ldots+\dfrac{1}{(2n-1)\times(2n+1)}=\dfrac{n}{2n+1}$, $\forall n\geq 1$.

Solution. Let $P(n)$: $\displaystyle\sum_{k=1}^{n}\frac{1}{(2k-1)\times(2k+1)}=\frac{n}{2n+1}$.

Base step. $n=1$, $\dfrac{1}{(2(1)-1)\times(2(1)+1)}=\dfrac{1}{3}=\dfrac{1}{2(1)+1}$, then $P(1)$ is true.

Inductive step. Assume $P(k)$ is true, and let us show that $P(k+1)$ is also true:

$$\begin{aligned}
P(k+1):\sum_{i=1}^{k+1}\frac{1}{(2i-1)\times(2i+1)} &= \sum_{i=1}^{k}\frac{1}{(2i-1)\times(2i+1)}+\frac{1}{(2k+1)\times(2k+3)},\\
&= \frac{k}{2k+1}+\frac{1}{(2k+1)\times(2k+3)},\\
&= \frac{1}{2k+1}\left[k+\frac{1}{2k+3}\right],\\
&= \frac{1}{2k+1}\left[\frac{k(2k+3)+1}{2k+3}\right],\\
&= \frac{2k^2+3k+1}{(2k+1)(2k+3)},\\
&= \frac{(2k+1)(k+1)}{(2k+1)(2k+3)}=\frac{k+1}{2k+3} \quad \Longrightarrow P(k+1) \text{ is also true.}
\end{aligned}$$

Thus $\sum_{k=1}^{n}\frac{1}{(2k-1)\times(2k+1)}=\frac{n}{2n+1}$ for any $n\geq 1$.

Question 47 Determine the sum $\sum_{k=1}^{n}(-1)^k k$.

Solution. The given sum can be re-expressed as follows:

$$S=\sum_{k=1}^{n}(-1)^k k=-1+2-3+4-5+6-\ldots\pm(n-1)\mp n.$$

Consider the two possible cases:

(i) $\exists p\in\mathbb{N}: n=2p$; the preceding sum is equivalent to:

$$S_{2p}=\sum_{k=1}^{2p}(-1)^k k=-1+2-3+4-5+6-\ldots-(2p-1)+2p=\sum_{k=1}^{p}2k-\sum_{k=1}^{p}(2k-1),$$

i.e.,

$$S_{2p}=\sum_{k=1}^{p}2k-\sum_{k=1}^{p}2k+\sum_{k=1}^{p}1=p.$$

(ii) $\exists p\in\mathbb{N}: n=2p+1$; the above sum is equivalent to:

$$S_{2p+1}=\sum_{k=1}^{2p+1}(-1)^k k=-1+2-3+4-5+6-\ldots+2p-(2p+1)=\sum_{k=1}^{p}2k-\sum_{k=1}^{p+1}(2k-1),$$

i.e., $S_{2p+1}=\sum_{k=1}^{p}2k-\sum_{k=1}^{p}2k-2(p+1)+\sum_{k=1}^{p+1}1=-2(p+1)+(p+1)=-(p+1).$

Finally,

$$\sum_{k=1}^{n}(-1)^k k\begin{cases}\dfrac{n}{2} & \text{if } \text{ n is even,}\\[2ex] \dfrac{-(n+1)}{2} & \text{if } \text{ n is odd.}\end{cases}$$

Question 48 Let n be a given integer. Prove that if there exists $p \in \mathbb{Z}:\ a-b=pn$, then there exists $q \in \mathbb{Z}:\ a^k - b^k = q\,n,$ for all positive integers k.

Solution.

- *Idea 1.* If

$$\begin{aligned}\exists\, p \in \mathbb{Z}:\ a-b=pn &\implies \exists\, p \in \mathbb{Z}:\ a=b+np\\ &\implies \exists\, p \in \mathbb{Z}:\ a^k=(b+np)^k,\ \forall\, k\geq 1,\quad \text{(by expanding)}\\ &\implies \exists\, p \in \mathbb{Z}:\ a^k=b^k+n\ \text{(expression in } b,\ p,\ n)\\ &\implies \exists\, m \in \mathbb{Z}:\ a^k=b^k+n\,m,\quad m=\text{expression in } b,\ p,\ n\in\mathbb{Z}.\end{aligned}$$

  To expand the expression in the second line of the proof, we use the binomial formula:

$$(b+np)^k=\sum_{i=0}^{k}\binom{k}{i}b^i\,(np)^{k-i}=\sum_{i=0}^{k-1}\binom{k}{i}b^i\,(np)^{k-i}+b^k=b^k+n\underbrace{\sum_{i=0}^{k-1}\binom{k}{i}b^i\,n^{k-i-1}p^{k-i}}_{\text{expression in } b,\ p,\ n}.$$

- *Idea 2.* If $n \mid (a-b)$, and according to the result in Question 41, we have $(a-b) \mid (a^k-b^k)$ for any $k \geq 1$, therefore $n \mid a^k - b^k$.

Question 49 Show that the nth derivative of x^n is $n!,\ \ n\geq 1$.

Solution. We will use the following notation from Calculus:

$$f'=f^{(1)},\ f''=f^{(2)},\ f'''=f^{(3)},\ldots,\ f^{(n)}\ \text{nth derivative of}\ f.$$

Let $P(n):\ (x^n)^{(n)}\ =n!$

Base step. $n=1,\ \ x'=1=1!$, then $P(1)$ is true.

Inductive step. Assume that $P(k)$ is true, and let us prove that $P(k+1)$ is also true.

$$(x^{k+1})^{(k+1)}\ =\ \left[(x^{k+1})^{(1)}\right]^{(k)}=\left[(k+1)\,x^k\right]^{(k)}\ =\ (k+1)\,(x^k)^{(k)}=(k+1)\,k!=(k+1)!$$

Thus $P(k+1)$ is true, and then $P(n)$ is true for any $n \geq 1$.

Question 50 Let S be the sum

$$S\ =\ 1+2+3+\ldots+n+(n+1)+n+(n-1)+(n-2)+\ldots+3+2+1.$$

Prove that S is a perfect square.

Solution. Directly by applying twice Gauss sum:

$$S=\frac{n(n+1)}{2}+(n+1)+\frac{n(n+1)}{2}\ =\ n(n+1)+(n+1)\ =\ (n+1)(n+1)\ =(n+1)^2.$$

Question 51 Prove by induction that

$$\left(1-\frac{1}{2}\right)\left(1-\frac{1}{3}\right)\left(1-\frac{1}{4}\right)\ldots\left(1-\frac{1}{n}\right)=\frac{1}{n},\ \ \forall n\geq 2.$$

Solution. Let $P(n):\ \displaystyle\prod_{k=2}^{n}\left(1-\frac{1}{k}\right)=\frac{1}{n}.$

Base step. $n=2,\ \left(1-\dfrac{1}{2}\right)=\dfrac{1}{2}$, so $P(2)$ is true.

Inductive step. Let us assume $P(m)$ and $P(m-1)$ are true, and let us prove that $P(m+1)$ is also true.

$$\prod_{k=2}^{m+1}\left(1-\frac{1}{k}\right)=\left(1-\frac{1}{m+1}\right)\prod_{k=2}^{m}\left(1-\frac{1}{k}\right)=\frac{m}{m+1}\times\frac{1}{m}=\frac{1}{m+1}.$$

Thus $P(m+1)$ is true, so $P(n)$ is true for any $n \geq 2$.

Question 52 Prove that for all $n \geq 2$

$$\sum_{k=1}^{n-1}\frac{1}{\sqrt{k}+\sqrt{k+1}} = \sqrt{n}-1.$$

Solution.

(i) *Idea 1.* Let $P(n): \sum_{k=1}^{n-1}\frac{1}{\sqrt{k}+\sqrt{k+1}} = \sqrt{n}-1.$

Base step. $n=2$, $\frac{1}{\sqrt{1}+\sqrt{2}}=\frac{\sqrt{2}-\sqrt{1}}{2-1}=\sqrt{2}-1$, so $P(2)$ is true.

Inductive step. Let us assume $P(m)$ is true, and let us prove that $P(m+1)$ is also true:

$$\sum_{k=1}^{m}\frac{1}{\sqrt{k}+\sqrt{k+1}}=\sum_{k=1}^{m-1}\frac{1}{\sqrt{k}+\sqrt{k+1}}+\frac{1}{\sqrt{m}+\sqrt{m+1}},$$

i.e.,

$$\sqrt{m}-1+\frac{\sqrt{m+1}-\sqrt{m}}{(m+1)-m}=\sqrt{m+1}-1 \implies P(m+1) \text{ is true.}$$

(ii) *Idea 2.* Using the conjugate of the denominators, we get:

$$\frac{1}{\sqrt{k}+\sqrt{k+1}}=\frac{\sqrt{k+1}-\sqrt{k}}{k+1-k}=\sqrt{k+1}-\sqrt{k}.$$

Thus

$$\sum_{k=1}^{n-1}\frac{1}{\sqrt{k}+\sqrt{k+1}}=\sum_{k=1}^{n-1}\sqrt{k+1}-\sqrt{k}$$

which by telescoping gives $\sum_{k=1}^{n-1}\frac{1}{\sqrt{k}+\sqrt{k+1}}=\sqrt{n}-\sqrt{1}=\sqrt{n}-1.$

Question 53 Show by induction that n^3+3n^2+2n is divisible by 6 for any $n \geq 0$.

Solution. Let $P(n): \exists\, k \in \mathbb{Z}: n^3+3n^2+2n=6k.$

Base step. $n=0$, $0=6(0)$, so $P(0)$ is true.

Inductive step. Let us assume $P(k)$ is true, and let us prove $P(k+1)$ is also true.

If we factor $k^3+3k^2+2k=k(k^2+3k+2)=k(k+1)(k+2)$, we will easily obtain the relation at rank $k+1$. In other words,

$$(k+1)(k+2)(k+3)=\underbrace{k(k+1)(k+2)}_{P(k)}+3\underbrace{(k+1)(k+2)}_{even}=6k+3(2p)=6(k+p),$$

for some integer p. Thus $6\,|\,(k+1)(k+2)(k+3)$, and finally, $6\,|\,n^3+3n^2+2n$ for any $n \geq 0$.

Question 54 Prove in two ways that 4^n-1 is divisible by 3 for any non-negative integer.

Solution.

(i) *Idea 1.* By induction, let $P(n): \exists\, k \in \mathbb{N}: \; 4^n - 1 = 3k$.
Base step. $n = 0, \;\; 4^0 - 1 = 0 = 3(0)$, so $P(0)$ is true.
Inductive step. Let us assume $P(m)$ is true, and let us prove $P(m+1)$ is also true;

$$4^{m+1} - 1 \;=\; 4 \times 4^m - 1 = (3+1) \times 4^m - 1 \;=\; 3 \times 4^m + (4^m - 1) = 3(4^m + k) = 3k',\; k' \in \mathbb{N}.$$

Thus $P(m+1)$ is true. Hence, $P(n)$ is true for any $n \geq 0$.

(i) *Idea 2.* It suffices to directly apply the identity $(a-b)\,|\,a^n - b^n$ in $\mathbb{Z}$, with $a = 4$ and $b = 1$.

Question 55 Show that $3^{2n+3} + 40n - 27$ is divisible by 64 for any non-negative integer.

Solution. Let $P(n): \exists\, k \in \mathbb{N}: \; 3^{2n+3} + 40n - 27 = 64k$.
Base step. $n = 0, \;\; 3^3 - 27 = 0 = 64(0)$, so $P(0)$ is true.
Inductive step. Let us assume $P(m)$ is true, and let us prove $P(m+1)$ is also true.
From $P(m)$, we have: $3^{2m+3} = 64k + 27 - 40m$, and then

$$\begin{aligned} 3^{2m+5} + 40(m+1) - 27 &= 9(3^{2m+3}) + 40m + 13 = 9(64k + 27 - 40m) + 40m + 13 \\ &= 64(9k+4) - 8 \times 40m = 64(9k + 4 - 5m) \\ &= 64k',\;\; k' \in \mathbb{Z} \Longrightarrow P(m+1) \text{ is true.} \end{aligned}$$

Hence, $P(n)$ is true for any $n \geq 0$.

Question 56 Prove that $2^n \; < \; n!, \;\; \forall\, n \geq 4$.

Solution. Let $P(n): \; 2^n \; < \; n!$
Base step. $n = 4, \; 2^4 = 16 < 4! = 24$, then $P(4)$ is true.
Inductive step. Assume $P(k)$ is true, and let us show $P(k+1)$ is also true.
Since $2^{k+1} = 2 \times 2^k \; < \; 2k!$, it suffices to prove that $2k! < (k+1)!$ to complete the induction. To this end, let us consider the difference:

$$2k! \;-\; (k+1)! \;=\; k!\,(\,2 - (k+1)\,) \;=\; (1-k)\,k! < 0, \;\; \forall\, k \geq 4 \;\Longrightarrow\; 2k! < (k+1)!$$

Thus $2^{k+1} = 2 \times 2^k \; < \; 2k! < (k+1)!$, and then $P(k+1)$ is satisfied. Hence, $P(n)$ is true for any $n \geq 4$.

Question 57 [**Bernoulli Inequality**] Show that for all $\alpha > -1, \; \alpha \neq 0$, and $n \geq 2$, we have

$$(1+\alpha)^n \; > \; 1 \; + \; n\,\alpha.$$

Solution. By induction, let $P(n): \;(1+\alpha)^n \; > \; 1 \; + \; n\,\alpha$.
Base step. $n = 2, \; (1+\alpha)^2 = 1 + 2\alpha + \alpha^2 \; > \; 1 + 2\alpha$, so $P(2)$ is true.
Inductive step. Assume $P(k)$ is true, and let us prove $P(k+1)$ is also true.
Since

$$\begin{aligned} (1+\alpha)^{k+1} &= (1+\alpha)(1+\alpha)^k \; > \; (1+\alpha)(1 \; + \; k\,\alpha), \\ &> 1 \; + \; k\,\alpha \; + \; \alpha \; + \; k\,\alpha^2, \\ &> 1 \; + \; (k+1)\,\alpha. \end{aligned}$$

Thus $P(k+1)$ is true, and finally, $(1+\alpha)^n \; > \; 1 \; + \; n\,\alpha, \;\; \forall\, \alpha > -1, \;\; \alpha \neq 0$, and $n \geq 2$.

Question 58 Given

$$\begin{cases} u_0 = 2, \; u_1 = 3 \\ u_{n+1} = 3u_n - 2u_{n-1}, \; n \geq 1. \end{cases}$$

Prove that $u_n = 1 + 2^n, \; \forall n \geq 0$.

Solution. We will use strong induction. Let $P(n): \ u_n = 1+2^n$.

Base step. $n=0,\ u_0 = 2 = 1+2^0$, then $P(0)$ is true

Inductive step. Let us assume that $P(k)$ and $P(k-1)$ are true, and let us prove that $P(k+1)$ is also true.

$$u_{k+1} = 3u_k - 2u_{k-1} \ = \ 3(1+2^k) - 2(1+2^{k-1}) \ = \ 1+3(2^k) - 2^k = 1+(2)2^k = 1+2^{k+1}.$$

Thus $P(k+1)$ is true, and then $u_n = 1+2^n$ for any $n \geq 0$.

5. More on Proof Techniques

Chapter Key Terms

(a) *proofs in sets*
(b) *proofs in calculus*
(c) *proofs in number theory*
(d) *Linear Algebra*
(e) *integral and fractional parts of real number*
(f) *classical sums*
(g) *telescoping sums*

5.1 Examples from Sets

In this subsection, we gather many proofs of the given results (theorems and propositions) in Chapter 3. These results are exposed here as solved examples. It is important for students to master how to prove that $z \in A$, where A is a given set. In general, we have to proceed as follows:

$$\text{to show that } z \in A = \{x\colon P(x)\} \text{ check that } P(z) \text{ is true,}$$

or

$$\text{to show that } z \in A = \{x \in U\colon P(x)\} \text{ check that } \begin{cases} z \in U, \\ P(z) \text{ is true.} \end{cases}$$

■ **Example 5.1** Check whether 20, $-100/3$, $-75 \in A = \{t \in \mathbb{Z}\colon 5 \text{ divides } t\}$.
Any element of set A needs to satisfy two conditions: (i) to be an integer, (ii) and be a multiple of 5:

- $t = 20 \in \mathbb{Z}$ and $t = 20 = 5\,(4) \implies 5\,|\,t \implies 20 \in A$.
- $t = -100/3 \notin \mathbb{Z} \implies -100/3 \notin A$.
- $t = -75 \in \mathbb{Z}$ and $t = -75 = 5\,(-15) \implies 5\,|\,t \implies -75 \in A$.

■

■ **Example 5.2** Check whether the following sets

$$M_1 = \{(x,y,z) : x+y+z=2\}, \quad M_2 = \left\{(x,y,\sqrt{\sin(xy)-2}) : x,y, \in \mathbb{R}\right\}$$

are empty.

Checking the non-emptiness of a set, requires providing an element from this set:

- A triple $(x,y,z) \in M_1$ needs to satisfy $x+y+z=2$, so it is easy to determine for instance, $(2,0,0),(0,2,0),(0,0,2)$, or $(1,1,0)$ are all in M_1. Thus we conclude that $M_1 \neq \emptyset$.
- For set M_2, we do not have any equation to check, but we have an expression under the radical symbol that needs to be non-negative. In other words,

  $$(x,y,z) \in M_2 \iff z = \sqrt{\sin(xy)-2}, \text{ and } \sin(xy)-2 \geq 0.$$

  The condition means that $\sin(xy) \geq 2$, which is impossible, since $\sin(xy) \leq 1$. Thus we conclude that $M_2 = \emptyset$.

■

■ **Example 5.3** Prove that

$$A \cap B = B \Longrightarrow B \subseteq A.$$

To show that $B \subseteq A$, we need to prove the following:

$$\boxed{x \in B \Longrightarrow x \in A.}$$

To this end, let us consider

$$\begin{aligned} x \in B \quad & \overset{hypothesis}{\Longleftrightarrow} \quad x \in A \cap B, \\ & \Longrightarrow \quad x \in A \text{ and } x \in B, \\ & \Longrightarrow \quad x \in A. \end{aligned}$$

which completes the proof. ■

■ **Example 5.4** Prove that

$$A \cap B = A \cup B \Longrightarrow A = B.$$

To show that $A = B$, we need to prove the following:

$$\boxed{A \subseteq B \text{ and } B \subseteq A.}$$

by assuming $A \cap B = A \cup B$ (hypothesis). To this end, let us consider

$$\begin{aligned} x \in A \quad & \Longrightarrow \quad x \in A \cup B, \\ & \overset{hypothesis}{\Longleftrightarrow} \quad x \in A \cap B, \\ & \Longrightarrow \quad x \in A \text{ and } x \in B, \\ & \Longrightarrow \quad x \in B. \end{aligned}$$

Thus $A \subseteq B$.

Similarly,

$$\begin{aligned} x \in B \quad & \Longrightarrow \quad x \in A \cup B, \\ & \overset{hypothesis}{\Longleftrightarrow} \quad x \in A \cap B, \\ & \Longrightarrow \quad x \in A \text{ and } x \in B, \\ & \Longrightarrow \quad x \in A. \end{aligned}$$

Thus $B \subseteq A$. which complete the proof. ■

■ **Example 5.5** Prove that
$$A = B \Longrightarrow A \cap B = A \cup B.$$
Here it is sufficient to replace B with A in the RHS:
$$A = B \Longrightarrow \mathbf{A} \cap B = A \cap A = A = A \cup A = \mathbf{A} \cup B.$$

■

(R) The two preceding examples are equivalent to the following result:
$$A \cap B = A \cup B \Longleftrightarrow A = B.$$
Usually, to prove the equivalence $P \Longleftrightarrow Q$, we prefer that students prove the two implications:
$$\boxed{P \Longrightarrow Q \quad \text{and} \quad Q \Longrightarrow P.}$$

■ **Example 5.6** *De Morgan's laws*

- *The complement of the union of two sets A and B is equal to the intersection of their complements.*
$$\overline{A \cup B} = \overline{A} \cap \overline{B}.$$
- *The complement of the intersection of two sets A and B is equal to the union of their complements.*
$$\overline{A \cap B} = \overline{A} \cup \overline{B}.$$

■

- $\overline{A \cup B} = \overline{A} \cap \overline{B}.$
$$\begin{aligned} \text{Let } x \in \overline{A \cup B} &\Longrightarrow x \notin (A \cup B), \quad \text{using Corollary, 3.5.3} \\ &\Longrightarrow x \notin A \text{ and } x \notin B, \\ &\Longrightarrow x \in \overline{A} \text{ and } x \in \overline{B}, \\ &\Longrightarrow x \in \overline{A} \cap \overline{B}. \end{aligned}$$
Thus $\overline{A \cup B} \subseteq \overline{A} \cap \overline{B}$.
Conversely,
$$\begin{aligned} \text{let } x \in \overline{A} \cap \overline{B} &\Longrightarrow x \in \overline{A} \text{ and } x \in \overline{B}, \\ &\Longrightarrow x \notin A \text{ and } x \notin B. \end{aligned}$$
Let us denote $p : x \in A, \;\; q : x \in B$, then
$$x \in \overline{A} \text{ and } x \in \overline{B} \equiv \overline{p} \wedge \overline{q} \equiv \overline{p \vee q} \equiv x \notin (A \cup B),$$
$$x \in \overline{A} \cap \overline{B} \Longrightarrow x \notin (A \cup B) \Longrightarrow x \in \overline{A \cup B}.$$
Thus we have $\overline{A} \cap \overline{B} \subseteq \overline{A \cup B}$ and, finally, $\overline{A \cup B} = \overline{A} \cap \overline{B}$.
- $\overline{A \cap B} = \overline{A} \cup \overline{B}$
$$\begin{aligned} \text{Let } x \in \overline{A \cap B} &\Longrightarrow x \notin (A \cap B), \quad \text{using Corollary 3.5.3} \\ &\Longrightarrow x \notin A \text{ or } x \notin B, \\ &\Longrightarrow x \in \overline{A} \text{ or } x \in \overline{B}, \\ &\Longrightarrow x \in \overline{A} \cup \overline{B}. \end{aligned}$$

Thus $\overline{A} \cup \overline{B} \subseteq \overline{A \cap B}$.
Conversely,

$$\begin{aligned} \text{let } x \in \overline{A} \cup \overline{B} &\implies x \in \overline{A} \text{ or } x \in \overline{B}, \\ &\implies x \notin A \text{ or } x \notin B. \end{aligned}$$

Let us denote $p : x \in A,\ \ q : x \in B$, then

$$x \notin A \text{ or } x \notin B \equiv \overline{p} \vee \overline{q} \equiv \overline{p \wedge q} \equiv x \notin (A \cap B).$$

So, $x \in \overline{A} \cup \overline{B} \implies x \notin (A \cap B) \implies x \in \overline{A \cap B}$. Hence, $\overline{A \cap B} = \overline{A} \cup \overline{B}$.

■ **Example 5.7** *Let A, B be two subsets of a universal set U. Then we have:*

$$A \subseteq B \iff \overline{B} \subseteq \overline{A}.$$

■

Let us use a proof by equivalences. To this end, we will use definitions and contraposition:

$$\begin{aligned} A \subseteq B &\iff [x \in A \implies x \in B,] \\ &\iff [x \notin B \implies x \notin A,] \qquad \text{(contraposition)} \\ &\iff [x \in \overline{B} \implies x \in \overline{A},] \\ &\iff \overline{B} \subseteq \overline{A}. \end{aligned}$$

One may proceed differently by proving the two implications (see Question 5.13).

■ **Example 5.8** *Let $(A_i)_{i \in I}$ be an indexed family of subsets of a universal set* **U** *and let B also be a subset of* **U**. *Then we have the following:*

1. $$B \cap \left(\bigcup_{i \in I} A_i\right) = \bigcup_{i \in I} (A_i \cap B).$$

2. $$B \cup \left(\bigcap_{i \in I} A_i\right) = \bigcap_{i \in I} (A_i \cup B).$$

3. $$\overline{\bigcap_{i \in I} A_i} = \bigcup_{i \in I} \overline{A_i}.$$

4. $$\overline{\bigcup_{i \in I} A_i} = \bigcap_{i \in I} \overline{A_i}.$$

■

1. $B \cap \left(\bigcup_{i \in I} A_i\right) = \bigcup_{i \in I} (A_i \cap B)$

$$\begin{aligned} \text{Let } x \in B \cap \left(\bigcup_{i \in I} A_i\right) &\implies (x \in B) \text{ and } (x \in \bigcup_{i \in I} A_i) \\ &\implies (x \in B) \text{ and } \exists j \in I :\ x \in A_j \\ &\implies x \in B \cap A_j \text{ for some } j \in I \\ &\implies x \in \bigcup_{i \in I} (B \cap A_i). \end{aligned}$$

Thus $B \cap \left(\bigcup_{i \in I} A_i\right) \subseteq \bigcup_{i \in I} (B \cap A_i)$.

Conversely,

$$\begin{aligned} \text{Let } x \in \bigcup_{i\in I}(B\cap A_i) &\implies \exists j \in I:\ x \in (B\cap A_j),\\ &\implies (x\in B) \text{ and } x\in A_j \text{ for some } j\in I,\\ &\implies (x\in B) \text{ and } x \in \bigcup_{i\in I} A_i,\\ &\implies x \in B\cap\Big(\bigcup_{i\in I} A_i\Big). \end{aligned}$$

Thus $\bigcup_{i\in I}(B\cap A_i)\subseteq B\cap\Big(\bigcup_{i\in I}A_i\Big)$, and finally, the proof is complete.

2. See Question 17.
3. $\overline{\bigcap_{i\in I} A_i} = \bigcup_{i\in I}\overline{A_i}$.
   (i)

$$\begin{aligned} \text{Let } x\in\overline{\bigcap_{i\in I}A_i} &\implies x\notin\bigcap_{i\in I}A_i\\ &\implies \exists j\in I:\ x\notin A_j\\ &\implies \exists j\in I:\ x\in\overline{A_j}\\ &\implies x\in\bigcup_{i\in I}\overline{A_i}. \end{aligned}$$

   Thus $\overline{\bigcap_{i\in I}A_i}\subseteq\bigcup_{i\in I}\overline{A_i}$.

   (ii)

$$\begin{aligned} \text{Let } x\in\bigcup_{i\in I}\overline{A_i} &\implies \exists j\in I:\ x\in\overline{A_j}\\ &\implies \exists j\in I:\ x\notin A_j\\ &\implies x\notin\bigcap_{i\in I}A_i\\ &\implies x\in\overline{\bigcap_{i\in I}A_i}. \end{aligned}$$

   Thus $\bigcup_{i\in I}\overline{A_i}\subseteq\overline{\bigcap_{i\in I}A_i}$, and finally the proof is complete.

4. See Question 18.

■ **Example 5.9** Let A, B, and C be sets. Prove that

$$A\subset B \implies A\cap C\subset B\cap C.$$

■

Assume that $A\subset B$. Let $x\in A\cap C$, then

$$\begin{aligned} x\in A\cap C &\implies x\in A \text{ and } x\in C,\\ &\underbrace{\overset{A\subset B}{\implies}} x\in B \text{ and } x\in C,\\ &\implies x\in B\cap C. \end{aligned}$$

Thus $A\cap C\subset B\cap C$.

■ **Example 5.10** For sets A and B, prove the following:

(i) $A=(A\cap B)\cup(A-B)$.

(ii) $A \cup B = (A - B) \cup (B - A) \cup (A \cap B)$.

■

Let **U** be the universal set.

(i) $A = (A \cap B) \cup (A - B)$.
Since $U = B \cup \overline{B}$ and $A = A \cap U$, we can write

$$A = A \cap U = A \cap (B \cup \overline{B}) = (A \cap B) \cup (A \cap \overline{B}).$$

The set $A - B = \{x \in \mathbf{U} : x \in A \text{ and } x \notin B\} = A \cap \overline{B}$; therefore, $A = (A \cap B) \cup (A - B)$.

(ii) $A \cup B = (A - B) \cup (B - A) \cup (A \cap B)$.

$$\begin{aligned}
(A - B) \cup (B - A) \cup (A \cap B) &= (A \cap \overline{B}) \cup (B \cap \overline{A}) \cup (A \cap B), \\
&= (A \cap \overline{B}) \cup \left[(B \cap \overline{A}) \cup (A \cap B)\right] \\
&= (A \cap \overline{B}) \cup \left[(B \cap \overline{A}) \cup (B \cap A)\right], \\
&= (A \cap \overline{B}) \cup \left[B \cap (\overline{A} \cup A)\right], \\
&= (A \cap \overline{B}) \cup \left[B \cap \mathbf{U}\right], \\
&= (A \cap \overline{B}) \cup B, \\
&= B \cup (A \cap \overline{B}), \\
&= (B \cup A) \cap (B \cup \overline{B}), \\
&= (B \cup A) \cap \mathbf{U}, \\
&= A \cup B.
\end{aligned}$$

■ **Example 5.11** Let A, B, and C be sets. Prove the following

(i) $A \cup B = A \iff B \subseteq A$.

(ii) $A \cap B = A \iff A \subseteq B$.

■

Let **U** be the universal set.

(i) $A \cup B = A \iff B \subseteq A$.
We prove the following implications:

(a) $A \cup B = A \Longrightarrow B \subseteq A$.
Let $x \in B$; then $x \in A \cup B = A$. Thus $x \in A$.

(b) $B \subseteq A \Longrightarrow A \cup B = A$.
Since $A \subseteq A \cup B$ is always true, it suffices to prove that $A \cup B \subseteq A$.
Let $x \in A \cup B$, then $x \in A$ or $x \in B \subseteq A$, which implies that $x \in A \cup A = A$.

Thus $A \cup B = A$.

(ii) $A \cap B = A \iff A \subseteq B$. We prove the following implications:

(a) $A \cap B = A \Longrightarrow A \subseteq B$.
Let $x \in A$, then $x \in A \cap B \subseteq B$. Thus $x \in B$.

(b) $A \subseteq B \Longrightarrow A \cap B = A$.
Since $A \cap B \subseteq A$ is always true, it suffices to prove that $A \subseteq A \cap B$.

Let $x \in A$, then $x \in B$, since $A \subseteq B$. This implies that $x \in A$ and $x \in B$; therefore, $x \in A \cap B$. Thus $A \cap B = A$.

■ **Example 5.12** Show that for each set A, B, and C we have:

$$(A \cup B) \cap C = (A \cap C) \cup (B \cap C).$$

■

Let us use equivalences. To this end, we denote $p : x \in A$, $q : x \in B$, and $r : x \in C$.

$$\begin{aligned}
x \in (A \cup B) \cap C &\iff x \in (A \cup B) \text{ and } (x \in C),\\
&\iff [(x \in A) \text{ or } (x \in B)] \text{ and } x \in C,\\
&\iff (p \vee q) \wedge r,\\
&\iff r \wedge (p \vee q),\\
&\iff (r \wedge p) \vee (r \wedge q),\\
&\iff (p \wedge r) \vee (q \wedge r),\\
&\iff [(x \in A) \text{ and } (x \in C)] \text{ or } [(x \in B) \text{ and } (x \in C)],\\
&\iff x \in (A \cap C) \text{ or } x \in (B \cap C),\\
&\iff x \in (A \cap C) \cup (B \cap C).
\end{aligned}$$

■ **Example 5.13** Prove that

$$A \subseteq B \iff \overline{B} \subseteq \overline{A}.$$

■

Let us prove both implications:

(i) $A \subseteq B \implies \overline{B} \subseteq \overline{A}$. Let

$$x \in \overline{B} \Longrightarrow x \notin B \underbrace{\overset{A \subseteq B}{\Longrightarrow}} x \notin A \Longrightarrow x \in \overline{A}.$$

Thus we have proven that $\overline{B} \subseteq \overline{A}$. It is worth student's time to pay attention to the following fact concerning $A \subseteq B$:

$$x \in A \implies x \in B \underbrace{\overset{\text{by contraposition}}{\iff}} x \notin B \Longrightarrow x \notin A.$$

(ii) $\overline{B} \subseteq \overline{A} \implies A \subseteq B$. Let us consider

$$x \in A \Longrightarrow x \notin \overline{A} \underbrace{\overset{\overline{B} \subseteq \overline{A}}{\Longrightarrow}} \Longrightarrow x \notin \overline{B} \Longrightarrow x \in B.$$

Thus we have proven that $A \subseteq B$.

■ **Example 5.14** Let $(A_i)_{i \in I}$ be a family of subsets of a universal set E, and let B also be a subset of E. Then we have

$$B \cup \left(\bigcap_{i \in I} A_i \right) = \bigcap_{i \in I} (A_i \cup B).$$

■

Let us prove the following two inclusions:

(i) Let

$$\begin{aligned}
x \in B \cup \left(\bigcap_{i \in I} A_i \right) &\implies x \in B \text{ or } x \in \bigcap_{i \in I} A_i,\\
&\implies x \in B \text{ or } \forall\, i \in I,\ x \in A_i,\\
&\implies \forall\, i \in I,\ x \in B \cup A_i,\\
&\implies x \in \bigcap_{i \in I} (A_i \cup B).
\end{aligned}$$

Thus we have proven that $B \cup \left(\bigcap_{i \in I} A_i \right) \subseteq \bigcap_{i \in I} (A_i \cup B)$.

(ii) Let

$$
\begin{aligned}
x \in \bigcap_{i\in I}(A_i \cup B) &\Longrightarrow \forall\, i \in I,\ x \in (B \cup A_i),\\
&\Longrightarrow \forall\, i \in I,\ x \in B \text{ or } x \in A_i,\\
&\Longrightarrow x \in B \text{ or } \forall\, i \in I,\ x \in A_i,\\
&\Longrightarrow x \in B \text{ or } x \in \bigcap_{i\in I} A_i,\\
&\Longrightarrow x \in B \cup \Big(\bigcap_{i\in I} A_i\Big).
\end{aligned}
$$

Thus we have proven that $\bigcap_{i\in I}(A_i \cup B) \subseteq B \cup \Big(\bigcap_{i\in I} A_i\Big)$.

■ **Example 5.15** Let $(A_i)_{i\in I}$ be a family of subsets of a universal set E. Then we have:

$$
\overline{\bigcup_{i\in I} A_i} = \bigcap_{i\in I} \overline{A_i}.
$$

■

Let us prove the two inclusions.

(i) Let us consider

$$
\begin{aligned}
x \in \overline{\bigcup_{i\in I} A_i} &\Longrightarrow x \notin \bigcup_{i\in I} A_i,\\
&\Longrightarrow \forall\, i \in I,\ x \notin A_i,\\
&\Longrightarrow \forall\, i \in I,\ x \in \overline{A_i},\\
&\Longrightarrow x \in \bigcap_{i\in I} \overline{A_i}.
\end{aligned}
$$

Thus we have proven that $\overline{\bigcup_{i\in I} A_i} \subseteq \bigcap_{i\in I} \overline{A_i}$.

(ii) Let us consider

$$
\begin{aligned}
x \in \bigcap_{i\in I} \overline{A_i} &\Longrightarrow \forall\, i \in I,\ x \in \overline{A_i},\\
&\Longrightarrow \forall\, i \in I,\ x \notin A_i,\\
&\Longrightarrow x \notin \bigcup_{i\in I} A_i,\\
&\Longrightarrow x \in \overline{\bigcup_{i\in I} A_i}.
\end{aligned}
$$

Thus we have proven that $\bigcap_{i\in I} \overline{A_i} \subseteq \overline{\bigcup_{i\in I} A_i}$.

■ **Example 5.16** Let A, B, C, and D be four sets, then

$$
(A \times B) \cup (C \times D) \subseteq (A \cup C) \times (B \cup D).
$$

■

Let us consider

$$
\begin{aligned}
(x,y) \in (A \times B) \cup (C \times D) &\Longrightarrow (x,y) \in (A \times B) \text{ or } (x,y) \in (C \times D),\\
&\Longrightarrow x \in A,\ y \in B \text{ or } x \in C,\ y \in D,\\
&\Longrightarrow x \in A \cup C,\ y \in B \cup D,\\
&\Longrightarrow (x,y) \in (A \cup C) \times (B \cup D).
\end{aligned}
$$

Thus we have proven that $(A \times B) \cup (C \times D) \subseteq (A \cup C) \times (B \cup D)$.

5.2 Examples from Calculus

Students will observe and experiment that direct proofs sometimes require particular ideas and unusual techniques, as it is the case in the next example from calculus I **[11]**. We recall the definitions of continuity and differentiability at $x = a \in D_f$.

- Let $a \in D_f$, f be continuous at a if and only if $\lim_{x\to a} f(x)$ exists and $\lim_{x\to a} f(x) = f(a)$.
- Let $a \in D_f$, f be differentiable at a if and only if

$$f'(a) = \lim_{h\to 0} \frac{f(a+h)-f(a)}{h}$$

exists and is finite.

- Let $a \in D_f$, f is differentiable at a if and only if

$$f'(a) = \lim_{x\to a} \frac{f(x)-f(a)}{x-a}$$

exists and is finite.

■ **Example 5.17** Prove that if f is a differentiable function at $a \in D_f$, then f is continuous at a. A clever idea used in Calculus I consists of expressing $f(x) - f(a)$ as follows:

$$\text{Let } x \neq a,\ f(x) = f(a) + \frac{f(x)-f(a)}{x-a}(x-a) \implies \lim_{x\to a} f(x) = f(a) + \lim_{x\to a} \frac{f(x)-f(a)}{x-a}(x-a).$$

Using the assumption that f is differentiable at a, $\lim_{x\to a} \frac{f(x)-f(a)}{x-a} = f'(a)$ which is finite and then:

$$\lim_{x\to a} f(x) = f(a) + f'(a) \lim_{x\to a}(x-a) = f(a) + f'(a) \times 0 = f(a) \implies \lim_{x\to a} f(x) = f(a).$$

Consequently, we have proved that f is continuous at a. ■

(R) Students will experiment and draw conclusions by themselves if direct proofs are easier to use than other proof techniques that will be studied in the coming sections.

(R) It is important to be aware that when using a direct proof, we are not obliged in many cases to start by the given hypothesis but we should use it during the proof (see Example 5.18).

■ **Example 5.18** Given a real-valued function f defined on $\mathbb{R}$, prove that if

$$\boxed{f \text{ is an even function} \implies \int_{-a}^{a} f(x)\,dx = 2\int_{0}^{a} f(x)\,dx}.$$

We will use a direct proof.

$$\int_{-a}^{a} f(x)\,dx = \int_{-a}^{0} f(x)\,dx + \int_{0}^{a} f(x)\,dx.$$

Using the following change of variable in the first definite integral[1]:

$$u = -x \implies du = -dx, \begin{cases} x = -a \\ x = 0 \end{cases} \implies \begin{cases} u = a \\ u = 0 \end{cases}$$

[1]It is well known in calculus I and II that $\int_a^b f(t)\,dt = -\int_b^a f(t)\,dt$.

$$\begin{aligned}\int_{-a}^{0} f(x)\,dx &= \int_{a}^{0} f(-u)\,(-du),\\ &= \int_{0}^{a} f(-u)\,du,\end{aligned}$$

and since f is an even function, $f(-u) = f(u)$, and thus we obtain[2]:

$$\int_{-a}^{0} f(x)\,dx = \int_{0}^{a} f(u)\,du = \int_{0}^{a} f(x)\,dx.$$

Finally,

$$\int_{-a}^{a} f(x)\,dx = \int_{0}^{a} f(x)\,dx + \int_{0}^{a} f(x)\,dx = 2\int_{0}^{a} f(x)\,dx,$$

which completes the proof. ■

■ **Example 5.19** Given a real-valued function f defined on $\mathbb{R}$, prove that if

$$\boxed{f \text{ is an odd function} \implies \int_{-a}^{a} f(x)\,dx = 0.}$$

This proof can be shown by following the steps shown in Example 5.18. The following example introduces a bright idea through which students will discover the usefulness of studying variations of a given function. Of course, the discussed question seems very far from the use of derivatives and variations. ■

■ **Example 5.20** Determine the domain of $f(x) = \ln\left(\sqrt{x - \ln(x)} - 1\right)$.
Let us see together whether this is an easy question:

$$x \in Dom(f) \iff \begin{cases} \sqrt{x-\ln(x)} - 1 > 0, \\ x - \ln(x) \geq 0, \\ x > 0 \end{cases} \iff \begin{cases} \sqrt{x-\ln(x)} > 1, \\ x - \ln(x) \geq 0, \\ x > 0 \end{cases} \iff \begin{cases} x-\ln(x) > 1^2 = 1, \\ x - \ln(x) \geq 0, \\ x > 0 \end{cases}$$

i.e.,

$$x \in Dom(f) \iff \begin{cases} x - \ln(x) > 1, \\ x > 0. \end{cases}$$

Thus to determine the domain it is sufficient to solve the inequality $x - \ln(x) - 1 > 0$ in $x > 0$.
To this end, we suggest:

- study the variations of $g(x) = x - \ln(x) - 1$
- and conclude the sign of $g(x)$ from the variations table.

Let us study the sign of the derivative of the function g for $x > 0$:

$$g'(x) = 1 - \frac{1}{x} = \frac{x-1}{x}, \quad \text{and then } g'(x) = 0 \iff x = 1,$$

then

x	0		1		∞
$g'(x)$		$-$	0	$+$	
$g(x)$		↘	g(1)	↗	

[2]Local variables don't matter: $\int_a^b f(u)\,du = \int_a^b f(x)\,dx = \int_a^b f(z)\,dz$.

Since $g(1) = 1 - \ln(1) - 1 = 0$, we conclude that $\forall\, x > 0,\ x \neq 1,\ \ g(x) > g(1) = 0$, and, finally

$$\boxed{Dom(f) \ = \ (0,\ 1) \cup (1,\ \infty).}$$

■

We see the usefulness of the variations of functions for studying the sign of a complex expression.

We close this section by proving an important theorem of Calculus: the *mean value theorem (M V T)*, also known as the *Lagrange Theorem*

■ **Example 5.21** Let $f\colon [a, b] \longrightarrow \mathbb{R}$ be a continuous function on $[a, b]$ and differentiable on $(a,\ b)$. Then there exists $c \in (a,\ b)$ such that

$$\frac{f(b) - f(a)}{b - a} = f'(c).$$

■

R Figure 5.2 provides an interpretation of the mean value theorem (MVT).

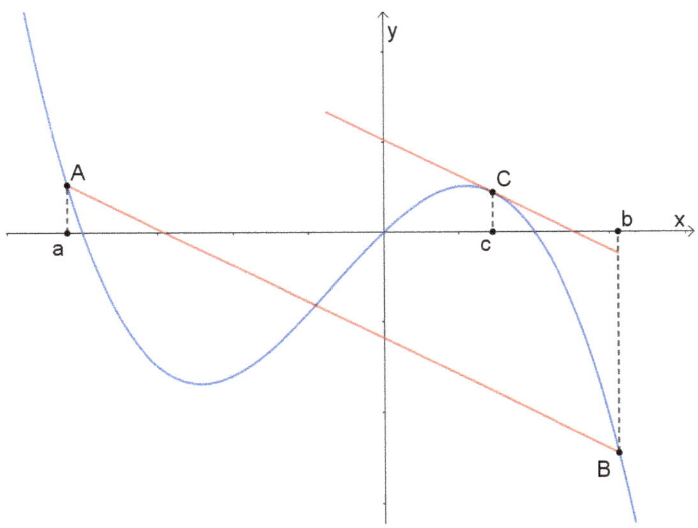

Figure 5.1: A graphical interpretation of MVT

It is clear that the ratio $\dfrac{f(b) - f(a)}{b - a}$ represents the slope of the straight line through $A\,(a, f(a))$ and $B\,(b, f(b))$. Thus MVT under the continuity and differentiability of f guarantees the existence of at least one point $(c, f(c))$ on the graph of f, where the tangent line to the curve $y = f(x)$ at $x = c$ is parallel[3] to the segment line $[AB]$.

R Readers can easily find out the importance of using MVT for finding the continuity over on $[0,\ 1]$ for

$$f(x) = \begin{cases} x & \text{if} \quad x \neq 1, \\ 0 & \text{if} \quad x = 1. \end{cases}$$

[3] $f'(c)$ is the slope of the tangent line to the curve $y = f(x)$ at $x = c$.

R Readers can also find out the importance of using MVT for finding the differentiability of $f(x) = |x|$ on $[-1, 1]$.

Proof. The continuity of f over $[a, b]$ assures the existence of $m, M \in \mathbb{R}$ such that $f([a, b]) = [m, M]$. We will proceed by cases:

- *Case 1.* $\boxed{f(a) = f(b)}$ Here also, we consider two cases:
  (a) If f is a constant function, then $f'(x) = 0$ for any $x \in (a, b)$; thus there is $c \in (a, b)$: $f'(c) = 0$, which completes the proof of the theorem.
  (b) If f is not a constant function, then there exists $\alpha \in (a, b)$: $f(\alpha) \neq f(a)$. We may assume that $f(a) > f(\alpha) \geq m$ [4] and since we have $f([a, b]) = [m, M]$, we conclude the existence of $c \in (a, b)$ such that $f(c) = m$. Thus we obtain

$$\begin{aligned} f(a) > f(\alpha) \geq f(c) = m &\implies m \text{ is a minimum over } [a, b] \text{ of } f \text{ at } x = c, \\ &\implies f'(c) = 0, \end{aligned}$$

  which completes the proof.
- *Case 2.* $\boxed{f(a) \neq f(b)}$.

  Let us define $h(x) = f(x) - \left(\frac{f(b) - f(a)}{b - a}\right) x$. It is easy to see that h is continuous on $[a, b]$ and differentiable on (a, b) as a difference of f and a linear function. Also, we have $h(a) = h(b)$. Indeed,

$$\begin{aligned} h(a) &= f(a) - \left(\frac{f(b) - f(a)}{b - a}\right) a, \\ &= \frac{bf(a) - af(a) - af(b) + af(a)}{b - a}, \\ &= \frac{bf(a) - af(b)}{b - a}; \end{aligned}$$

  and

$$\begin{aligned} h(b) &= f(b) - \left(\frac{f(b) - f(a)}{b - a}\right) b, \\ &= \frac{bf(b) - af(b) - bf(b) + bf(a)}{b - a}, \\ &= \frac{-af(b) + bf(a)}{b - a}, \\ h(b) &= \frac{bf(a) - af(b)}{b - a}. \end{aligned}$$

  Hence, $h(b) = h(a)$, and the assumptions of the theorem for Case 1 are satisfied; then we conclude the existence of $c \in (a, b)$: $h'(c) = 0$. Here,

$$h'(x) = f'(x) - \frac{f(b) - f(a)}{b - a} \implies h'(c) = f'(c) - \frac{f(b) - f(a)}{b - a} = 0.$$

  Finally, there exists $c \in (a, b): \; f'(c) = \dfrac{f(b) - f(a)}{b - a}$, which completes the proof.

[4] We can also assume that $f(a) < f(\alpha) \leq M$ and reach the same result. Also, it is easy to see that $c \neq a$ and $c \neq b$.

■

■ **Example 5.22** Use the mean value theorem (MVT) to prove that

$$\forall x \in \mathbb{R}: e^x \geq x+1.$$

It is clear that at $x=0$, the equality case holds (i.e., we have: $e^0 = 1 = 0+1$). To use MVT, we set $f(t) = e^t$, which is continuous and differentiable everywhere on $\mathbb{R}$. Let us split the discussion as follows:

(i) $\boxed{x > 0.}$

According to the continuity and differentiability of f on $\mathbb{R}$, MVT applied to f on $[0, x]$ guarantees

$$\exists\, c \in (0, x): f'(c) = \frac{f(x)-f(0)}{x-0} = \frac{e^x - 1}{x} \implies e^c = \frac{e^x-1}{x}.$$

Since $c > 0 \implies e^c > e^0 = 1$, and then we obtain

$$\frac{e^x-1}{x} = e^c > 1 \implies e^x - 1 > x \implies e^x > x+1.$$

(ii) $\boxed{x < 0.}$

Similarly, MVT applied to f on $[x, 0]$ guarantees

$$\exists\, c' \in (x, 0): f'(c') = \frac{f(0)-f(x)}{0-x} = \frac{1-e^x}{-x} \implies e^{c'} = \frac{e^x-1}{x}.$$

Since $c' < 0 \implies e^{c'} < e^0 = 1$ and then we obtain

$$\frac{e^x-1}{x} = e^{c'} < 1 \implies e^x - 1 > x \implies e^x > x+1.$$

[5] Finally, we proved that $\forall x \in \mathbb{R}$, so we have $e^x \geq x+1$.

■

(R) **Rolle's Theorem**: Let $f\colon [a, b] \longrightarrow \mathbb{R}$ be a continuous function on $[a, b]$ and differentiable on (a, b) with $f(a) = f(b)$. Then there exists $c \in (a, b)$ such that $f'(c) = 0$.

Thus Rolle's theorem under the continuity and differentiability of f guarantees the existence of at least one point $(c, f(c))$ on the graph of f, where the tangent line to the curve $y = f(x)$ at $x = c$ is horizontal (parallel to the horizontal segment line $[AB]$).

■ **Example 5.23** Show that the equation $x^3 - x^2 + x + 1/2 = 0$ has a unique solution.
Let us denote $f(x) = x^3 - x^2 + x + 1/2$ satisfying $f(0) = 1/2$ and $f(-1) = -5/2$.

Since

1. f is a continuous polynomial function on $\mathbb{R}$ and particularly on $[-1, 0]$,
2. $f(0) \times f(-1) = -5/4 < 0$,

---

[5] Here we will multiply both terms of the inequality by $\mathbf{x} < \mathbf{0}$.

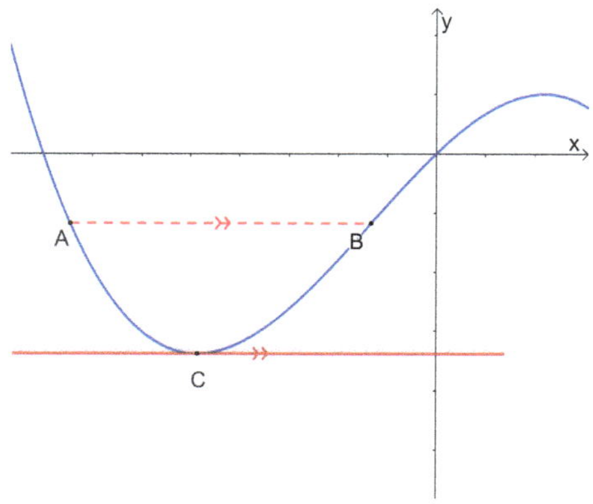

Figure 5.2: graphical interpretation of Rolle's theorem.

by applying the *intermediate value theorem* (IVT) to f on $[-1, 0]$, there exists $c \in (-1, 0)$: $f(c) = 0$. Thus we proved the existence of at least one solution to $f(x) = 0$, and we need to prove the uniqueness of the root. To this end, let us proceed by contradiction and assume that

$$\exists\, c_1, c_2 \in (-1, 0),\ c_1 < c_2\colon f(c_1) = f(c_2) = 0.$$

Since f is a polynomial function, it is continuous on $[c_1, c_2]$ and differentiable on (c_1, c_2) and $f(c_1) = f(c_2)$, so Rolle's theorem implies that

$$\exists\, d \in (c_1, c_2)\colon f'(d) = 0.$$

But $f'(x) = 3x^2 - 2x + 1 = 0$ has no real root because its discriminant $\Delta = 4 - 4(3)(+1) = -8 < 0$, then what we assumed is wrong and the solution of $f(x) = 0$ is unique. ∎

▪ Example 5.24 Does there exist a differentiable function $f\colon [0, 2] \to \mathbb{R}$ such that

$$f(0) = -1,\quad f(2) = 4,\quad f'(x) \leq 2\ \forall\, x \in [0, 2]\,?$$

Let us proceed by contradiction and assume that there exists a differentiable function f (which is then continuous) satisfying the preceding assumptions. Thus by applying MVT to f on $[0, 2]$, we obtain:

$$\exists\, c \in (0, 2) :\ f'(c) = \frac{f(2) - f(0)}{2 - 0} = \frac{4 - (-1)}{2} = \frac{5}{2} \Longrightarrow \exists\, c \in (0, 2) :\ f'(c) > 2,$$

which contradicts the fact that $f'(x) \leq 2$ for all $x \in [0, 2]$. Thus what we assumed is wrong, and such function does not exist. ∎

5.3 Examples from Number Theory

In this subsection, we will work on some proofs from number theory.

▪ Example 5.25 Find the number of integers between 11 and 2022 for which $1 + 2^n$ is a multiple of 3.

Since $2^n + 1 = 2^n - (-1)^n + 1 + (-1)^n$, and using the well-known identity

$$x^n - y^n = (x - y) \sum_{k=0}^{n-1} x^{n-k-1} y^k,$$

we get $2^n - (-1)^n = (2-(-1)) \sum_{k=0}^{n-1} 2^{n-k-1}(-1)^k = 3\,m, \quad m \in \mathbb{Z}$. Thus we obtain:

$$2^n + 1 = 2^n - (-1)^n + 1 + (-1)^n = 3\,m + (1 + (-1)^n) = \begin{cases} 3m & \text{if } n \text{ is odd,} \\ 3m+2 & \text{if } n \text{ is even.} \end{cases}$$

To complete the answer, it is sufficient to determine the number of odd integers between 11 and 2022, which is equal to $(2022 - 11 + 1)/2 = 1006$. ■

■ **Example 5.26** Find all positive integers satisfying $n \mid n+11$.
Since $n|n$ and to obtain $n \mid n+11$, we have $n \mid (n+11) - n = 11$. Thus $n \in \{1,\ 11\}$. ■

■ **Example 5.27** Find all positive integers satisfying $n \mid n+16$.
Since $n|n$ and to obtain $n \mid n+16$, we have $n \mid (n+16) - n = 16$. Thus $n \in \{1,2,4,8\ 16\}$. ■

■ **Example 5.28** Find all positive integers satisfying $n^2+1 \mid n+1$.
Since $n^2+1 \,|\, n+1$, $n^2+1 \leq n+1$, which is satisfied only if $n = 1$, and then $n \in \{1\}$. ■

■ **Example 5.29** Show that if for $x, y \in \mathbb{Z}$: $7 \,|\, 3x - 5y$ if and only if $7|4y - x$.
The general idea is to combine $3x - 5y$ with $7(ax + by)$ to reach $4y - x$ with integers x and y. In other words, $2(3x - 5y) - 7(x - 2y) = -x + 4y$.

Conversely, let us assume that $7 \,|\, 4y - x$, then, similarly, $4(-x + 4y) + 7(x - 3y) = 3x - 5y$, and the proof is complete.

Of course, through solving a linear system we can find a, b, and c such that

$$a(-x+4y) + 7(bx + cy) = 3x - 5y, \quad \text{we can get } a = 4,\ b = 1, \text{and } c = -3$$

or $a = 11,\ b = 2,\ c = -7$. ■

Theorem 5.3.1 **Euclid's theorem** The set of prime numbers is infinite.

Proof. Let us proceed by contradiction and assume there are a finite number of prime numbers denoted by $p_1, p_2, \cdots, p_N$, and let p_N denotes the greatest prime number.

If we define $P = 1 + p_1 \times p_2 \times p_3 \times \cdots \times p_N$, it is clear that $P > p_N$, implies either P is prime greater than p_N, which contradicts our assumption, or P is a composite integer with other prime factors distinct from all $p_1, p_2, \cdots, p_N$, because $P \equiv 1 \ (mod\ p_i)$, $i \in \{1, 2, \cdots, N\}$, which is also in contradiction with our assumption. Thus the set of primes is infinite. ■

R There is a simpler proof for this theorem given by the eminent mathematician Kummer. Let us assume that there are a finite number of prime numbers denoted by $p_1, p_2, \cdots, p_n$, and let $N > 2$ be the product of all these primes. It is clear that $N - 1 > p_n$, so it is not prime, and then there exists k such that p_k divides $N - 1$. Thus as p_k divides N, it implies that p_k divides the difference $N - (N-1) = 1$, which is impossible. Thus the number of prime numbers is infinite.

Theorem 5.3.2 **Fermat numbers.**

(a) The Fermat numbers of order m: $F_m = 2^{2^m} + 1$ can be expressed as

$$F_{n+1} = 2 + F_0 F_1 \cdots F_{n-2} F_{n-1} F_n.$$

(b) Any two distinct Fermat numbers F_p and F_q $p > q$ are relatively prime[a]

$$GCD(F_p, F_q) = 1.$$

[a]GCD stands for greatest common divisor

Proof.
(a) Let us consider the Fermat number of order $m = n+1$,

$$\begin{aligned}
F_{n+1} - 2 &= 2^{2^{n+1}} - 1 = \left(2^{2^n} - 1\right)\left(2^{2^n} + 1\right), \\
&= \left(2^{2^{n-1}} - 1\right)\left(2^{2^{n-1}} + 1\right)\left(2^{2^n} + 1\right), \\
&= \left(2^{2^{n-1}} - 1\right) F_{n-1} F_n. \\
&= \left(2^{2^{n-2}} - 1\right)\left(2^{2^{n-2}} + 1\right) F_{p-1} F_n, \\
&= \left(2^{2^{n-2}} - 1\right) F_{n-2} F_{n-1} F_n, \\
&= \vdots \\
&= \left(2^{2^0} - 1\right) F_0 F_1 \ldots F_{n-2} F_{n-1} F_n, \\
&= F_0 F_1 \ldots F_{n-2} F_{n-1} F_n.
\end{aligned}$$

(b) Let us denote $d = GCD(F_p, F_q)$, where $p < q$ and using (a), we conclude that $d \mid F_q - 2$ and then $d \mid F_q - (F_q - 2) = 2 \Longrightarrow d \in \{\pm 1, \pm 2\}$. But, since all Fermat numbers are odd, we conclude that $GCD(F_p, F_q) = d = 1$.

■

Theorem 5.3.3 **Cataldi-Fermat** If the number $2^m - 1$ is prime, then m is itself prime.

Proof. Solve by contraposition. We assume that m is not prime, then there exists $p, q > 1 \in \mathbb{N}$ such that $m = pq$, and then $2^m - 1 = 2^{pq} - 1 = (2^p)^q - 1 = (2^p - 1) \sum_{k=0}^{q-1} (2^p)^{q-1-k}$. This implies that $(2^p - 1) \mid 2^m - 1$ so $(2^m - 1)$ is not prime. ■

(R) If n is prime, the number $M_n = 2^n - 1$ is called a Mersenne number.

Theorem 5.3.4 If $X_m = 2^m + 1$ is prime, then m is a power of 2.

Proof. By contradiction. We assume that m is not a power of 2. In other words, there exists $k \geq 0$, and an odd integer $n \in \mathbb{N}$: $m = 2^k n$. We have, then,

$$X_m = 2^m + 1 = 2^{2^k n} + 1$$

. Using the fact that n is odd, we obtain the following:

$$X_m = \left(2^{2^k}\right)^n + 1 = (2^{2^k} + 1) \sum_{i=1}^{n} (-1)^{i-1} \left(2^{2^k}\right)^{n-i}$$

. But this implies that F_m is composite and it is not prime. Thus n must be equal to 1, which implies that what we supposed is wrong, and then m is a power of 2. ■

In the next examples, we will focus on divisibility tests. We aim to let readers learn how to explore and enjoy the proof techniques. We will prove the divisibility test for 2, 3, 4, 7, and 11.

Let us consider a given integer[6]:

$$N = \overline{d_1 d_2 d_3 \cdots d_k}, \quad \text{with } k \text{ digits, and } d_i \in \{0,1,2,3,4,5,6,7,8,9\} \ \forall\, i = 1,2,\cdots,k.$$

Let us first start by explaining the notation.
For example $N = 4576$ $(d_1 = 4,\ d_2 = 5,\ d_3 = 7$ and $d_4 = 6)$ and N can be expressed as follows:

$$N = 4576 \ = \ 4\times 1000 \ + \ 5\times 100 \ + 7\times 10 \ + \ 6\times 1 = 4\times 10^3 + \ 5\times 10^2 \ + 7\times 10^1 \ + \ 6\times 10^0,$$

thus in general for $k \geq 1$,

$$N = \overline{d_1 d_2 d_3 \cdots d_k} \ = \ \sum_{i=1}^{k} d_i \times 10^{k-i}.$$

Definition 5.3.1 Divisibility Rule by 2 An integer with an even unit place is divisible by 2.

In other words, if the unit digit of a number n is an even number, then n is divisible by 2. Let us consider the integer $N = \overline{d_1 d_2 d_3 ... d_k}$, $k \geq 1$, and since for any $i = 1$ to $i = k-1$, 10^{k-i} is even, therefore, we conclude that

$$\underbrace{N}_{even} = \sum_{i=1}^{k} d_i \times 10^{k-i} = \underbrace{\sum_{i=1}^{k-1} d_i \times 10^{k-i}}_{even} \ + \ d_k.$$

Finally, N is divisible by 2 if and only if the last digit d_k is even, which is equivalent to d_k being divisible by 2.

Definition 5.3.2 Divisibility Rule by 3 A given integer is divisible by 3 if the sum of its digits is also divisible by 3.

We first suggest proving this result when $N = \overline{abcd}$, where $a,b,c,d \in \{0,1,2,3,4,5,6,7,8,9\}$. Since

$$N = \overline{abcd} \ = \ 1000a + 100b + 10c + d \ = \ (999a + a) + (99b + b) + (9c + c) + d,$$

i.e.,

$$\underbrace{N}_{divisible\ by\ 3} = \overline{abcd} \ = \ \underbrace{3(333a + 33b + 3c)}_{divisible\ by\ 3} \ + (a + b + c).$$

Thus we have proven that $3 \mid N \Longleftrightarrow 3 | a + b + c$.

Let us consider the integer $N = \overline{d_1 d_2 d_3 \cdots d_k}$, $k \geq 1$ and using the fact that

$$\begin{aligned} 10 = 3(3) + 1 \ &\Longrightarrow \ 10^{k-i} = (3(3) + 1)^{k-i}, \\ &\Longrightarrow \ \exists\, m \in \mathbb{N}:\ 10^{k-i} = 3m + 1, \\ &\Longrightarrow \ d_i \times 10^{k-i} = (3m + 1)d_i,\ 1 \leq i \leq (k-1). \end{aligned}$$

[6]In $N = \overline{d_1 d_2 d_3 \cdots d_k}$ the notation does not mean the products of the d_is, but the digits that form the number N.

Then

$$\underbrace{N}_{divisible\ by\ 3} = \sum_{i=1}^{k} d_i \times 10^{k-i} = \sum_{i=1}^{k-1} d_i \times 10^{k-i} + d_k,$$
$$= \sum_{i=1}^{k-1} (3m+1)d_i + d_k,$$
$$= 3m\sum_{i=1}^{k-1} d_i + \sum_{i=1}^{k-1} d_i + d_k,$$
$$= \underbrace{3m\sum_{i=1}^{k-1} d_i}_{divisible\ by\ 3} + \sum_{i=1}^{k} d_i.$$

Finally, N is divisible by 3 if and only if the sum of its digits $\sum_{i=1}^{k} d_i$ is divisible by 3.

For example, $N = 20360014722 = 3(6786671574)$ is a multiple of 3, but we can reach the same conclusion by considering

$$2+0+3+6+0+0+1+4+7+2+2 = 27 = 3\times 9.$$

Also, we advise canceling from the sums:

$$2+\mathbf{0}+\mathbf{3}+\mathbf{6}+\mathbf{0}+\mathbf{0}+1+4+\mathbf{7}+\mathbf{2}+2 = 2+1+4+2 = 9.$$

Definition 5.3.3 — Divisibility Rule by 4. An integer is divisible by 4 if the number formed by the last two digits of the number is divisible by 4.

For example, 1436 is divisible by 4 because $36 = 4\times 9$ is a multiple of 4. That is, $1436 = 4\times 359$. The number 2021 is not a multiple of 4 because $2021 = 2020+1 = 4(505)+1$, and the divisibility rule by 4, is not satisfied, since $21 = 4\times 5+1$.
Let us consider an integer $N = \overline{d_1 d_2 d_3 ... d_k}$, $k \geq 3$, then using the fact that

$$10^2 = 4\times 25 \Longrightarrow 10^{k-i} = 10^2 \times 10^{k-2-i} \text{ is a multiple of 4}, \ \forall\, i = 1,2,\cdots,(k-2),$$

we have

$$\underbrace{N}_{multiple\ of\ 4} = \sum_{i=1}^{k} d_i \times 10^{k-i} = \underbrace{\sum_{i=1}^{k-2} d_i \times 10^{k-i}}_{multiple\ of\ 4} + 10d_{k-1} + d_k.$$

Thus N is a multiple of 4 if and only if $10d_{k-1} + d_k = \overline{d_{k-1}d_k}$ is also a multiple of 4.

Definition 5.3.4 — Divisibility Rule by 7. An integer is divisible by 7 if the number formed by dropping the unit place and then subtracting twice of it from the number is also divisible by 7.

For example, 406 is divisible by 7. According to the divisibility rule by 7, we have
$40-2(6) = 40-12 = 28 = 4\times 7$, thus $7 \mid 406$.

Another example is 1931. Since $1931-2(3) = 1925 \hookrightarrow 192-2(5) = 182$, then $18-2(2) = 14 = 7\times 2$. Thus $7 \mid 19313$.

(a) Let $N = \overline{d_1 d_2 d_3 \cdots d_k}$ and let us drop the unit digit d_k. Assume $\exists\, p \in \mathbb{Z}$ such that the difference is a multiple of 7. For example,

$$\overline{d_1 d_2 d_3 \cdots d_{k-1}} - 2(d_k) = d_1\, 10^{k-2} + d_2\, 10^{k-3} + \cdots + d_{k-2}\, 10 + d_{k-1} - 2(d_k) = 7p.$$

Then $d_{k-1} \equiv 7p+2(d_k) - d_1\,10^{k-2} - d_2\,10^{k-3} - \cdots - d_{k-2}\,10$, and by plugging in the original expression of N, we obtain:

$$N = d_1\,10^{k-1} + d_2\,10^{k-2} + \cdots + d_{k-2}\,10^2 + 10\left[7p + 2(d_k) - d_1\,10^{k-2} - d_2\,10^{k-3} \cdots d_{k-2}\,10\right] + d_k,$$

which can be simplified to $N = 10\,[7p+2(d_k)] + d_k = 7(10p+3\,d_k)$, a multiple of 7.

(b) To prove the converse, let us assume N is a multiple of 7. Then, $\exists\, q \in \mathbb{Z}$: $N = 7q$, and let M be the integer obtained by applying the divisibility rule by 7 that is,

$$M = \overline{d_1 d_2 d_3 \cdots d_{k-1}} - 2(d_k) = d_1\,10^{k-2} + d_2\,10^{k-3} + \cdots + d_{k-2}\,10 + d_{k-1} \; - \; 2(d_k).$$

Thus $d_{k-1} = M - d_1\,10^{k-2} + d_2\,10^{k-3} + \cdots + d_{k-2}\,10 \; + 2(d_k)$.
Now, let us consider

$$N = d_1\,10^{k-1} + d_2\,10^{k-2} + \cdots + d_{k-2}\,10^2 + d_{k-1}\,10 \; + d_k \; = \; 7q.$$

Let us substitute d_{k-1} in the preceding expression of N we have

$$N = d_1\,10^{k-1} + d_2\,10^{k-2} + \cdots + d_{k-2}\,10^2 + 10\left(M - d_1\,10^{k-2} - - \cdots - d_{k-2}\,10 \; + 2d_k\right) + d_k.$$

Thus

$$N = 10M \; + 20(d_k) + d_k \; = \; 7q \implies 10M \text{ is divisible by } 7.$$

Since $GCD(10,7) = 1$, we conclude that M is a multiple of 7.

Definition 5.3.5 — Divisibility Rule by 11. An integer is divisible by 11 if the number formed by dropping the unit place and then subtracting it from the number is also divisible by 11.

For example, 121 is divisible by 11. According to the divisibility by 11 rule, $12 - 1 = 11$ which is a multiple of 11.
Let us check whether 22242 is divisible by 11. Following the divisibility by 11 rule, we have

$$2224 - 2 = 2222 \hookrightarrow \; 222 - 2 = 220 \; \hookrightarrow \; 22 - 0 = 22 = 11 \times 2.$$

Thus 22242 is divisible by 11.

(a) Let us drop the unit digit d_k from $N = \overline{d_1 d_2 d_3 \cdots d_k}$ and assume $\exists\, p \in \mathbb{Z}$ such that the difference is a multiple of 11. For example

$$\overline{d_1 d_2 d_3 \cdots d_{k-1}} - d_k = d_1\,10^{k-2} + d_2\,10^{k-3} + \cdots + d_{k-2}\,10 + d_{k-1} \; - \; d_k = 11\,p.$$

Then $d_{k-1} = \; 11\,p + d_k - d_1\,10^{k-2} - d_2\,10^{k-3} - \cdots - d_{k-2}\,10$, and by plugging in the original expression of N, we obtain

$$N = d_1\,10^{k-1} + d_2\,10^{k-2} + \cdots + d_{k-2}\,10^2 + 10\left[11p + d_k - d_1\,10^{k-2} - d_2\,10^{k-3} ... d_{k-2}\,10\right] + d_k,$$

which can be simplified to $N = 10\,[11p + d_k] + d_k = 11(10p + d_k)$, a multiple of 11.

(b) To prove the converse, let us assume that N is a multiple of 11. That is, $\exists\, q \in \mathbb{Z}: \; N = 11\,q$, and by writing the resulting expression M after applying the divisibility rule by 11, we have

$$M = \overline{d_1 d_2 d_3 \cdots d_{k-1}} - 2(d_k) = d_1\,10^{k-2} + d_2\,10^{k-3} + \cdots + d_{k-2}\,10 + d_{k-1} \; - \; d_k.$$

So,

$$d_{k-1} = M - d_1\,10^{k-2} + d_2\,10^{k-3} + \cdots + d_{k-2}\,10\, + d_k.$$

Then,

$$N = d_1\,10^{k-1} + d_2\,10^{k-2} + \cdots + d_{k-2}\,10^2 + d_{k-1}\,10\, + d_k \;=\; 11q,$$

and by substituting d_{k-1} in the preceding expression of N, we have

$$N = d_1\,10^{k-1} + d_2\,10^{k-2} + \cdots + d_{k-2}\,10^2 + 10\left(M - d_1\,10^{k-2} - \cdots - d_{k-2}\,10\, + d_k\right) + d_k.$$

Thus $10M\, + 10(d_k) + d_k \;=\; 11q \implies 10M$ is divisible by 11, and since $GCD(10,11) = 1$, M is a multiple of 11.

We finish this section by proving a very well-known result called the **binomial theorem**, also known as binomial formula. The following Remark 5.3 contains a few basic identities needed for the proof of binomial theorem.

(R) The following identities exist for a non-negative integer n

1. $\binom{n}{0} = \binom{n+1}{0}$
2. $\binom{n}{n} = \binom{n+1}{n+1}$
3. $\binom{n}{k} + \binom{n}{k+1} = \binom{n+1}{k+1}$ Pascal formula
4. $$\binom{n}{0} = \frac{n!}{0!\,\times n!} = 1 = \frac{(n+1)!}{0!\,\times(n+1)!} = \binom{n+1}{0}$$
5. $$\binom{n}{n} = \frac{n!}{n!\,\times 0!} = 1 = \frac{(n+1)!}{(n+1)!\,\times 0!} = \binom{n+1}{n+1}$$
6. Since $(m+1)! \;=\; (m+1)\,m!$,

$$\begin{aligned}
\binom{n}{k} + \binom{n}{k+1} &= \frac{n!}{k!\,\times(n-k)!} + \frac{n!}{(k+1)!\,\times(n-k-1)!,} \\
&= \frac{(k+1)\,n!}{(k+1)\,k!\,\times(n-k)!} + \frac{(n-k)\,n!}{(k+1)!\,\times(n-k-1)!\,(n-k),} \\
&= \frac{(k+1)\,n!}{(k+1)!\times(n-k)!} + \frac{(n-k)\,n!}{(k+1)!\,\times(n-k)!,} \\
&= \frac{[(k+1)+(n-k)]\;n!}{(k+1)!\times(n-k)!,} \\
&= \frac{(n+1)\,n!}{(k+1)!\times(n-k)!}, \quad \text{since } n-k = (n+1)-(k+1) \\
&= \frac{(n+1)!}{(k+1)!\times[(n+1)-(k+1)]!.} \\
Therefore, \binom{n}{k} + \binom{n}{k+1} &= \binom{n+1}{k+1}
\end{aligned}$$

5.3.1 Binomial Theorem

A Greek mathematician named Euclid first mentioned the **binomial theorem** in the 4th century BC. In the binomial theorem, the algebraic expression $(x+y)^n$ is expanded as a sum of the terms involving the exponents of variables x and y. The coefficients of a binomial expansion are numerical values that are associated with each term.

■ **Example 5.30** The binomial theorem formula is used in the expansion of any power of a binomial in the form of a series. Let us prove the binomial expansion formula in one variable, $(x+1)^n$ by mathematical induction.

$$(x+1)^n = \sum_{k=0}^{n} \binom{n}{k} x^k.$$

Let $P(n)$: $(x+1)^n = \sum_{k=0}^{n} \binom{n}{k} x^k.$

base step: For $n=0$, $1 = \binom{0}{0} = 1$, then $P(0)$ is true.

inductive step: Assume now that $P(k)$ is true, and let us show that

$$P(k+1): (x+1)^{k+1} = \sum_{m=0}^{k+1} \binom{k+1}{m} x^m$$

is also true.

$$\begin{aligned}
(x+1)^{k+1} &= (x+1)\,(x+1)^k, \\
&= (x+1)\underbrace{\sum_{m=0}^{k} \binom{k}{m} x^m}_{P(k)}, \\
&= x\sum_{m=0}^{k} \binom{k}{m} x^m + 1\sum_{m=0}^{k} \binom{k}{m} x^m, \\
&= \sum_{m=0}^{k} \binom{k}{m} x^{m+1} + \sum_{m=0}^{k} \binom{k}{m} x^m, \\
&= \sum_{m=0}^{k-1} \binom{k}{m} x^{m+1} + \binom{k}{k} x^{k+1} + \binom{k}{0} x^0 + \sum_{m=1}^{k} \binom{k}{m} x^m.
\end{aligned}$$

After re-indexing the second summation, we have

$$\sum_{m=1}^{k} \binom{k}{m} x^m = \sum_{m=0}^{k-1} \binom{k}{m+1} x^{m+1},$$

then

$$\begin{aligned}
(x+1)^{k+1} &= \sum_{m=0}^{k-1} \binom{k}{m} x^{m+1} + \binom{k}{k} x^{k+1} + \binom{k}{0} x^0 + \sum_{m=0}^{k-1} \binom{k}{m+1} x^{m+1}, \\
&= \sum_{m=0}^{k-1} \left[\binom{k}{m} + \binom{k}{m+1}\right] x^{m+1} + \binom{k+1}{k+1} x^{k+1} + \binom{k+1}{0} x^0,
\end{aligned}$$

and according to (3), we obtain

$$(x+1)^{k+1} = \sum_{m=0}^{k-1} \binom{k+1}{m+1} x^{m+1} + \binom{k+1}{0} x^0 + \binom{k+1}{k+1} x^{k+1},$$

$$(x+1)^{k+1} = \sum_{m=1}^{k} \binom{k+1}{m} x^{m} + \binom{k+1}{0} x^0 + \binom{k+1}{k+1} x^{k+1},$$

$$(x+1)^{k+1} = \sum_{k=0}^{k+1} \binom{k+1}{m} x^{m}.$$

Thus $P(k+1)$ is true, which proves the binomial formula. ■

■ **Example 5.31** Prove the binomial formula.

$$(x+y)^n = \sum_{k=0}^{n} \binom{n}{k} x^k y^{n-k}, \quad x, y \in \mathbb{R}.$$

We will consider the following three cases:

- $\mathbf{x = y = 0}$ (proof is trivial).
- $\mathbf{x \neq 0,\ y = 0}$.

$$(x+0)^n = x^n = \binom{n}{n} x^n = \sum_{k=0}^{n} \binom{n}{k} x^k\, 0^{n-k}.$$

- $\mathbf{xy \neq 0}$,

$$(x+y)^n = y^n \left(\frac{x}{y}+1\right)^n = y^n \sum_{k=0}^{n} \binom{n}{k} \left(\frac{x}{y}\right)^k.$$

Thus

$$(x+y)^n = \sum_{k=0}^{n} \binom{n}{k} x^k y^{n-k}.$$

■

5.4 Integer Part

This section concerns a nice part of mathematics, where we focus on the **integer part** of a given real number and as well as its fractional part. The subject is based on simple definitions and properties but students usually find this part complicated. Of course, it is not an easy or direct part but it needs a solid background in high-school mathematics.

Definition 5.4.1 The **integer part** of a given real number x is the greatest integer less than or equal to x, denoted $\lfloor x \rfloor$.

For example,

$$\lfloor 3.4 \rfloor = 3, \quad \lfloor 3/2 \rfloor = 1, \quad \lfloor 3 \rfloor = 3, \quad \lfloor 3.99 \rfloor = 3, \ \lfloor 3.0001 \rfloor = 3,$$
$$\lfloor \sqrt{7} \rfloor = 2, \quad \lfloor 2\pi \rfloor = 6, \quad \lfloor -1,9 \rfloor = -2, \quad \lfloor -1.001 \rfloor = -2,$$
$$\lfloor -2.5 \rfloor = -3, \quad \lfloor -\sqrt{3} \rfloor = -2, \quad \lfloor -4 \rfloor = -4.$$

In other words, the integer part of a given real number x can be defined as follows:

$$\lfloor x \rfloor = m \Longleftrightarrow m \leq x < m+1, \quad x \in \mathbb{R},\ m \in \mathbb{Z}.$$

Definition 5.4.2 fractional part The **fractional or decimal part** of a real number x is a number in $\lfloor 0, 1)$ that is equal to the difference $x - \lfloor x \rfloor$ and is denoted as

$$\{x\} = x - \lfloor x \rfloor.$$

For example,

$$\{3.4\} = 0.4, \quad \{3/2\} = 0.5, \quad \{3\} = 0, \quad \{3.99\} = 0.99,$$

$$\{3.0001\} = 0.0001, \quad \{2\pi\} = 2\pi - 6, \quad \{-1,9\} = 0.9,$$

$$\{-1.001\} = 0.001, \quad \{-2.5\} = 0.5, \quad \{-4\} = 0, \ \{\sqrt{7}\} = \sqrt{7} - 2.$$

Direct conclusions drawn from these definitions are described in Theorem 5.4.1.

Theorem 5.4.1 — basic properties.

(a) $\lfloor x \rfloor = x \Longleftrightarrow x \in \mathbb{Z}, \ \forall x \in \mathbb{R}.$

(b) $\{x\} = 0 \Longleftrightarrow x \in \mathbb{Z}, \ \forall x \in \mathbb{R}.$

(c) $\{x\} \in [0, \ 1).$

(d) $\lfloor \lfloor x \rfloor \rfloor = \lfloor x \rfloor$ and $\{ \{x\} \} = \{x\}.$

(e) $\lfloor \{x\} \rfloor = 0$ and $\{ \lfloor x \rfloor \} = 0.$

(f) $x \geq 0 \Longleftrightarrow \lfloor x \rfloor \geq 0$ and $x < 0 \Longleftrightarrow \lfloor x \rfloor < 0.$

(g) $\forall x \in \mathbb{R}, \ x - 1 < \lfloor x \rfloor \ \leq x.$

Proof. $(a) - (e)$ follow from the definitions.
To prove (f), we use the definition of $\lfloor x \rfloor$: $\lfloor x \rfloor \leq x < \lfloor x \rfloor + 1$. Since $0 \leq x < \lfloor x \rfloor + 1$, this implies that $\lfloor x \rfloor + 1 > 0$, and then $\lfloor x \rfloor > -1$. Using the fact that $\lfloor x \rfloor$ is an integer, we conclude that $\lfloor x \rfloor \geq 0$. Now, for $x < 0$, we have $\lfloor x \rfloor \leq x < 0$, which completes the proof.

To prove (g), we need to write the definition:

$$\lfloor x \rfloor \leq x < \lfloor x \rfloor + 1 \Longleftrightarrow \lfloor x \rfloor \leq x \text{ and } x < \lfloor x \rfloor + 1,$$

which is equivalent to

$$x - 1 < \lfloor x \rfloor \text{ and } \lfloor x \rfloor \leq x,$$

and the proof is completed. ■

▪ **Example 5.32** Let x and y be two real numbers. Show that

$$x \leq y \Longleftrightarrow \lfloor x \rfloor \leq \lfloor y \rfloor.$$

Since $\lfloor x \rfloor \leq x \leq y$, then $\lfloor x \rfloor \leq y < \lfloor y \rfloor + 1$ which implies that $\lfloor x \rfloor - 1 < \lfloor y \rfloor$ and thus $\lfloor x \rfloor \leq \lfloor y \rfloor$ because $\lfloor y \rfloor$ and $\lfloor x \rfloor - 1$ are integers. ▪

▪ **Example 5.33** Let x and y be two real numbers, show that

$$\lfloor x \rfloor = \lfloor y \rfloor \Longleftrightarrow |x - y| < 1.$$

According to the definition of an integer part, we have

$$\lfloor x \rfloor \leq x < \lfloor x \rfloor + 1 \text{ and } \lfloor y \rfloor \leq y < \lfloor y \rfloor + 1,$$

which is equivalent to

$$\lfloor x \rfloor \leq x < \lfloor x \rfloor + 1, \quad -\lfloor y \rfloor - 1 \leq -y < -\lfloor y \rfloor.$$

Adding both inequalities, we obtain $\lfloor x \rfloor - \lfloor y \rfloor - 1 \leq x - y < -\lfloor y \rfloor + \lfloor x \rfloor + 1$. This implies that $-1 < x - y < +1$, since $\lfloor x \rfloor = \lfloor y \rfloor$. ∎

∎ **Example 5.34** Let x and y be two real numbers. Show that

$$\{x\} = \{y\} \Longrightarrow x - y \in \mathbb{Z}.$$

It is obvious using the definition.

$$\{x\} = \{y\} \Longleftrightarrow x - \lfloor x \rfloor = y - \lfloor y \rfloor \Longleftrightarrow x - y = \lfloor x \rfloor - \lfloor y \rfloor \in \mathbb{Z}.$$

∎

∎ **Example 5.35** Let $x \in \mathbb{R}$ and $k \in \mathbb{Z}$. Show that

$$\lfloor x + k \rfloor = k + \lfloor x \rfloor \text{ and} \{x + k\} = \{x\}.$$

∎

∎ **Example 5.36** Let $x \in \mathbb{R}$. Show that

$$\lfloor -x \rfloor = -\lfloor x \rfloor - 1 \text{ if } x \text{ is not an integer, and } \lfloor -x \rfloor = -\lfloor x \rfloor \text{ if } x \in \mathbb{Z}.$$

∎

∎ **Example 5.37** Let $x \in \mathbb{R}$. Show that

$$\{-x\} = 1 - \{x\}, \text{ if } x \text{ is not an integer and } \{-x\} = \{x\} \text{ if } x \in \mathbb{Z}.$$

∎

∎ **Example 5.38** Let $n \in \mathbb{Z}$. Show that

$$\left\lfloor \frac{n-1}{2} \right\rfloor + \left\lfloor \frac{n+2}{4} \right\rfloor + \left\lfloor \frac{n+4}{4} \right\rfloor = n.$$

(Hint: use the integer division of n by 4.)

The integer n can be expressed as follows: $n \in \{4p, 4p+1, 4p+2, 4p+3, \; p \in \mathbb{Z}\}$. By considering the 4 cases the conclusion is obvious. ∎

∎ **Example 5.39** Show the following for any $x, \; y \in \mathbb{R}$:

- $\lfloor x + y \rfloor \geq \lfloor x \rfloor + \lfloor y \rfloor$.
- $\lfloor xy \rfloor \geq \lfloor x \rfloor \times \lfloor y \rfloor, \quad \forall x, y \geq 0.$
- $\left\lfloor \frac{x}{n} \right\rfloor = \left\lfloor \frac{\lfloor x \rfloor}{n} \right\rfloor, \quad n \in \mathbb{N}, \; n \neq 0.$
- $\lfloor x \rfloor = \left\lfloor \frac{\lfloor nx \rfloor}{n} \right\rfloor, \quad n \in \mathbb{N}, \; n \neq 0.$

- For any x, y, we write $x = \lfloor x \rfloor + \{x\}$ and $y = \lfloor y \rfloor + \{y\}$. Then using $\{x\}, \{y\} \in (0,1)$, we obtain $x+y = \lfloor x \rfloor + \{x\} + \lfloor y \rfloor + \{y\} \geq \lfloor x \rfloor + \lfloor y \rfloor$. Since $\lfloor x+y \rfloor \leq x+y$, $\lfloor x+y \rfloor \geq \lfloor x \rfloor + \lfloor y \rfloor$.

- By using the non-negativity of x and y and the following two inequalities, $\lfloor x \rfloor \leq x$, $\lfloor y \rfloor \leq y$, we obtain the product term as the following:

$$\lfloor x \rfloor \lfloor y \rfloor \leq xy \text{ which implies that } \lfloor x \rfloor \lfloor y \rfloor \leq \lfloor xy \rfloor.$$

- Let $m = \left\lfloor \frac{x}{n} \right\rfloor$, then $m \leq \frac{x}{n} < m+1$, implies that $mn \leq x < mn+n$, which is equivalent to $mn \leq \lfloor x \rfloor < mn+n$ (because mn and $mn+n$ are integers). Then, $m \leq \frac{\lfloor x \rfloor}{n} < m+1$ implies that $\left\lfloor \frac{\lfloor x \rfloor}{n} \right\rfloor = \left\lfloor \frac{x}{n} \right\rfloor$.

- Just take $x = nm$ in the previous relation.

■

■ **Example 5.40** Show that for any $a \in \mathbb{R}$ and $n \in \mathbb{N}$,

$$\sum_{k=0}^{n-1} \left\lfloor \frac{a+k}{n} \right\rfloor = \lfloor a \rfloor.$$

(Hint: consider the Euclidean division of $\lfloor a \rfloor$ by n, or use the periodicity of the given expression.)

1. Following the Hint, we have: $\lfloor a \rfloor = q \times n + r$, where q is the quotient and the remainder$r$, where $0 \leq r < n$. Consider

$$nq+r \leq a < nq+r+1 \Longleftrightarrow \frac{nq+r+k}{n} \leq \frac{a+k}{n} < \frac{nq+r+1+k}{n},$$

i.e.,

$$q+\frac{r+k}{n} \leq \frac{a+k}{n} < q+\frac{r+1+k}{n}.$$

Now that it is easy to continue. We wish to separately study the following two cases: $\frac{r+1+k}{n} \leq 1$ and $\frac{r+1+k}{n} > 1$.

- If $\frac{r+1+k}{n} \leq 1$, we have $q \leq q+\frac{r+k}{n} \leq \frac{a+k}{n} < q+\frac{r+1+k}{n} \leq q+1$, then $\left\lfloor \frac{a+k}{n} \right\rfloor = q$.

- If

$$\frac{r+1+k}{n} > 1 \Longrightarrow r+k+1 > n \Longrightarrow r+k \geq n \Longrightarrow \frac{r+k}{n} \geq 1 \Longrightarrow q+\frac{r+k}{n}+q \geq q+1.$$

Also,

$$k \leq n-1,\ r < n \Longrightarrow r+k+1 < 2n \Longrightarrow 1 < \frac{r+1+k}{n} < 2 \Longrightarrow q+\frac{r+1+k}{n} < q+2.$$

Thus

$$q+1 \leq q+\frac{r+k}{n} \leq \frac{a+k}{n} < q+\frac{r+1+k}{n} < q+2,$$

then $q+1 \leq \frac{a+k}{n} < q+2$, which implies: $\left\lfloor \frac{a+k}{n} \right\rfloor = q+1$.

This gives the following

$$\sum_{k=0}^{n-1} \left\lfloor \frac{a+k}{n} \right\rfloor = \sum_{k=0}^{n-1-r} q + \sum_{k=n-r}^{n-1} (q+1).$$

The first sum equals $\sum_{k=0}^{n-1-r} q = q\lfloor (n-r-1)-0+1 \rfloor = qn - qr$, and the second one equals $\sum_{k=n-r}^{n-1} (q+1) = (q+1)\lfloor (n-1)-(n-r)+1 \rfloor = qr+r$. Then we have

$$\sum_{k=0}^{n-1} \left\lfloor \frac{a+k}{n} \right\rfloor = qn+r = \lfloor a \rfloor.$$

2. Let us denote $f(a) = \sum_{k=0}^{n-1} \left\lfloor \frac{a+k}{n} \right\rfloor$ the question itself, asks to prove that f coincides with the integer part function. To this end, let us consider

$$\begin{aligned} f(a+1) &= \left\lfloor \frac{a+1}{n} \right\rfloor + \left\lfloor \frac{a+2}{n} \right\rfloor + \ldots + \left\lfloor \frac{a+n-1}{n} \right\rfloor + \left\lfloor \frac{a+n}{n} \right\rfloor. \\ &= \left\lfloor \frac{a+1}{n} \right\rfloor + \left\lfloor \frac{a+2}{n} \right\rfloor + \ldots + \left\lfloor \frac{a+n-1}{n} \right\rfloor + \left\lfloor \frac{a}{n} \right\rfloor + 1. \\ &= \left\lfloor \frac{a}{n} \right\rfloor + \left\lfloor \frac{a+1}{n} \right\rfloor + \ldots + \left\lfloor \frac{a+n-1}{n} \right\rfloor + 1. \\ &= f(a)+1. \end{aligned}$$

Thus for any $a \in \mathbb{R}$: $f(a+1) = f(a)+1 \Longrightarrow f$ seems equivalent to the integer part function. Further, for ant $a \in \lfloor 0,1)$, it is easy to affirm that $f(a) = 0$. Indeed,

$$0 \le a < 1,\ 0 \le k \le n-1 \Longrightarrow 0 \le a+k < n \Longrightarrow 0 \le \frac{a+k}{n} < 1 \Longrightarrow \left\lfloor \frac{a+k}{n} \right\rfloor = 0 \Longrightarrow f(a) = 0.$$

Now, we can affirm that $f(a) = \lfloor a \rfloor,\ \forall a \in \mathbb{R}$ and finally we proved the Hermite equation:

$$\sum_{k=0}^{n-1} \left\lfloor \frac{a+k}{n} \right\rfloor = \lfloor a \rfloor.$$

■

■ **Example 5.41** Solve

$$\left\lfloor \frac{x}{2} \right\rfloor + \left\lfloor \frac{2x}{3} \right\rfloor = x.$$

It is clear that $x \in \mathbb{Z}$. The presence of the denominators 3 and 2 suggests we should consider the Euclidean division of x by $2 \times 3 = 6$. Thus $x = 6q+r$, where the remainder $r \in \{0,1,2,3,4,5\}$. The preceding equation becomes

$$\left\lfloor \frac{6q+r}{2} \right\rfloor + \left\lfloor \frac{2(6q+r)}{3} \right\rfloor = 6q+r \Longleftrightarrow \left\lfloor 3q + \frac{r}{2} \right\rfloor + \left\lfloor 4q + \frac{r}{3} \right\rfloor = 6q+r,$$

i.e.,

$$q + \left\lfloor \frac{r}{2} \right\rfloor + \left\lfloor \frac{2r}{3} \right\rfloor = r.$$

And now let us consider the six possible situations:

0	$q=0 \Longrightarrow x=0,$
1	$q=1 \Longrightarrow x=7,$
2	$q=0 \Longrightarrow x=2,$
3	$q=0 \Longrightarrow x=3,$
4	$q=0 \Longrightarrow x=4,$
5	$q=0 \Longrightarrow x=5.$

Finally, $S=\{0,2,3,4,5,7\}$. ■

■ **Example 5.42** Solve the following system:

$$\begin{aligned} x+\lfloor y\rfloor+\{z\} &= 10.6, && (5.1)\\ \lfloor x\rfloor+\{y\}+z &= 11.7, && (5.2)\\ \{x\}+y+\lfloor z\rfloor &= 16.5. && (5.3)\end{aligned}$$

The solution is based on some basic operations between the equations.

Since in general, $\lfloor a\rfloor+\{a\}=a$, and the sum of equations: (5.1), (5.2), and (5.3) can be simplified to

$$x+y+z=19.4, \quad (5.4)$$

and now differance of equation (5.4) and equation (5.1) gives

$$\{y\}+\lfloor z\rfloor=8.8 \Longrightarrow \{y\}=0.8,\ \lfloor z\rfloor=8.$$

According to (5.2) we have

$$\lfloor x\rfloor+z=10.9 \Longrightarrow \lfloor x\rfloor+\lfloor z\rfloor=10 \text{ and } \{z\}=0.9 \Longrightarrow z=8.9$$

and then $\lfloor x\rfloor=2$. Now (5.1) gives $\lfloor y\rfloor=7$, $\{x\}=0.7$ and then $x=2.7$. Thus $S=\{(2.7,\ 7.8,\ 8.9)\}$.
■

5.5 More about Sums

In this section, we prove some classical sums needed for many advanced topics in mathematics. The student will also find a subsection devoted to telescoping sums. It is important to notice that mastering factorization of algebraic expressions, polynomials, and identities are the main tools required to this topic.

5.5.1 Classical Sums

$$\boxed{GS=\sum_{k=1}^{n} k = 1+2+3+\ldots+n=\frac{n(1+n)}{2} \hookrightarrow \text{Gauss sum.}}$$

$$\boxed{\sum_{k=0}^{n} 2k+1 = 1+3+5+7+\ldots+2n+1=(n+1)^2 \hookrightarrow \text{Sum of odd numbers.}}$$

$$\boxed{\sum_{k=1}^{n} 2k = 2+4+6+8+\ldots+2n=n(n+1) \hookrightarrow \text{Sum of even numbers.}}$$

In the next two examples, we show how to construct formulae for the sum of squares and the sum of cubes.

■ **Example 5.43** Determine the sum of the first n squares

$$SS = 1^2+2^2+3^2+\ldots+n^2.$$

■

The method we will expose can be found in many old and recent French textbooks. The method is based on the identity $(n+1)^3 = n^3+3n^2+3n+1$ and the Gauss sum.

$$\begin{array}{ccccccccccc}
\text{Identity} & & (k+1)^3 & = & k^3 & + & 3k^2 & + & 3k & & +1 \\
\hline
k=1 & \hookrightarrow & 2^3 & = & 1^3 & + & 3\times 1^2 & + & 3\times 1 & & +1 \\
k=2 & \hookrightarrow & 3^3 & = & 2^3 & + & 3\times 2^2 & + & 3\times 2 & & +1 \\
k=3 & \hookrightarrow & 4^3 & = & 3^3 & + & 3\times 3^2 & + & 3\times 3 & & +1 \\
\vdots & & & & \vdots & \vdots & \vdots & \vdots & \vdots & & \\
k=n-2 & \hookrightarrow & (n-1)^3 & = & (n-2)^3 & + & 3(n-2)^2 & + & 3(n-2) & & +1 \\
k=n-1 & \hookrightarrow & n^3 & = & (n-1)^3 & + & 3(n-1)^2 & + & 3(n-1) & & +1 \\
k=n & \hookrightarrow & (n+1)^3 & = & n^3 & + & 3n^2 & + & 3n & & +1, \\
\hline
\text{Sums} & \hookrightarrow & (n+1)^3 & = & 1^3 & + & 3\,\sum_{k=1}^{n} k^2 & + & 3\,GS & & +\sum_{k=1}^{n} 1
\end{array}$$

i.e.,

$$(n+1)^3 = 1 \,+\, 3\,SS+3\,GS\,+n \,=\, 3\,SS+3\left(\frac{n(n+1)}{2}\right)+(n+1),$$

and then

$$3\,SS = (n+1)^3 \,-\, 3\left(\frac{n(n+1)}{2}\right) - (n+1) \,=\, (n+1)\left[(n+1)^2-\frac{3n}{2}-1,\right]$$

i.e.,

$$SS = \frac{(n+1)}{3}\left[\frac{2(n+1)^2}{2}-\frac{3n}{2}-\frac{2}{2},\right] = \frac{(n+1)}{6}\left[2(n+1)^2-3n-2\right]$$

and finally

$$SS = \frac{(n+1)}{6}\left[2n^2+4n+2-3n-2\right] = \frac{(n+1)}{6}(2n^2+n) = \frac{n(n+1)(2n+1)}{6}.$$

Thus

$$\boxed{SS \,=\, \sum_{k=1}^{n} k^2 = 1^2+2^2+3^2+\ldots+n^2 \,=\, \frac{n(n+1)(2n+1)}{6},\ \forall n\geq 1}.$$

■ **Example 5.44** Determine the sum of the first n cubes

$$SC = 1^3+2^3+3^3+\ldots+n^3.$$

■

We will again apply the same approach as in the previous example. The method is based on the identity $(n+1)^4 = n^4+4n^3+6n^2+4n+1$ – the sum of squares and the Gauss sum.

Identity		$(k+1)^4$	=	k^4	+	$4k^3$	+	$6k^2$	+	$4k$	$+1$
$k=1$	$\hookrightarrow$	2^4	=	1^4	+	4×1^3	+	6×1^2	+	4×1	$+1$
$k=2$	$\hookrightarrow$	3^4	=	2^4	+	4×2^3	+	6×2^2	+	4×2	$+1$
$k=3$	$\hookrightarrow$	4^4	=	3^4	+	4×3^3	+	6×3^2	+	4×3	$+1$
$\vdots$						$\vdots$	$\vdots$	$\vdots$	$\vdots$	$\vdots$	
$k=n-2$	$\hookrightarrow$	$(n-1)^4$	=	$(n-2)^4$	+	$4(n-2)^3$	+	$6(n-2)^2$	+	$4\times(n-2)$	$+1$
$k=n-1$	$\hookrightarrow$	n^4	=	$(n-1)^4$	+	$4(n-1)^3$	+	$6(n-1)^2$	+	$4\times(n-1)$	$+1$
$k=n$	$\hookrightarrow$	$(n+1)^4$	=	n^4	+	$4n^3$	+	$6n^2$	+	$4\times n$	$+1$
Sums	$\hookrightarrow$	$(n+1)^4$	=	1^4	+	$4\sum_{k=1}^n k^3$	+	$6\sum_{k=1}^n k^2$	+	$\sum_{k=1}^n k$	$+n,$

i.e.,

$$(n+1)^4 = 4\,SC + 6SS + 4GS + (n+1)$$

and then

$$4\,SC = (n+1)\left[(n+1)^3 - 6\frac{n(2n+1)}{6} - 4\frac{n}{2} - 1\right],$$

i.e.,

$$SC = \frac{(n+1)}{4}\left[(n+1)^3 - n(2n+1) - 2n - 1\right] = \frac{(n+1)}{4}\left[(n+1)^3 - (2n+1)(n+1)\right],$$

$$SC = \frac{(n+1)^2}{4}\left[(n+1)^2 - (2n+1)\right],$$

and then

$$SC = \frac{n^2(n+1)^2}{4}.$$

Thus

$$\boxed{\sum_{k=1}^{n} k^3 = 1^3 + 2^3 + 3^3 + \ldots + n^3 = \frac{n^2(n+1)^2}{4} = (GS)^2 = (1+2+3+\ldots+n)^2}.$$

■ **Example 5.45** Determine the sum

$$1^4 + 2^4 + 3^4 + \ldots + n^4, \quad 1^5 + 2^5 + 3^5 + \ldots + n^5.$$

■

Solution. (Hint: use the same approach as in the previous examples).

R James Bernoulli proposed a class of polynomials, called Bernoulli polynomials, of degree n, defined as follows:

$$B_n(x+1) - B_n(x) = n\,x^{n-1}.$$

These polynomials can be used to build a general rule for the sum

$$1^p + 2^p + 3^p + 4^p + \ldots + n^p = \sum_{k=1}^{n} k^p = \frac{B_{p+1}(n+1) - B_{p+1}(0)}{p+1}.$$

We can check the Gauss sum and the sum of squares:

- $p=1 \hookrightarrow GS = \frac{B_2(n+1)-B_2(0)}{2}$. These polynomials satisfy $B_2(x+1)-B_2(x)=2x$, and by a simple identification for $B_2(x)=ax^2+bx+c$, we have $B_2(x)=x^2-x+c$, and then the above sum becomes:

$$GS = \frac{B_2(n+1)-B_2(0)}{2} = \frac{(n+1)^2-(n+1)-c}{2} = \frac{n^2+n}{2} = \frac{n(n+1)}{2}.$$

- $p=2$ (already done).

■ Example 5.46 Find a formula for

$$T(n) = 1\times 2\times 3+2\times 4\times 5+\ldots+n(n+1)(n+2).$$

■

The required sum is equivalent to $\sum_{k=1}^{n} k(k+1)(k+2)$, and direct expansion we obtain

$$T(n) = \sum_{k=1}^{n} k(k+1)(k+2) = \sum_{k=1}^{n} \{k(k^2+3k+2)\} = \sum_{k=1}^{n} \{k^3+3k^2+2k\},$$

hence

$$T(n) = \sum_{k=1}^{n} k^3 + 3\sum_{k=1}^{n} k^2 + 2\sum_{k=1}^{n} k = SC+3\times SS+2\times GS.$$

Thus

$$T(n) = \frac{n^2(n+1)^2}{4} + \frac{n(n+1)(2n+1)}{2} + n(n+1) = \frac{n(n+1)}{4}\left[n(n+1)+2(2n+1)+4\right].$$

Finally,

$$T(n) = \frac{n(n+1)}{4}(n^2+5n+6) = \frac{n(n+1)(n+2)(n+3)}{4}.$$

R This induces a nice result:

$$1\times 2\times 3+2\times 4\times 5+\ldots+n(n+1)(n+2) = \frac{n(n+1)(n+2)(n+3)}{4} = \frac{n(n+1)}{2}\times\frac{(n+2)(n+3)}{2},$$

which is equivalent to

$$1\times 2\times 3+2\times 4\times 5+\ldots+n(n+1)(n+2) = [1+2+3+\ldots+n]\,[1+2+3+\ldots+(n+2)].$$

■ Example 5.47 Give a formula for the sum

$$\sum_{j=1}^{n} jx^{j-1} = 1+2x+3x^2+\ldots+nx^{n-1}.$$

■

Solution. Let S denote the required sum. If $x=1$ we obtain the Gauss sum, and if $x=0$, the sum $S=1$. Now, we assume that $x(1-x)\neq 0$. If we multiply S by x we obtain

$$xS = x\times\sum_{j=1}^{n} jx^{j-1} = \sum_{j=1}^{n} jx^j.$$

Thus

$$S-xS = \sum_{j=1}^{n} jx^{j-1} - \sum_{j=1}^{n} jx^j = 1+x+x^2+\ldots+x^{n-1}-nx^n = (1-nx^n)+x\sum_{k=0}^{n-2} x^k,$$

i.e.,

$$S = \frac{1-nx^n}{1-x} + x\times\frac{1-x^{n-1}}{(1-x)^2} = \frac{1-nx^n}{1-x} + \frac{x-x^n}{(1-x)^2}.$$

5.5.2 Telescoping Sums

Telescoping Sum is a magic method that allows us to determine several very complex sums, where almost everything cancels out. The technique is obvious, but it is useful to know for engineering and mathematics students. The main idea can be exposed as follows: Let (u_n) and (v_n) be two given sequences such that

$$v_n = u_{n+1} - u_n \hookrightarrow \sum_{i=0}^{n} v_i = (u_1 - u_0) + (u_2 - u_1) + (u_3 - u_2) + \ldots + (u_n - u_{n-1}) + (u_{n+1} - u_n).$$

Thus after simplifying we obtain

$$\sum_{i=0}^{n} v_i = -u_0 + u_{n+1} = u_{n+1} - u_0 = \text{last term} \ - \ \text{first term},$$

or we have another situation where

$$v_n = u_n - u_{n+1} \hookrightarrow \sum_{i=0}^{n} v_i = (u_0 - u_1) + (u_1 - u_2) + (u_2 - u_3) + \ldots + (u_{n-1} - u_n) + (u_n - u_{n+1}),$$

and then

$$\sum_{i=0}^{n} v_i = u_0 - u_{n+1} = \text{first term} \ - \ \text{last term}.$$

■ **Example 5.48** Evaluate $\frac{1}{1\times 2} + \frac{1}{2\times 3} + \ldots + \frac{1}{n(n+1)}$. ■

Solution. If we denote $v_k = \frac{1}{k(k+1)}$, we need to find a sequence (u_n) such that $v_k = u_k - u_{k+1}$ or $v_k = u_{k+1} - u_k$. Since we have a product in the denominator, one may immediately think about splitting the fraction into the difference of two fractions:

$$\frac{1}{k(k+1)} = \frac{a}{k} - \frac{b}{k+1} = \frac{a(k+1) - bk}{k(k+1)} = \frac{k(a-b)+a}{k(k+1)}.$$

Thus the two numerators are equal for any value of $k \geq 1$, which implies by identification we have: $\begin{cases} a-b=0, \\ a=1 \end{cases} \Longrightarrow a=b=1.$ Then $v_k = \frac{1}{k} - \frac{1}{k+1}$, i.e., $u_k = \frac{1}{k}$. Thus we have $v_k = u_k - u_{k+1}$, and the above sum is equal to first term – last term, i.e.,

$$\sum_{k=1}^{n} \frac{1}{k(k+1)} = \frac{1}{1\times 2} + \frac{1}{2\times 3} + \ldots + \frac{1}{n(n+1)} \frac{1}{1} - \frac{1}{n+1} = \frac{n}{n+1}.$$

■ **Example 5.49** We aim to find the sum $S_n = \frac{1}{1\times 2} + \frac{1}{2\times 3} + \ldots + \frac{1}{n(n+1)}$:

(a) Evaluate S_1, S_2, S_3, and S_4.
(b) Propose a formula for S_n.
(c) Prove your formula by induction.

■

(a)

$$S_1 = \frac{1}{1\times 2} = \frac{1}{2}, \quad S_2 = S_1 + \frac{1}{2\times 3} = \frac{2}{3}, \quad S_3 = S_2 + \frac{1}{3\times 4} = \frac{3}{4}, \text{ and } S_4 = S_3 + \frac{1}{4\times 5} = \frac{4}{5}.$$

(b) One can observe:

$$S_1 = \frac{1}{1+1}, \quad S_2 = \frac{2}{3} = \frac{2}{2+1}, \quad S_3 = \frac{3}{3+1}, \quad S_4 = \frac{4}{4+1}.$$

Thus we can propose $S_n = \dfrac{n}{n+1}$ for $n \geq 1$.

(c) Let us prove the formula by induction. We set $P(n): \; S_n = \dfrac{n}{n+1}$.
For $n = 1$, $P(1)$ is true, and we assume that $P(n)$ is true. Let us prove that $P(n+1)$ remains true. To this end, let us consider

$$S_{n+1} = S_n + \frac{1}{(n+1)(n+2)} = \frac{n}{n+1} + \frac{1}{(n+1)(n+2)} = \frac{n(n+2)+1}{(n+1)(n+2)} = \frac{n^2+2n+1}{(n+1)(n+2)},$$

i.e.,

$$S_{n+1} = \frac{(n+1)^2}{(n+1)(n+2)} = \frac{n+1}{n+2}.$$

Thus, $P(n+1)$ is true, and the formula is true for all integers $n \geq 1$.

(R) By manipulating the sigma sums correctly, we can solve many problems related to the telescoping technique. This is the case for the following sum:

$$\sum_{k=2}^{n} \frac{1}{(k-1)(k+1)}.$$

■ **Example 5.50** Find the sum $T = \dfrac{1}{1\times 3} + \dfrac{1}{2\times 4} + \ldots + \dfrac{1}{2011\times 2013}$. ■

If we rewrite $\dfrac{1}{(k-1)(k+1)} = \dfrac{1}{2}\left(\dfrac{1}{k-1} - \dfrac{1}{k+1}\right)$ directly, then

$$2T_n = \sum_{k=2}^{n} \frac{1}{k-1} - \sum_{k=2}^{n} \frac{1}{k+1}.$$

Now, when k runs from 2 to n, the denominator $k-1$ starts at 1 and ends at $n-1$. Thus we can write $\displaystyle\sum_{k=2}^{n} \frac{1}{k-1} = \sum_{i=1}^{n-1} \frac{1}{i}$, and the same thing for the second term, i.e.,

$$\sum_{k=2}^{n} \frac{1}{k+1} = \sum_{i=3}^{n+1} \frac{1}{i}$$

. Hence, we obtain

$$2T_n = \sum_{i=1}^{n-1} \frac{1}{i} - \sum_{i=3}^{n+1} \frac{1}{i} = \frac{1}{1} + \frac{1}{2} + \sum_{i=3}^{n-1} \frac{1}{i} - \sum_{i=3}^{n-1} \frac{1}{i} - \frac{1}{n} - \frac{1}{n+1} = \frac{1}{1} + \frac{1}{2} - \frac{1}{n} - \frac{1}{n+1},$$

i.e.,

$$\sum_{k=2}^{n} \frac{1}{(k-1)(k+1)} = \frac{1}{2} + \frac{1}{4} - \frac{1}{2n} - \frac{1}{2(n+1)} = \frac{3}{4} - \frac{2n+1}{2n(n+1)} = \frac{(3n+2)(n-1)}{4n(n+1)}.$$

Finally, for $n = 2012$, $T = \dfrac{(3\times 2012+2)\times 2011}{4\times 2012\times 2013} = \dfrac{1589}{2000}$.

■ **Example 5.51** Find the sum $T_n = \frac{1}{1\times 2\times 3} + \frac{1}{2\times 3\times 4} + \ldots + \frac{1}{99\times 100\times 100}$. ■

Since the general term is $\frac{1}{(k-1)k(k+1)}$, we need to try to express it as a sum of simpler fractions as follows:

$$\frac{1}{(k-1)k(k+1)} = \frac{1}{k-1} + \frac{b}{k} + \frac{c}{k+1} \Longleftrightarrow (a+b+c)k^2 + (a-c)k - b = 1\ \forall k \geq 2,$$

which implies that $a = c = 1/2$ and $b = -1$. Thus the required sum can be evaluated as follows (for $n = 100$):

$$\sum_{k=2}^{n} \frac{1}{(k-1)k(k+1)} = \sum_{k=2}^{n} \frac{1}{2(k-1)} - \sum_{k=2}^{n} \frac{1}{k} + \sum_{k=2}^{n} \frac{1}{2(k+1)}.$$

Since

$$k=2 \hookrightarrow k-1=1,\ k=n \hookrightarrow k-1=n-1 \Longrightarrow \sum_{k=2}^{n} \frac{1}{2(k-1)} = \sum_{i=1}^{n-1} \frac{1}{2i} = \frac{1}{2} + \frac{1}{4} + \sum_{i=3}^{n-1} \frac{1}{2i},$$

$$k=2 \hookrightarrow k+1=3,\ k=n \hookrightarrow k+1=n+1 \Longrightarrow \sum_{k=2}^{n} \frac{1}{2(k+1)} = \sum_{i=3}^{n+1} \frac{1}{2i} = \sum_{i=3}^{n-1} \frac{1}{2i} + \frac{1}{2n} + \frac{1}{2(n+1)},$$

and then we can write

$$\sum_{k=2}^{n} \frac{1}{(k-1)k(k+1)} = \sum_{i=1}^{n-1} \frac{1}{2i} - \sum_{i=2}^{n} \frac{1}{i} + \sum_{i=3}^{n+1} \frac{1}{2i},$$

which can be simplified

$$\sum_{k=2}^{n} \frac{1}{(k-1)k(k+1)} = \frac{1}{2} + \frac{1}{4} + 2\sum_{i=3}^{n-1} \frac{1}{2i} - \sum_{i=3}^{n-1} \frac{1}{i} - \frac{1}{2} - \frac{1}{n} + \frac{1}{2n} + \frac{1}{2(n+1)},$$

i.e.,

$$\sum_{k=2}^{n} \frac{1}{(k-1)k(k+1)} = \frac{1}{4} - \frac{1}{n} + \frac{1}{2n} + \frac{1}{2(n+1)} = \frac{1}{4} - \frac{1}{2n} + \frac{1}{2(n+1)} = \frac{1}{4} - \frac{1}{2n(n+1)},$$

and, finally, for $n = 100$, the sum is equal to $\frac{1}{4} - \frac{1}{200\times 101} = \frac{5049}{20200}$.

Notably, the instructor can introduce the notion of limit, for example, after such questions. What happens to the sum if n becomes larger and larger. Using calculators, Excel, or any math software, students can evaluate the preceding sum for different large values of n. If we denote $S_n = \sum_{k=2}^{n} 1/(k-1)k(k+1)$,

n	10	25	50	100	250	1000	10,000
S_n	0.2454545	0.2492308	0.2498039	0.2499505	0.2499920	0.2499995	0.2499999

Here it is clear that as n is getting larger and larger, the sum is getting closer and closer from 0.25. In this situation we can say that the sum (S_n) converges to $1/4$, and we can write

$$\sum_{k=2}^{\infty} \frac{1}{(k-1)k(k+1)} = \frac{1}{4}.$$

■ **Example 5.52** Use telescoping to determine the sum of squares.

$$SS = 1^2 + 2^2 + 3^2 + ... + n^2.$$

■

The method we used in the beginning of the section uses telescoping. Indeed, if we consider $v_n = u_{n+1} - u_n$ where $u_n = n^3$, we obtain the following:

$$v_k = (k+1)^3 - k^3 = 3k^2 + 3k + 1 \Longrightarrow \sum_{k=0}^{n} v_k = u_{n+1} - u_0 = (n+1)^3 = 3SS + 3\,GS + (n+1),$$

which will produce the same result as in Example 5.43.

■ **Example 5.53** Simplify the following product (for $t \neq 0,\ t \neq 1$):

$$\Pi = \left(1 + \frac{1}{t}\right)\left(1 + \frac{1}{t^2}\right)\left(1 + \frac{1}{t^4}\right)...\left(1 + \frac{1}{t^{2^{2022}}}\right).$$

■

This question can be considered a complicated one, but if someone uses the formula of the difference of two squares, they will solve the question in a record time. Here, we have $\left(1 + \frac{1}{t}\right)$, and the following term is $\left(1 + \frac{1}{t^2}\right)$, which suggests multiply the first one by its conjugate and obtaining the conjugate of the second term and so on. In other words

$$\left(1 - \frac{1}{t}\right)\left(1 + \frac{1}{t}\right) = \left(1 - \frac{1}{t^2}\right) \Longrightarrow \left(1 - \frac{1}{t^2}\right)\left(1 + \frac{1}{t^2}\right) = \left(1 - \frac{1}{t^4}\right)\cdots$$

Thus our product multiplied by $\left(1 - \frac{1}{t}\right)$ will produce the following:

$$\begin{aligned} \left(1 - \tfrac{1}{t}\right)\Pi &= \left(1 - \tfrac{1}{t^2}\right)\left(1 + \tfrac{1}{t^2}\right)\left(1 + \tfrac{1}{t^4}\right)...\left(1 + \tfrac{1}{t^{2^{2012}}}\right), \\ &= \left(1 - \tfrac{1}{t^4}\right)\left(1 + \tfrac{1}{t^4}\right)...\left(1 + \tfrac{1}{t^{2^{2022}}}\right), \\ \vdots \quad & \quad \vdots \\ \left(1 - \tfrac{1}{t}\right)\Pi &= \left(1 - \tfrac{1}{t^{2^{2023}}}\right). \end{aligned}$$

Thus $\Pi = \dfrac{1 - \frac{1}{t^{2^{2023}}}}{1 - \frac{1}{t}}.$

■ **Example 5.54** Find the sum

$$\sum_{k=1}^{n} k \times k! = 1 \times 1! + 2 \times 2! + ... + n \times n!.$$

■

Let us denote $v_n = n \times n!$; if we can find a sequence u_n that induces telescoping, the problem is solved. Looking for such sequences is itself a problem. Here, a nice property of the factorial is the key. $(n+1)! = (n+1) \times n!$, and thus we can take $u_n = n!$ to obtain:

$$u_{n+1} - u_n = (n+1)! - n! = n!(n+1-1) = v_n$$

and then

$$\sum_{k=1}^{n} k \times k! = \sum_{k=1}^{n} \{u_{k+1} - u_k\} = u_{n+1} - u_1 = (n+1)! - 1! = (n+1)! - 1.$$

■ **Example 5.55** Find a formula for the sum

$$\sum_{k=1}^{n} \frac{k}{(k+1)!}.$$

■

Since

$$\frac{k+1}{(k+1)!} = \frac{1}{k!} \Longrightarrow \frac{k}{(k+1)!} + \frac{1}{(k+1)!} = \frac{1}{k!},$$

and then

$$\frac{k}{(k+1)!} = \frac{1}{k!} - \frac{1}{(k+1)!} = u_k - u_{k+1} \Longrightarrow \sum_{k=1}^{n} \frac{k}{(k+1)!} = u_1 - u_{n+1},$$

i.e.,

$$\sum_{k=1}^{n} \frac{k}{(k+1)!} = \frac{1}{1!} - \frac{1}{(n+1)!} = 1 - \frac{1}{(n+1)!}.$$

■ **Example 5.56** Evaluate the sum:

$$\sum_{n=1}^{p} \frac{1}{n\sqrt{n+1}+(n+1)\sqrt{n}}.$$

■

Of course, one must think about the conjugate to obtain $\dfrac{n\sqrt{n+1}-(n+1)\sqrt{n}}{n^2(n+1)-(n+1)^2n}$ and by factoring we get

$$\frac{1}{n\sqrt{n+1}+(n+1)\sqrt{n}} = \frac{n\sqrt{n+1}-(n+1)\sqrt{n}}{n(n+1)(n-(n+1))} = \frac{-n\sqrt{n+1}+(n+1)\sqrt{n}}{n(n+1)} = \frac{1}{\sqrt{n}} - \frac{1}{\sqrt{n+1}}.$$

Thus we get

$$\sum_{n=1}^{p} \frac{1}{n\sqrt{n+1}+(n+1)\sqrt{n}} = \sum_{n=1}^{p} \left(\frac{1}{\sqrt{n}} - \frac{1}{\sqrt{n+1}}\right),$$

and by telescoping, we get the first term minus the last one:

$$\sum_{n=1}^{p} \frac{1}{n\sqrt{n+1}+(n+1)\sqrt{n}} = \frac{1}{\sqrt{1}} - \frac{1}{\sqrt{p+1}} = 1 - \frac{1}{\sqrt{p+1}}.$$

■ **Example 5.57** Evaluate the sum:

$$S = \frac{1}{\sqrt{15}+\sqrt{13}} + \frac{1}{\sqrt{13}+\sqrt{11}} + \frac{1}{\sqrt{11}+3} + \frac{1}{3+\sqrt{7}} + \frac{1}{\sqrt{7}+\sqrt{5}}.$$

■

Using the conjugate is the key. All the terms will have the same denominator, which is 2, and then we obtain

$$2S = \left(\sqrt{15}-\sqrt{13}\right) + \left(\sqrt{13}-\sqrt{11}\right) + \left(\sqrt{11}-3\right) + \left(3-\sqrt{7}\right) + \left(\sqrt{7}-\sqrt{5}\right).$$

Then we simplify the sum to $S = \frac{1}{2}\left(\sqrt{15}-\sqrt{7}\right)$.

■ **Example 5.58** Evaluate

$$\sum_{k=1}^{2022} \sqrt{1+\frac{1}{k^2}+\frac{1}{(k+1)^2}}.$$

■

We will give two solutions[7]:

- Is $1+\frac{1}{k^2}+\frac{1}{(k+1)^2}$ a perfect square or not? Since we have three squares, we thought about the identity of the square of three variables. In other words

$$(a+b+c)^2 = a^2+b^2+c^2+2(ab+bc+ca),$$

where the last double terms must vanish. Our first attempt was

$$\left(1+\frac{1}{k}+\frac{1}{k+1}\right)^2 = 1+\frac{1}{k^2}+\frac{1}{(k+1)^2}+\frac{4}{k},$$

but after this, using the fact that $\frac{1}{k(k+1)} = \frac{1}{k}-\frac{1}{k+1}$, we thought that we needed a minus sign.

Thus we considered

$$\left(1+\frac{1}{k}-\frac{1}{k+1}\right)^2 = 1+\frac{1}{k^2}+\frac{1}{(k+1)^2}+\frac{2}{k}-\frac{2}{k(k+1)}-\frac{2}{k+1},$$

and since $1+\frac{1}{k}-\frac{1}{k+1} = 1+\frac{1}{k(k+1)} > 0$, we have:

$$\sqrt{1+\frac{1}{k^2}+\frac{1}{(k+1)^2}} = 1+\frac{1}{k}-\frac{1}{k+1},$$

and by telescoping

$$\sum_{k=1}^{2022}\left(1+\frac{1}{k}-\frac{1}{k+1}\right) = 2022+1+\sum_{k=1}^{2022}\left(\frac{1}{k}-\frac{1}{k+1}\right) = 2022+1-\frac{1}{2023}.$$

Finally,

$$\sum_{k=1}^{2022}\sqrt{1+\frac{1}{k^2}+\frac{1}{(k+1)^2}} = 2023-\frac{1}{2023}.$$

- By reducing to the same denominator we get:

$$1+\frac{1}{k^2}+\frac{1}{(k+1)^2} = \frac{k^4+2k^3+3k^2+2k+1}{k^2(k+1)^2},$$

and here we need to try to find a perfect square in the numerator, since the denominator is a square. Thus we can rewrite the numerator as follows:

$$k^4+2k^3+3k^2+2k+1 = 1^2+k^2+(k^2)^2+2k+2k^2+2k^3 = (1+k+k^2)^2$$

then

$$1+\frac{1}{k^2}+\frac{1}{(k+1)^2} = \left(\frac{1+k+k^2}{k(k+1)}\right)^2.$$

And since $\left(\frac{1+k+k^2}{k(k+1)}\right) > 0$, we get:

$$\sqrt{1+\frac{1}{k^2}+\frac{1}{(k+1)^2}} = \frac{1+k+k^2}{k(k+1)} = 1+\frac{1}{k}-\frac{1}{k+1},$$

[7] The second is provided by Dr. Dorin Andrica.

■ **Example 5.59** Evaluate the following sum

$$S=\sum_{k=1968}^{2022}\frac{1}{k^{2/3}+(k^2+k)^{1/3}+(1+k^2+2k)^{2/3}}.$$

■

It is direct by using the fact that $a^{1/3}=\sqrt[3]{a}$ and the following identity

$$(\sqrt[3]{a^2}+\sqrt[3]{ab}+\sqrt[3]{b^2})(\sqrt[3]{a}-\sqrt[3]{b})=a-b.$$

The final answer is $S=\sqrt[3]{2022}-\sqrt[3]{1968}$.

■ **Example 5.60** Evaluate the following sum[8]:

$$S=\sum_{k=1}^{40}\frac{4k+\sqrt{4k^2-1}}{\sqrt{2k+1}+\sqrt{2k-1}}.$$

■

Multiplying and dividing by the conjugate

$$S=\sum_{k=1}^{40}\frac{(4k+\sqrt{4k^2-1})(\sqrt{2k+1}-\sqrt{2k-1})}{2}.$$

Since $4k=(\sqrt{2k+1})^2+(\sqrt{2k-1})^2$, $4k^2-1=(2k-1)(2k+1)$, and using the identity: $a^3-b^3=(a-b)(a^2+ab+b^2)$, we get

$$S=\sum_{k=1}^{40}\frac{\left[(\sqrt{2k+1})^2+(\sqrt{2k-1})^2+\sqrt{2k-1}\sqrt{2k+1}\right](\sqrt{2k+1}-\sqrt{2k-1})}{2},$$

i.e.,

$$S=\frac{1}{2}\sum_{k=1}^{40}(\sqrt{2k+1})^3-(\sqrt{2k-1})^3,$$

Thus

$$S=\frac{1}{2}\left[(\sqrt{3})^3-(\sqrt{1})^3+(\sqrt{5})^3-(\sqrt{3})^3+\ldots+(\sqrt{81})^3-(\sqrt{79})^3\right].$$

And by telescoping we obtain

$$\sum_{k=1}^{40}\frac{4k+\sqrt{4k^2-1}}{\sqrt{2k+1}+\sqrt{2k-1}}=\frac{1}{2}(\sqrt{81})^3-\sqrt{1^3})=\frac{1}{2}(9^3-1)=364.$$

5.6 Examples from Linear Algebra

In this subsection, we will work on some proofs from linear algebra. Before we begin, let us recall some results:

---

[8]The sum was given in the 2001 Junior Balkan Mathematical Olympiad competition 2001.

1. A is `symmetric` if and only if $A^\top = A$, A is a square matrix.
2. A is `skew-symmetric` if and only if $A^\top = -A$, A is a square matrix.
3. $(AB)^\top = B^\top A^\top$, $\forall A\ (m\times n)$, and $B\ (n\times p)$.
4. For any square matrix A, we have $det(A^\top) = det(A)$.
5. Let A, B be two square matrices, then
$$\text{If } AB = I \Longrightarrow B = A^{-1}.$$
6. For any square matrices A and B, we have
$$det(AB) = det(A)\,det(B) = det(B)\,det(A) = det(BA).$$
7. $det\,(\alpha A) = \alpha^n\,det(A)$, A is $n\times n$.
8. If A is a square matrix of order n, then we have:
$$A \text{ invertible} \Longleftrightarrow A^{-1} \text{ exists} \Longleftrightarrow det(A) \neq 0 \Longleftrightarrow AA^{-1} = A^{-1}A = I_n.$$
9. For any square matrix A, we have
$$A \text{ is not invertible} \Longleftrightarrow \exists x \neq 0:\ Ax = 0.$$
10. λ is an eigenvalue to $A \Longleftrightarrow \exists\,\bar{x}\neq 0:\ A\bar{x} = \lambda\,\bar{x}$.
11. λ is an eigenvalue to $A \Longleftrightarrow det(A-\lambda I) = 0$.

■ **Example 5.61** Let A and B, two invertible n by n matrices, show the following
1. AB is also invertible and $(AB)^{-1} = B^{-1}A^{-1}$.
2. $\left(A^\top\right)^{-1} = \left(A^{-1}\right)^\top$.
3. A is symmetric then A^{-1} is symmetric.

■

Assume that A and B are invertible.
1. Since
$$\begin{aligned} A,\ B \text{ are invertible} &\Longrightarrow det(A)\neq 0,\ det(B)\neq 0,\\ &\Longrightarrow det(A)\times det(B) = det(AB) \neq 0,\\ &\Longrightarrow AB \text{ is also invertible.}\end{aligned}$$
Now,
$$AB\times B^{-1}A^{-1} = A\left(BB^{-1}\right)A = AIA^{-1} = AA^{-1} = I,$$
and
$$B^{-1}A^{-1}\times AB = B^{-1}\left(AA^{-1}\right)B = B^{-1}IB = BB^{-1} = I.$$
Thus $(AB)^{-1} = B^{-1}A^{-1}$.
2. Since $I^\top = I$, and using the first result in this example
$$\begin{aligned} I = A\times A^{-1} &\Longrightarrow I^\top = \left(A\times A^{-1}\right)^\top,\\ &\Longrightarrow I = \left(A^{-1}\right)^\top\times A^\top,\\ &\Longrightarrow \left(A^\top\right)^{-1} = \left(A^{-1}\right)^\top.\end{aligned}$$
3. Using the second result in this example,
$$A \text{ is symmetric} \Longrightarrow A = A^\top \Longrightarrow A^{-1} = \left(A^\top\right)^{-1} = \left(A^{-1}\right)^\top \Longrightarrow A^{-1} \text{ is symmetric.}$$

■ **Example 5.62** Let A be an $n \times n$ invertible matrix. Prove that $det(A^{-1}) = \dfrac{1}{det(A)}$.
It is obvious using the fact that $det(A^{-1}A) = det(A^{-1})det(A) = det(I_n) = 1$. ■

■ **Example 5.63** Let A, B, and P be $n \times n$ matrices, where at least P is invertible. Prove that if

$$A = P^{-1}BP \implies det(A) = det(B).$$

Using these results, we have

$$\begin{aligned} A = P^{-1}BP &\implies det(A) = det\left(P^{-1}BP\right), \\ &\implies det(A) = det\left(P^{-1}B\right)det(P), \\ &\implies det(A) = det\left(P^{-1}\right)det(B)det(P), \qquad det(P) \neq 0 \\ &\implies det(A) = det(B)\frac{1}{det(P)}det(P), \\ &\implies det(A) = det(B). \end{aligned}$$

■

■ **Example 5.64** Let A and B be two $n \times n$ matrices.
1. If A and B are symmetric, then AB is also symmetric.
2. If n is odd, then $det\left(AB - AB^{\top}\right) = 0$.

■

Here, we advise students to provide all arguments that support their choices (T/F).
1. False
   Indeed, since A and B are symmetric, $A^{\top} = A$, and$B^{\top} = B$, and then we have

$$(AB)^{\top} = B^{\top}A^{\top} = BA \neq AB \text{ in general.}$$

2. True
   It is not an easy question:

$$det\left(AB - AB^{\top}\right) = det\left(A\left(B - B^{\top}\right)\right) = det(A) \times det\left(B - B^{\top}\right).$$

One may observe the presence of the matrix $B - B^{\top}$, satisfying, in general,

$$\left(B - B^{\top}\right)^{\top} = B^{\top} - \left(B^{\top}\right)^{\top} = B^{\top} - B = -\left(B - B^{\top}\right) \implies B - B^{\top},$$

is skew-symmetric. Thus

$$\begin{aligned} det(B - B^{\top}) &= det\left[(B - B^{\top})^{\top}\right], \\ &= det\left[-(B - B^{\top})\right], \\ &= (-1)^n det(B - B^{\top}), \quad n \text{ is odd} \\ &= -det(B - B^{\top}) \implies 2det(B - B^{\top}) = 0 \implies det(B - B^{\top}) = 0. \end{aligned}$$

Finally, $det\left(AB - AB^{\top}\right) = det(A) \times det\left(B - B^{\top}\right) = 0$.

■ **Example 5.65** Let A be a square matrix with eigenvalue λ. Prove that for $\alpha \in \mathbb{R}$, $\alpha \neq 0$, $\lambda - \alpha$ is an eigenvalue for $A - \alpha I$.

$$\begin{aligned}
\lambda \text{ is an eigenvalue to } A &\implies \exists\, \bar{x} \neq 0 \colon A\bar{x} = \lambda\bar{x},\\
&\implies \exists\, \bar{x} \neq 0 \colon A\bar{x} - \alpha\bar{x} = \lambda\bar{x} - \alpha\bar{x},\\
&\implies \exists\, \bar{x} \neq 0 \colon A\bar{x} - \alpha I\bar{x} = \lambda\bar{x} - \alpha\bar{x},\\
&\implies \exists\, \bar{x} \neq 0 \colon (A - \alpha I)\,\bar{x} = (\lambda - \alpha)\bar{x},\\
&\implies (\lambda - \alpha) \text{ is an eigenvalue for } A - \alpha I.
\end{aligned}$$

■

■ **Example 5.66** Let A be a square matrix with eigenvalue λ, prove that λ is an eigenvalue for $A^{\top}$. Here, we will show how some students proceed without reaching the correct answer.

$$\begin{aligned}
\lambda \text{ is an eigenvalue to } A &\implies \exists\, \bar{x} \neq 0 : A\bar{x} = \lambda\bar{x},\\
&\implies \exists\, \bar{x} \neq 0 : (A\bar{x})^{\top} = (\lambda\bar{x})^{\top},\\
&\implies \exists\, \bar{x} \neq 0 : A^{\top}\bar{x} = \lambda\,\bar{x},\\
&\implies \lambda \text{ is also an eigenvalue for } A^{\top}.
\end{aligned}$$

But $(A\bar{x})^{\top} = (\bar{x})^{\top} A^{\top} \neq A^{\top}\bar{x}$ and $(\lambda\,\bar{x})^{\top} = \lambda\,(\bar{x})^{\top} \neq \lambda\,\bar{x}$. Thus the conclusion is wrong. Let us proceed differently by using the property $det(M^{\top}) = det(M)$ and $I^{\top} = I$.

$$\begin{aligned}
\lambda \text{ is an eigenvalue to } A &\implies det(A - \lambda I) = 0,\\
&\implies det\left[(A - \lambda I)^{\top}\right] = 0,\\
&\implies det\left[A^{\top} - (\lambda I)^{\top}\right] = 0,\\
&\implies det\left[A^{\top} - \lambda I^{\top}\right] = 0,\\
&\implies det\left[A^{\top} - \lambda I\right] = 0,\\
&\implies \lambda \text{ is also an eigenvalue for } A^{\top}.
\end{aligned}$$

■

5.7 Chapter 5 Practice Problems

Question 1 Show in two ways that $x^3 - 15x + \alpha = 0$ has at most one real root on $[-2,\ 2]$.

Question 2 Let $a, b,$ and c be positive real numbers such that $a \geq b \geq c$. Prove that

$$\frac{a^2 - b^2}{c} + \frac{b^2 - c^2}{a} + \frac{a^2 - c^2}{b} \geq 3a - 4b + c.$$

Question 3 Let m, n be odd integers and $p \in \mathbb{N}$. Prove that

$$2^p \text{ divides } (m - n) \text{ if and only if } 2^p \text{ divides } m^3 - n^3.$$

Question 4 Prove that if p is a prime greater than 3, then $p^2 - 1$ is divisible by 24.

Question 5 Prove the following:

1. $a \mid 0$ for any integer a.
2. $1 \mid a$ and $-1 \mid a$ for any integer a.
3. $0 \mid a \implies a = 0$.

Question 6 Show that for any nonnegative integer p,

$$\frac{p}{3} + \frac{p^2}{2} + \frac{p^3}{6} \in \mathbb{N} \cup \{0\}.$$

Question 7 Let r be a root of $x^3 + bx + c = 0$. Prove that $b^2 - 4rc \geq 0$. (Hint: use contradiction.)

Question 8 Let $X_1 = 2$, and $X_{n+1} = X_n^2 - X_n + 1$ for any $n \geq 1$. Prove that

$$X_{n+1} = X_1 X_2 X_3 \cdots X_{n-1} X_n + 1.$$

Question 9 Prove that all Fermat numbers of order higher than 1 have 7 as the last digit.

Question 10 Use proof by contradiction to show that if (D) and (D') are two straight lines perpendicular to a third line (d), then (D) and (D') are parallel.

Question 11 Find three different examples of three integers, where the first integer divides the second integer, the second integer divides the third integer, and third one divides the first one.

Question 12 Disprove the following: for every positive integer x and every integer $n \geq 2$, the equation $x^n + (x+1)^n = (x+2)^n$ has no solution.

Question 13 Each of the following statements is either true or false. If a statement is true, prove it. If a statement is false, disprove it:

(a) If $C \subseteq A \cup B$, then $C \subseteq A$ or $C \subseteq B$.

(b) If A and B are sets, then $\mathscr{P}(A) \cap \mathscr{P}(B) = \mathscr{P}(A \cap B)$.

Question 14 Prove that a cube of any positive integer could be expressed as a difference of two squares.

Question 15 An integer n will be called good if we can write $n = a_1 + a_2 + \ldots + a_k$, where $a_1, a_2, \ldots, a_k$ are positive integers (not necessarily distinct), satisfying $\sum_{i=1}^{k} \frac{1}{a_i} = 1$.

(a) Check that the integers 33 and 73 are good.

(b) Show that if n is good, $2n+8$ and $2n+9$ are also good.

Question 16 Show that for any $x, y \in \mathbb{R}$:

- $\lfloor x+y \rfloor \leq \lfloor x \rfloor + \lfloor y \rfloor + 1$.
- $2\lfloor x \rfloor \leq \lfloor 2x \rfloor \leq 2\lfloor x \rfloor + 1$.

Question 17 Show that for any $x, y \in \mathbb{R}_+$:

- $\lfloor xy \rfloor \leq \lfloor x \rfloor \times \lfloor y \rfloor + \lfloor x \rfloor + \lfloor y \rfloor$.
- $\lfloor x \rfloor^2 \leq \lfloor x^2 \rfloor \leq \lfloor x \rfloor^2 + 2\lfloor x \rfloor$.

Question 18 Show that for any $x, y \in \mathbb{R}$:

- $\{x\} + \{y\} - 1 \leq \{x+y\} \leq \{x\} + \{y\}$.
- $2\{x\} - 1 \leq \{2x\} \leq 2\{x\}$.

Question 19 Show that for any $x \in \mathbb{R}$, $m \in \mathbb{Z}$, and $n \in \mathbb{N}$:

$$\frac{m}{n} \leq x < \frac{m+1}{n} \Longrightarrow \lfloor x \rfloor = \left[\frac{m}{n}\right].$$

Denote $q = \left[\frac{m}{n}\right]$, and use the fact that $m = nq + r$, $0 \leq r < n$.

Question 20 Show that for any $x \in \mathbb{R}$, $n \in \mathbb{N}^*$, $0 \leq \lfloor nx \rfloor - n\lfloor x \rfloor \leq n-1$.

Question 21 If A is a given set with a finite number of elements, $n(A)$ is bounded, then

$$n(\mathscr{P}(A)) = 2^{n(A)}.$$

5.8 Chapter 5 Additional Practice Problems with Solutions

Question 1 Show that $\sqrt{x+1} < 1+x/2$ if $x > 0$. (Hint: use MVT.)

Solution. Define $f(t) = \sqrt{t+1}$, and consider the interval $[0,\ x]$ for $x > 0$. The composition of the differentiable functions $t \longmapsto t+1$ and $t \longmapsto \sqrt{t}$ on $[0,\ \infty)$ are also continuous on $[0,x]$ and differentiable on $(0,\ x)$; thus by MVT there exists $c \in (0,\ x)$ such that

$$f'(c) = \frac{f(x)-f(0)}{x-0} \Longrightarrow \frac{1}{2\sqrt{c+1}} = \frac{\sqrt{x+1}-1}{x}.$$

Since

$$c > 0 \Longrightarrow c+1 > 1 \Longrightarrow 2\sqrt{c+1} > 2 \Longrightarrow \frac{1}{2\sqrt{c+1}} < \frac{1}{2}.$$

Therefore,

$$\frac{\sqrt{x+1}-1}{x} < \frac{1}{2} \Longrightarrow \sqrt{x+1} < 1+\frac{x}{2}.$$

Question 2 Show that $1+2x+x^3+4x^5 = 0$ has exactly one real root.

Solution. Define $f(x) = 1+2x+x^3+4x^5$, and consider the interval $[-1,0]$. We have $f(-1) = -6, f(0) = 1$. Since the polynomial function f is continuous and $f(-1)f(0) < 0$, by the IVT, there exists $c \in (-1,0)$ such that $f(c) = 0$, implying that the equation $f(x) = 0$ has at least one root.

To prove that it has exactly one root, let us suppose by contradiction that there exist at least two distinct roots – namely, x_1 and x_2. Consider the interval $[x_1,\ x_2]$. f is continuous on $[x_1,\ x_2]$ and differentiable on $(x_1,\ x_2)$. Then by MVT, there exists $x \in (x_1,\ x_2)$ such that $f'(x) = 2+3x^2+20x^4 = 0$. Since $f'(x) > 0$ for all $x \in \mathbb{R}$, this is a contradiction; thus $f(x) = 0$ has exactly one root.

Question 3 If $f(1) = 10$ and $f'(x) \geq 2$ for $x \in [1,\ 4]$, how small can $f(4)$ possibly be?

Solution. By MVT to the continuous on $[1,\ 4]$ and differentiable on $(1,\ 4)$ function, there exists $x \in (1,\ 4)$ such that

$$f'(x) = \frac{f(4)-f(1)}{4-1} \Longrightarrow f(4) = 3f'(x)+10 \geq 3(2)+16 \Longrightarrow f(4) \geq 16.$$

Question 4 Show that $2x - \sin(x) = 1$ has exactly one real root.

Solution. Define $f(x) = 2x - \sin(x) - 1$, and consider the interval $[0,\ 1]$. We have $f(0) = -1$ and $f(1) = 2 - \sin(1) - 1 = 1 - \sin(1) > 0$. Since f is continuous and $f(0)f(1) < 0$, the IVT guarantees that the equation $f(x) = 0$ has at least one root in $(0,\ 1)$.
To prove that it has exactly one root, let us suppose by contradiction that it has at least two distinct roots, namely, x_1 and x_2. Consider the interval $[x_1,\ x_2]$. f is continuous on $[x_1,\ x_2]$ and differentiable on $(x_1,\ x_2)$. Thus by MVT, there exists $c \in (x_1,\ x_2)$ such that $f'(c) = 2 - cos(c) = 0$, which is a contradiction. Thus what we assumed is wrong, and the given equation has exactly one root.

Question 5 Given a twice differentiable function f satisfying $f''(x) \geq 0,\ \ \forall\, x \in [a,\ b]$, prove that

$$\forall\, x,\ y \in [a,\ b],\ \ f((1-t)x+ty)\ \leq\ (1-t)f(x)\ +\ tf(y),\ \ \ t \in [0,\ 1].$$

Solution. Let $h(x) = (1-t)f(x)+tf(y) - f((1-t)x+ty),\ x,\ y \in [a,\ b],\ t \in [0,\ 1]$. Since f is differentiable, we have

$$h'(x) = (1-t)f'(x) - \ (1-t)f'((1-t)x+ty)\ =\ (1-t)\left[f'(x) - \ f'((1-t)x+ty)\right].$$

Using our knowledge that

- $t \in [0,\ 1]$ implies that $(1-t) \geq 0$, and
- $f''(x) \geq 0$, and $\forall\, x \in [a,\ b]$, f' is increasing on $[a,\ b]$,

$(1-t)x+ty\ = x+t(y-x)$, we have the following:

$$x < y \implies (1-t)x+ty \geq x, \quad \forall\, t \in [0,\ 1].$$

Then $f'(x) \leq f'\left((1-t)x+ty\right)$, and Thus we obtain

$$\boxed{x\ <\ y \implies h'(x)\ \leq 0}.$$

Now, we consider the following case:

$$x\ >\ y \implies (1-t)x+ty\ =\ x+t(y-x) \leq\ x, \quad \forall\, t \in [0,\ 1].$$

Then $f'(x) \geq f'\left((1-t)x+ty\right)$, and Thus we obtain

$$\boxed{x\ >\ y \implies h'(x)\ \geq 0}.$$

Also, since $h'(y) = (1-t)\left[f'(y) - f'\left((1-t)y+ty\right)\right]\ =\ f'(y) - f'(y)\ =\ 0$ and $h(y) = (1-t)\,f(y) + t\,f(y) - f\left((1-t)y+ty\right) = f(y) - f(y)\ =\ 0$, we conclude

x	a		y		b
$h'(x)$		$-$	0	$+$	
h	$h(a)$	↘	0	↗	$h(b)$

Finally, $h(x) \geq 0$ on $[a,\ b]$, and then

$$f\left((1-t)x+ty\right)\ \leq\ (1-t)\,f(x)\ +\ t\,f(y), \quad \forall x \in [a,\ b],\ \ t \in [0,\ 1].$$

Question 6 Determine the following sums:

(i) $\displaystyle\sum_{k=1}^{n} k\binom{n}{k}$.

(ii) $\displaystyle\sum_{k=1}^{n} \frac{1}{k+1}\binom{n}{k}$.

(iii) $\displaystyle\sum_{k=1}^{n} \binom{n}{k}^2$.

Solution.

(i) Since

$$\sum_{k=1}^{n} k\binom{n}{k} = \sum_{k=1}^{n} k\frac{n!}{k!(n-k)!} = \sum_{k=1}^{n} \frac{n(n-1)!}{(k-1)!((n-1)-(k-1))!},$$
$$= \sum_{k=1}^{n} n\binom{n-1}{k-1},$$
$$= n\sum_{k'=0}^{n-1}\binom{n-1}{k'},$$
$$= n\sum_{k'=0}^{n-1}\binom{n-1}{k'}1^{k'}1^{n-1-k'},$$
$$= n(1+1)^{n-1}.$$

Thus $\sum_{k=1}^{n} k\binom{n}{k} = n2^{n-1}$.

(ii) Since

$$\sum_{k=1}^{n}\frac{1}{k+1}\binom{n}{k} = \sum_{k=1}^{n}\frac{1}{k+1}\frac{n!}{k!(n-k)!},$$
$$= \sum_{k=1}^{n}\frac{1}{(n+1)}\frac{(n+1)!}{(k+1)!(n-k)!},$$
$$= \sum_{k=1}^{n}\frac{1}{(n+1)}\frac{(n+1)!}{(k+1)![(n+1)-(k+1)]!},$$
$$= \sum_{k=1}^{n}\frac{1}{n+1}\binom{n+1}{k+1},$$
$$= \frac{1}{n+1}\sum_{k'=2}^{n+1}\binom{n+1}{k'},$$
$$= \frac{1}{n+1}\left[\sum_{k'=0}^{n+1}\binom{n+1}{k'} - \binom{n+1}{0} - \binom{n+1}{1}\right],$$
$$= \frac{1}{n+1}\left[2^{n+1}-1-(n+1)\right] = \frac{2^{n+1}-1}{n+1} - 1.$$

Thus

$$\sum_{k=1}^{n}\frac{1}{k+1}\binom{n}{k} = \frac{2^{n+1}-1}{n+1} - 1.$$

(iii) First of all, it is easy to show that $\binom{n}{k} = \binom{n-k}{k}$. This obtained equality suggests two ways:

$$\sum_{k=1}^{n}\binom{n}{k}\times\binom{n-k}{k} = \sum_{k=1}^{n}\binom{n}{k}^2$$

. The second way of forming a committee of n members out of $2n$ second point of second-level list $\binom{2n}{n}$. Thus we obtain

$$\sum_{k=1}^{n}\binom{n}{k}^2 = \binom{2n}{n}.$$

Question 7 Suppose that $f\colon \mathbb{R} \longrightarrow \mathbb{R}$ is differentiable with $f''(x) \geq 0$ for all x. Prove that for all a and b, $a < b$,

$$f\left(\frac{a+b}{2}\right) \leq \frac{f(a)+f(b)}{2}.$$

Solution. The solution is direct application of the result of Question 5 using $t = 1/2$, $x = a$ and $y = b$.

Question 8 Let a, b, and c be three real numbers such that their sum is zero. Prove that the equation $3ax^2 + 2bx + c = 0$ has at least one real solution.

Solution. We will provide two different ideas to tackle this question.

Idea 1. Using MVT

To this end, let us denote[9] $h(x) = ax^3 + bx^2 + cx + d$ satisfying $h(0) = d$ and $h(1) = a + b + c + d = 0 + d = d$. Since h is a polynomial function, then it is continuous on $[0,\ 1]$, differentiable on $(0,\ 1)$, satisfying $h(0) = h(1)$. Then by applying Rolle's theorem, we obtain

$$\exists\, \alpha \in (0,\ 1):\ h'(\alpha) = 0 = 3a\alpha^2 + 2b\alpha + c.$$

Then the equation has at least one real root.

Idea 2. Using Δ.

The requested equation will have at least one real root, when its discriminant $\Delta_2 = 4b^2 - 12ac \geq 0$, thus $b^2 \geq 3ac$.

Let us proceed by contradiction and assume that $b^2 < 3ac$. The condition $a + b + c = 0$ suggests we consider the equation $ax^2 + bx + c = 0$ with $x = 1$ as a root, which implies that $\Delta_1 = b^2 - 4ac \geq 0 \Longrightarrow b^2 \geq 4ac$. Thus using the fact that $ac > 0$ from our assumption $b^2 < 3ac$, we will obtain $4ac \leq b^2 < 3ac$ which is impossible, so what we assumed is wrong. Finally, $\Delta_2 \geq 0$.

Question 9 Does there exist a quadratic polynomial P with integer coefficients satisfying $P(P(\sqrt{2})) = 0$?

Solution. Set $P(x) = x^2 - 4$. Then $P(P(x)) = (x^2 - 4)^2 - 4$. Therefore, $P(P(\sqrt{2})) = 0$.

Question 10 Prove that if a, b, and c are odd integers such that $a + b + c = 0$, then $abc < 0$.

Solution. Since the sum of three odd integers is odd, the statement $a + b + c = 0$ is false; thus the implication is always true, whether or not $abc < 0$ is true.

Question 11 Prove that if x, y, z are three real numbers such that $x^2 + y^2 + z^2 < xy + yz + zx$, then $x + y + z > 0$.

Solution. If we multiply by 2 the LHS and we complete the squares, we end up with $LHS = (x-y)^2 + (y-z)^2 + (z-x)^2 < 0$, which is always false; thus the implication is always true, $x + y + z > 0$ is true.

Question 12 Let $a,\ b,\ c,\ d \in \mathbb{R}$, prove that

$$a + b + c + d = 4 \implies ab + bc + cd + da \leq 4.$$

Solution. We know $ab + bc + cd + da = (a+c)(b+d)$. Set $p = (a+c)(b+d)$ and $s = (a+c) + (b+d)$. Then $(a+c)$ and $(b+d)$ are roots of the quadratic equation $x^2 - sx + p = 0$. Therefore, $\Delta = s^2 - 4p \geq 0$. Since $s = 4$, we obtain $p \leq 4$ completes the proof.

[9] Why this choice particularly? It is important students to be curious and ask about such ideas. Looking carefully at the given equation, we observe that it is a derivative of the function h.

Question 13 Prove that for any integers a and b

$$\left[\,(a\,|\,b)\ \wedge\ (b\,|\,a)\,\right]\ \Longrightarrow a=\pm b.$$

Solution. Since $a|b$, there exists $k\in\mathbb{Z}$ such that $b=ak$. Moreover, since $b|a$, there exists $k'\in\mathbb{Z}$ such that $a=bk'$. Therefore, we get $b=bkk'$, implying that $kk'=1$. Since k and k' are integers, we obtain either $k=k'=1$ or $k=k'=-1$ completes the proof.

Question 14 Prove that for any integers a, b, and c

$$\left[\,(a\,|\,b)\wedge(b\,|\,c)\right]\ \Longrightarrow\ a\,|\,c.$$

Solution. Sine $a|b$, there exists $k\in\mathbb{Z}$ such that $b=ak$. Moreover, since $b|c$, there exists $k'\in\mathbb{Z}$ such that $c=bk'$. Therefore, $c=akk'$ implies $a|c$.

Question 15 Show that if $|mn|\neq 1$, then m^4+4n^4 is not prime.

Solution. The question consists proving that m^4+4n^4 is composite (for non zero integers m,n). To this end, we need to factor this sum of two squares. We typically accept that a^2+b^2 cannot be factored and the sum of two squares cannot be factored. Sophie Germain, a French mathematician, did not accept this conclusion, and she proposed the following:

$$m^4+4n^4=(m^2)^2+(2n^2)^2\ =\ (m^2)^2+(2n^2)^2\ +\ 2\,(m^2)(2n^2)\ -\ (2mn)^2=(m^2+2n^2)^2-(2mn)^2.$$

Thus

$$m^4+4n^4\ =\ (m^2+2n^2-2mn)(m^2+2n^2+2mn)=\left((m-n)^2+n^2\right)\left((m+n)^2+n^2\right).$$

Since the integers m and n are not 0, the factor $\left((m+n)^2+n^2\right)$ is always greater than 1, and the other factor $\left((m-n)^2+n^2\right)$ can be 1 only if $m=n=1$ but this case is not possible because of the hypothesis $|mn|\neq 1$. Finally, m^4+4n^4 is not prime.

Question 16 Show that for any non-negative integer m,

$$\frac{m}{6}+\frac{m^2}{2}+\frac{m^3}{3}\ \in\ \mathbb{N}\cup\{0\}.$$

Solution. The statement is true for $m=0$. Next, let $m\neq 0$. Then we have

$$\frac{m}{6}+\frac{m^2}{2}+\frac{m^3}{3}=\frac{m+3m^2+2m^3}{6}=\frac{m(2m^2+3m+1)}{6}=\frac{m(m+1)(2m+1)}{6}=\frac{6\sum_{k=1}^{m}k^2}{6}=\sum_{k=1}^{m}k^2\in\mathbb{N}.$$

Question 17 Show that if $A\cap B=\emptyset$, $A\ \subseteq\ (B\cup C)$, then $A\ \subseteq\ C$.

Solution. Let $x\in A$. Since $A\subseteq(B\cup C)$, then $x\in B$ or $x\in C$. Furthermore, since $A\cap B=\Phi$, we get $x\in C$.

Question 18 Show that $\forall\,x\in\ (0,\ \pi/2),\ \sin(x)\ +\cos(x)\ >\ 1$.

Solution. Suppose by contradiction that $\sin(x)+\cos(x)\leq 1$. Then $(\sin(x)+\cos(x))^2\leq 1$. Therefore,

$$\sin^2(x)+\cos^2(x)+2\sin(x)\cos(x)\leq 1\Longrightarrow\sin(2x)\leq 0,$$

which is a contradiction, since if $x\in(0,\frac{\pi}{2})$, then $2x\in(0,\pi)$, and thus $\sin(2x)>0$. Therefore, what we assumed is wrong.

Question 19 For every $x \in \left[\frac{\pi}{2}, \pi\right]$, $\sin(x) - \cos(x) \geq 1$.

Solution. Suppose by contradiction that $\sin(x) - \cos(x) < 1$. Then $(\sin(x) - \cos(x))^2 < 1$. Therefore,

$$\sin^2(x) + \cos^2(x) - 2\sin(x)\cos(x) < 1 \Longrightarrow \sin(2x) > 0,$$

which is a contradiction, since if $x \in [\frac{\pi}{2}, \pi]$, then $2x \in [\pi, 2\pi]$, and thus $\sin(2x) \leq 0$. Therefore, what we assumed is wrong.

Question 20 Use proof by contradiction to show that if x, y are non-negative real numbers, then

$$\sqrt{xy} \geq \frac{2xy}{x+y}.$$

Solution. Suppose by contradiction that for $x, y \geq 0$ (not both equal to zero) $\sqrt{xy} < \dfrac{2xy}{x+y}$. Then we have

$$(x+y)\sqrt{xy} < 2xy = 2\sqrt{xy}\sqrt{xy} \Longrightarrow x+y < 2\sqrt{x}\sqrt{y}.$$

The last inequality is equivalent to $(\sqrt{x})^2 + (\sqrt{y})^2 - 2\sqrt{x}\sqrt{y} < 0$, which induce the following contradiction $(\sqrt{x} - \sqrt{y})^2 < 0$; thus what we assumed is wrong.

Question 21 Given a quadratic equation $ax^2 + bx + c = 0$ where a, b, and c are odd integers. Prove that the quadratic equation does not have rational roots.

Solution. Suppose by contradiction that $\bar{x} = \frac{p}{q}$ is a root of the equation $ax^2 + bx + c = 0$, where $q \neq 0, p, q \in \mathbb{Z}$, and $GCD(p, q) = 1$. Then we have $ap^2 + bqp + cq^2 = 0$. Let us consider the following possible cases: (i) p and q both are odd; (ii) p is odd and q is even; and (iii) p is even and q is odd. It is easy to see that for all cases, $ap^2 + bqp + cq^2$ is odd, and thus it can not be equal to zero. Therefore, what we assumed is wrong.

Question 22 The sum of the cubes of three consecutive positive integers is divisible by 9.

Solution. Let n be a positive integer. The sum of the cubes of

$$(n-1)^3 + n^3 + (n+1)^3 = 3n^3 + 6n = 3n(n^2+2) = 3n\left[(n^2-1)+3\right] = 3n(n-1)(n+1) + 9n,$$

and since the product of three consecutive integers is divisible by 3, the proof is complete.

6. Sequences

Chapter Key Terms

(a) *sequences*
(b) *monotone sequence variations*
(c) *bounded sequence*
(d) *convergence and divergence*
(e) *arithmetic progression*
(f) *geometric progression*
(g) *Fibonacci sequence*

6.1 Sequences

Definition 6.1.1 A sequence is a function whose domain is the set $\mathbb{N}$ of natural numbers. If S is a sequence, we usually denote it as (s_n) or $\{s_n\}$ instead of $S(n)$. The image of n is the nth term of the sequence S, denoted by $s(n)$ or s_n, also called the general term of the sequence.

■ **Example 6.1** Consider the sequence (a_n) given by $a_n = \dfrac{1}{n}$. Writing out a few terms of the sequence, we obtain

$$(a_n) = \left(1, \frac{1}{2}, \frac{1}{3}, \ldots, \frac{1}{200} \ldots\right).$$

Similarly, $(b_n) = (1,2,3,4,\ldots,100,101,\ldots)$ is the sequence of positive integers. We write its general term as $b_n = n$. ■

(R) As above, a sequence is a list of numbers following a given pattern. We may, wish to change the domain of a sequence from $\mathbb{N}$ to $\mathbb{N} \cup \{0\}$ or $\{n \in \mathbb{N} : l \leqslant n,$ for some integer $l\}$. So we write the sequences as $(s_n)_{n=0}^{\infty}$ or $(s_n)_{n=l}^{\infty}$, respectively.

■ **Example 6.2** The following sequences exemplify the definition by describing the first few terms.

1. Let $(u_n)_{n=0}^{\infty}$ be a sequence given by $u_n = \dfrac{n+1}{2n+1}$.

   Then the first term of $(u_n)_{n=0}^{\infty}$ is $u_0 = 1$, and the tenth term is, $u_9 = \dfrac{9+1}{2(9)+1} = \dfrac{10}{19}$.

2. Let (v_n) be a sequence defined by $v_n = 1 + \dfrac{1}{n}$, $n \neq 0$.

   So, $v_1 = 2$, $v_2 = \dfrac{3}{2}$, $v_3 = \dfrac{4}{3}$, $v_4 = \dfrac{5}{4}$, and the nth term is $v_n = \dfrac{n+1}{n}$.

3. The sequence $(w_n)_{n=0}^{\infty}$ has nth term given by $w_n = (-1)^n$. So the first five terms are:

   $$w_0 = +1,\ w_1 = -1,\ w_2 = +1,\ w_3 = -1, \text{ and } w_4 = +1.$$

4. The sequence (a_n) defined as $a_n = 20$ is a constant sequence (for all n, $a_n = 20$).

■

■ **Example 6.3** Let the sequence $(t_n)_{n=0}^{\infty}$ be defined as

$$t_n = n! = 1 \times 2 \times 3 \times \ldots \times n.$$

Then, $t_0 = 0! = 1$, $t_1 = 1! = 1$, $t_2 = 2! = 1 \times 2 = 2$, $t_3 = 3! = 1 \times 2 \times 3 = 6, \ldots$.

The **factorial** of a number n, denoted by $n!$ is defined as

$$n! = 1 \times 2 \times 3 \times \ldots \times n.$$

Moreover,
(i) $n! = n \times (n-1)!$,
(ii) $(n!)^2 \neq n^2!$,
(iii) $(2n)! \neq 2 \times n!$.

■

■ **Example 6.4** Let (u_n) be a sequence defined by

$$u_n = \frac{n(n-1)(n-2)(n-3)(n-4)(n-5)(n-6)+1}{n}, \quad n \geq 1.$$

n	1	2	3	4	5	6	7	8	9	10
u_n	1	0.5	0.33	0.25	0.20	0.17	720.14	5040.13	20160.11	60480.10

Clearly, if we know few terms of this sequence, we can not guess its behavior. ■

Definition 6.1.2 The sequence (u_n) is said to be **explicitly** defined when a real-valued function f defined on $[0, \infty)$ is used to generate the sequence terms, that is

$$u_n = f(n), \quad \forall n \in \mathbb{N} \cup \{0\}.$$

For example, a sequence (u_n) can be defined from f as follows:

$$u_n = f(n) = \frac{2n(n^2-1)}{n^4+1}, \quad u_0 = 0,\ u_1 = 0, \text{ and } u_2 = \frac{12}{17}.$$

A recursive or iterative process can also be used to define a sequence. This process is based on an initial term of the sequence so that each subsequent term is defined as a function of the prior term. For example,

$$\begin{cases} u_0 = 1, \\ u_{n+1} = 2 + u_n, \quad n \geq 0 \end{cases}, \quad \begin{cases} v_0 = \sqrt{2}, \\ v_{n+1} = \dfrac{2}{1+v_n}, \quad n \geq 0 \end{cases}, \text{ and } \begin{cases} w_0 = -2, \\ w_n = \sqrt[3]{1+w_{n-1}^2}, \quad n \geq 0. \end{cases}$$

(R) When two sequences share the same initial and general terms, they are equal. If they have different iterations, they are not equal sequences. The sequences (u_n) and (v_n) defined here, for example, are different sequences.

$$\begin{cases} u_0 = 1, \\ u_{n+1} = f(u_n), \quad n \geq 0 \end{cases}, \quad \begin{cases} v_0 = 2, \\ v_{n+1} = f(v_n), \quad n \geq 0. \end{cases}$$

6.2 Variations

Definition 6.2.1 Let (u_n) be a real sequence, we say that

- (u_n) is increasing when $u_n \leq u_{n+1}$ for all n:

$$u_0 \leq u_1 \leq u_2 \leq \ldots \leq u_n \leq u_{n+1} \leq \ldots.$$

- (u_n) is strictly increasing when $u_n < u_{n+1}$ for all n:

$$u_0 < u_1 < u_2 < \ldots < u_n < u_{n+1} < \ldots.$$

- (u_n) is decreasing when $u_n \geq u_{n+1}$ for all n:

$$u_0 \geq u_1 \geq u_2 \geq \ldots \geq u_n \geq u_{n+1} \geq \ldots.$$

- (u_n) is strictly decreasing when $u_n > u_{n+1}$ for all n.

$$u_0 > u_1 > u_2 > \ldots > u_n > u_{n+1} > \ldots.$$

- (u_n) is monotone when (u_n) is either increasing or decreasing.

Methods to study variations of a sequence (u_n).

1. The difference term $d_n = u_{n+1} - u_n$.
   If $d_n \leq 0$ for any n then (u_n) is decreasing, and if $d_n \geqslant 0$, then (u_n) is increasing.

2. The ratio term $\dfrac{u_{n+1}}{u_n}$, where (u_n) is a sequence of all positive terms.
   If $\dfrac{u_{n+1}}{u_n} \geq 1$, then (u_n) is increasing, and if $\dfrac{u_{n+1}}{u_n} \leq 1$, then (u_n) is decreasing.
   However, if for all $n: \; v_n < 0$, we can set $u_n = -v_n > 0$, and apply the ratio term test.

3. The function term: $u_n = f(n)$.
   If the sequence (u_n) is defined by a function f as $u_n = f(n)$, then (u_n) has the same variations as f on $[0, \infty)$.

■ **Example 6.5** Determine whether the following are increasing or decreasing sequences:

$$a_n = 3n - 1, \; b_n = \frac{1}{n}, \; c_n = n^3, \; g_n = 2^n, \; e_n = 2^n - n, \; f_n = \frac{n}{2^n}, \; n \geq 1.$$

1. $a_n = 3n - 1$
Let us consider the difference term $d_n = a_{n+1} - a_n = 3(n+1) - 1 - 3n + 1 = 3 > 0,\ \forall\, n$. Then, $a_{n+1} \geqslant a_n$.
Thus (a_n) is an increasing sequence.

2. $b_n = \frac{1}{n} = f(n)$, where $f : x \to \frac{1}{x}$, $x > 0$ is a decreasing function because it is the reciprocal function. Thus (b_n) is decreasing.
An alternative, we may also consider

$$d_n = b_{n+1} - b_n = \frac{1}{n+1} - \frac{1}{n} = \frac{n-n-1}{n(n+1)} = \frac{-1}{n(n+1)} < 0,\ \forall\, n \geq 1.$$

Thus (b_n) is decreasing.

3. $c_{n+1} - c_n = (n+1)^3 - n^3 = 3n(n+1) + 1 > 0,\ \forall n.$
Thus (c_n) is an increasing sequence.

4. Since the terms of the sequence (g_n) are all positive, let us compare $\frac{g_{n+1}}{g_n} = \frac{2^{n+1}}{2^n} = 2 > 1.$
Thus (g_n) is an increasing sequence.

5. Consider the difference term

$$d_n = e_{n+1} - e_n = 2^{n+1} - (n+1) - 2^n + n = 2^n(2-1) - 1 = 2^n - 1 = 2^n - 2^0.$$

Since the sequence (2^n) is increasing, $2^n \geq 2^0$. We conclude that $d_n \geq 0$. Thus (e_n) is an increasing sequence.

6. Since f_n is positive, let us compare $\frac{f_{n+1}}{f_n}$ to 1 as follows:

$$\frac{f_{n+1}}{f_n} = \frac{n+1}{2^{n+1}} \times \frac{2^n}{n} = \frac{n+1}{2n}.$$

Now, for

$$n \geq 1 \Longrightarrow 2n = n + n \geq n + 1 > 0 \Longrightarrow \frac{n+1}{2n} \leq 1.$$

Thus (f_n) is an increasing sequence.
As an alternative, we may also consider

$$f_{n+1} - f_n = \frac{n+1}{2^{n+1}} - \frac{n}{2^n} = \frac{1-n}{2^{n+1}} \leq 0,\ \ \forall\, n \geq 1.$$

Thus (f_n) is increasing sequence.

■

(R) A non constant sequence may be neither increasing nor decreasing. For instance, $u_n = (-1)^n$.

■ **Example 6.6** Let $z_{n+1} = \frac{z_n^2}{16} + 4$, $z_0 = 4$. Study the monotony of (z_n).
We will expose three ways of studying the variations of (z_n). It is clear that $z_n > 0$ for any

non-negative integer n and if we consider the function f generating the given sequence. For example,

$$f(x) = \frac{x^2}{16} + 4, \; x > 0.$$

- We use the variations of f on $[0, \infty)$ to conclude the variations of (z_n). The square function $x \mapsto x^2$ is increasing on $[0, \infty)$, and then $x \mapsto f(x)$ is increasing on $[0, \infty)$. An easy and direct induction shows that $(z_n) \nearrow$. Indeed, if $P(n) : z_n \geq z_{n-1}, \;\; n \geq 1$, we have
Base step. For $n = 1$, $z_1 = 5 > 4 = z_0$, then $P(1)$ is true.
Inductive step. Assume now that $P(k)$ is true for an integer $k \geq 1$, and let us show that $P(k+1) : \; z_{k+1} \geq z_k$ is also true.
According to $P(k)$, we have: $z_k \geq z_{k-1}$ and since f is increasing, then
$$z_{k+1} = f(z_k) \geq f(z_{k-1}) = z_k$$
. Thus $P(k+1)$ is true, and (z_n) is increasing.

- We propose to show that $f(x) \geq x$ for any x. Indeed, it is easy through a trivial completing of the square. Let $x \in \mathbb{R}$:
$$f(x) - x = \frac{x^2}{16} + 4 - x = \frac{x^2 + 64 - 16x}{16} = \frac{(x-8)^2}{16} \geq 0, \quad \forall x \in \mathbb{R} \implies f(x) \geq x.$$
Thus we conclude that $z_{n+1} = f(z_n) \geq z_n$, so (z_n) is increasing.

- Use the direct simple technique:
$$z_{n+1} - z_n = \frac{z_n^2}{16} + 4 - z_n = \frac{z_n^2 - 46 - 16z_n}{16} = \frac{(z_n - 8)^2}{16} \geq 0 \implies (z_n) \nearrow.$$

■

6.3 Bounded Sequence

Definition 6.3.1 Let (u_n) be a sequence of real numbers. We say (u_n) is:

- **bounded above** if there exists a real number M such that for $n \in \mathbb{N}, \;\; u_n \leq M$ or
- **bounded below** if there exists a real number m such that for $n \in \mathbb{N}, \;\; u_n \geq m$.

Definition 6.3.2 A sequence is **bounded** if it is bounded above and bounded below.
A sequence (u_n), bounded above by M and bounded below by m is bounded if for $n \in \mathbb{N}$,

$$m \leq u_n \leq M.$$

Alternatively, we say (u_n) is a **bounded sequence** if there exists $A > 0$ such that

$$|u_n| \leq A, \; \forall n \in \mathbb{N}.$$

R Sometimes, the boundedness occurs after a certain term $n_j \geq 1$. For example, the sequence (δ_n) with the nth term $\delta_n = 2^n - n!$, $n \geq 0$ is not bounded above by 0 for all $n \geq 1$. Indeed,

$$\delta_1 = 2^1 - 1! > 0, \;\; \delta_2 = 2^2 - 2! > 0, \;\; \delta_3 = 2^3 - 3! > 0, \text{ and } \delta_4 = 2^4 - 4! \; = \; 16 - 24 \; < \; 0.$$

However, it can be shown that (δ_n) is bounded above by 0 for any $n \geq 4$.

■ **Example 6.7** Let us determine the boundedness of the following sequences:

$$u_n = \frac{1}{n},\ n \geq 1, \quad v_n = \frac{n-1}{n+1}, \text{ and } w_n = 2 - 2n + n^2.$$

1. $u_n = \dfrac{1}{n},\ n \geq 1.$

   It is clear that for all $n \geq 1$ we have $0 < u_n = \dfrac{1}{n} \leq 1$, thus (u_n) is bounded.

2. $v_n = \dfrac{n-1}{n+1}.$

   Observe that the numerator of v_n is always less than or equal to the denominator, which is positive. This implies that $v_n \leq 1$ in other words, (v_n) is bounded from above by 1.

   On the other hand, $v_n = \dfrac{n-1}{n+1} = \dfrac{n+1-2}{n+1} = 1 - \dfrac{2}{n+1}$. For $n > 0$, we can write:

   $$n+1 \geq 1 \Longrightarrow \frac{1}{n+1} \leq 1 \Longrightarrow \frac{-2}{n+1} \geq -2 \Longrightarrow 1 - \frac{2}{n+1} \geq -1.$$

   Thus (v_n) is bounded below by -1. Therefore, $-1 \leq v_n \leq 1$, $\forall n \geq 0$. This proves that (v_n) is a bounded sequence.

3. Let f be the function so that $w_n = 2 - 2n + n^2 = f(n)$. Note that the graph of f is a parabola that opens upward, and it has a minimum value at $f(-b/2a) = f(1) = 1$ but no maximum value. Thus we conclude that (w_n) is bounded below by 1.

   Another way to prove this is by noticing that

   $$w_n = 2 - 2n + n^2 = 1 + 1 - 2n + n^2 = 1 + (1-n)^2 \geq 1, \forall n \in \mathbb{N}, \text{ because } (1-n)^2 \geq 0.$$

■

■ **Example 6.8** Let

$$z_{n+1} = \frac{z_n^2}{16} + 4,\ z_0 = 4.$$

Show that (z_n) is bounded from above by 8.

We will use induction. Let $P(n)$: $z_n < 8$.

Base step. For $n = 0$, $z_0 = 4 < 8$, then $P(0)$ is true.

Inductive step. Assume that $P(k)$ is true for some arbitrary positive integer k; and we wish to show that $P(k+1)$: $z_{k+1} < 8$ is also true.

$$0 < z_k < 8 \Longrightarrow z_k^2 < 64 \Longrightarrow z_{k+1} = \frac{z_k^2}{16} + 4 < \frac{64}{16} + 4 = 8 \Longrightarrow z_{k+1} < 8.$$

Thus $P(k+1)$ is true, and then $z_n < 8$ for any $n \geq 0$. Finally, (z_n) is bounded from above by 8. ■

6.4 Convergence and Divergence

In this section, we will introduce the convergence of a sequence and then discuss several theorems that will help us find the limit of a sequence in a more efficient manner.

Definition 6.4.1 [Convergent sequences] A sequence (u_n) is said to **converge** to the real number L provided that for every $\varepsilon > 0$, there exists a natural number n_0 such that for all $n \in \mathbb{N}$, $n \geq n_0$ implies $|u_n - L| < \varepsilon$.
If (u_n) converges to L, we say L is the **limit** of the sequence (u_n), and we write $\lim_{n\to\infty} u_n = L$ or $u_n \to L$. That is,

$$\lim_{n\to\infty} u_n = L \iff \forall\, \varepsilon > 0,\ \exists n_0 \in \mathbb{N} : \forall\, n,\ n \geq n_0 \implies |u_n - L| < \varepsilon.$$

(R) Intuitively, It can be deduced from the definition that a convergent sequence with limit L is a sequence in which after a certain index $n_0 \geq 1$, all the terms cluster around the limit. In other words, the sequence terms after n_0 belong to a particular interval $(L - \varepsilon,\ L + \varepsilon)$, where ε represents a tiny positive value.[1]

■ **Example 6.9** Prove that

$$\lim_{n\to\infty} \frac{1}{n} = 0.$$

Let us consider any arbitrary $\varepsilon > 0$. We want to show the existence of certain n_0 that makes the following implication true:

$$\forall\, n \in \mathbb{N},\ n \geq n_0 \Longrightarrow \left|\frac{1}{n} - 0\right| < \varepsilon.$$

Symbolically,

$$\forall\, \varepsilon > 0,\ ?\,\exists\, n_0 \in \mathbb{N} : \forall\, n \geq n_0 \Longrightarrow \left|\frac{1}{n} - 0\right| < \varepsilon.$$

We know that:

$$\left|\frac{1}{n} - 0\right| < \varepsilon \Longleftrightarrow \frac{1}{n} < \varepsilon \Longleftrightarrow n > \frac{1}{\varepsilon}.$$

So, we may take $n_0 = \left\lfloor \frac{1}{\varepsilon} \right\rfloor + 1$. Therefore,

$$\forall\, \varepsilon > 0,\ \exists n_0 = \left\lfloor \frac{1}{\varepsilon} \right\rfloor + 1,\ n \geq n_0 \implies n \geq \left\lfloor \frac{1}{\varepsilon} \right\rfloor + 1 > \frac{1}{\varepsilon} \Longrightarrow \frac{1}{n} < \varepsilon.$$

We can summarize the convergence of $\left(\frac{1}{n}\right)$ to 0, as follows:

$$\text{Given } \varepsilon > 0,\ \left|\frac{1}{n} - 0\right| = \frac{1}{n} < \varepsilon \text{ if } n > \left\lfloor \frac{1}{\varepsilon} \right\rfloor + 1.$$

■

■ **Example 6.10** Prove that

$$\lim_{n\to\infty} \left(\sqrt{n+1} - \sqrt{n-1}\right) = 0.$$

By the definition of convergence, we have to prove

$$\forall\, \varepsilon > 0,\ ?\,\exists n_0 \in \mathbb{N} : \forall\, n \geq n_0 \implies \left|\sqrt{n+1} - \sqrt{n-1}\right| < \varepsilon.$$

[1] In general, ε is always less than 1.

We wish to show the existence of that n_0 which makes the implication true. Since

$$\left|\sqrt{n+1}-\sqrt{n-1}\right| = \left|\frac{2}{\sqrt{n+1}+\sqrt{n-1}}\right|,$$

and using the inequality $\sqrt{n+1} > \sqrt{n-1} \Longrightarrow \sqrt{n+1}+\sqrt{n-1} > 2\sqrt{n-1}$, we get:

$$\left|\sqrt{n+1}-\sqrt{n-1}\right| = \left|\frac{2}{\sqrt{n+1}+\sqrt{n-1}}\right| < \frac{2}{2\sqrt{n-1}} = \frac{1}{\sqrt{n-1}}.$$

Thus

$$\frac{1}{\sqrt{n-1}} < \varepsilon \Longrightarrow \sqrt{n-1} > \frac{1}{\varepsilon} \Longrightarrow n-1 > \frac{1}{\varepsilon^2}.$$

Now, if we take $n_0 = \left\lfloor \frac{1}{\varepsilon^2}+1 \right\rfloor + 1$, we obtain

$$\forall\, n \geq n_0 = \left\lfloor \frac{1}{\varepsilon^2}+1 \right\rfloor + 1 > \frac{1}{\varepsilon^2}+1, \;\; i.e., \quad n-1 > \frac{1}{\varepsilon^2}.$$

i.e.,

$$\sqrt{n-1} > \frac{1}{\varepsilon} \Longrightarrow \frac{1}{\sqrt{n-1}} < \varepsilon.$$

Hence:

$$\forall\, \varepsilon > 0, \;\exists\, n_0 = \left\lfloor \frac{1}{\varepsilon^2}+1 \right\rfloor + 1 : \;\forall\, n \geq n_0 \Longrightarrow \left|\sqrt{n+1}-\sqrt{n-1}-0\right| < \frac{1}{\sqrt{n-1}} < \varepsilon.$$

We can summarise the convergence of $\left(\sqrt{n+1}-\sqrt{n-1}\right)$ to 0 as follows:

$$\text{Given } \varepsilon > 0, \; \left|\sqrt{n+1}-\sqrt{n-1}-0\right| < \varepsilon \text{ if } \; n > \left\lfloor \frac{1}{\varepsilon^2+1} \right\rfloor + 1.$$

■

(R) The following sequences are defined as reference functions:

$$\boxed{\lim_{n\to\infty} n = +\infty, \quad \lim_{n\to\infty} n^2 = +\infty, \quad \lim_{n\to\infty} n^3 = +\infty, \quad \lim_{n\to\infty} \sqrt{n} = +\infty}.$$

$$\boxed{For \;\; t > 0, \; \lim_{n\to\infty} n^t = +\infty, \qquad for \;\; a > 1, \; \lim_{n\to\infty} a^n = +\infty}.$$[2]

$$\boxed{\lim_{n\to\infty} \ln(n) = +\infty, \qquad \lim_{n\to\infty} e^n = +\infty}$$

Definition 6.4.2 Let (u_n) be a sequence of real numbers, if $\lim\limits_{n\to\infty} u_n = +\infty$ $(or\ -\infty)$ then (u_n) is divergent.

$$\lim_{n\to\infty} u_n = +\infty \Longleftrightarrow \forall\, M > 0, \;\exists\, n_0 \in \mathbb{N} : \;\forall\, n \geq n_0 \Longrightarrow u_n > M,$$

and

$$\lim_{n\to\infty} u_n = -\infty \Longleftrightarrow \forall\, M > 0, \;\exists\, n_0 \in \mathbb{N} : \;\forall\, n \geq n_0 \Longrightarrow u_n < -M.$$

[2] See proof in Example 6.13

R Intuitively, a divergent sequence to $+\infty$ could be viewed as one in which after a certain index $n_0 > 0$, all the terms are greater than any given real number:

■ **Example 6.11** We use the definition of divergence to show that $\lim_{n\to\infty} \ln(n) = +\infty$:

$$\forall M > 0, \ ?\exists n_0 \in \mathbb{N}: \text{ if } n \geq n_0 \implies u_n > M.$$

Since the function $x \mapsto \ln(x)$ is increasing, we have $n > e^M \implies \ln(n) > M$, and then we take $n_0 = \lfloor e^M \rfloor + 1$. Thus we will have:

$$n \geq n_0 = \lfloor e^M \rfloor + 1 > e^M \implies \ln(n) > \ln(e^M) = M.$$

Then,

$$\forall M > 0, \ \exists n_0 = \lfloor e^M \rfloor + 1 \ : \ \forall n \geq n_0 \implies \ln(n) > M,$$

which proves that $\lim_{n\to\infty} \ln(n) = +\infty$. ■

Theorem 6.4.1 Let $f: [0, \infty) \longrightarrow \mathbb{R}$ be a function and (u_n) be a sequence defined by $u_n = f(n)$:

1. If $\lim_{x\to\infty} f(x) = +\infty$, then $\lim_{n\to\infty} u_n = +\infty$.
2. If $\lim_{x\to\infty} f(x) = -\infty$, then $\lim_{n\to\infty} u_n = -\infty$.

Proof. See [8]. ■

Theorem 6.4.2 Let (u_n) and (v_n) be two sequences such that for all $n \geq n_0$, $u_n \leq v_n$. Then:

1. (v_n) diverges to $+\infty$ if (u_n) diverges to $+\infty$,
2. (u_n) diverges to $-\infty$ if (v_n) diverges to $-\infty$.

Proof. See [8]. ■

■ **Example 6.12** Prove that $\lim_{n\to\infty} \dfrac{6n^2+9n+5}{2n+3} = \infty$.

Let $v_n = \dfrac{6n^2+9n+5}{2n+3}$, then

$$v_n = \frac{3n(2n+3)+5}{2n+3} = 3n + \frac{5}{2n+3} > 3n.$$

Since $(3n) \to \infty$ as $n \to \infty$; therefore, by Theorem 6.4.2 with $u_n = 3n$, (v_n) diverges to ∞. Thus $\lim_{n\to\infty} \dfrac{6n^2+9n+5}{2n+3} = \infty$. ■

■ **Example 6.13** For $a > 1$,

$$\lim_{n\to\infty} a^n = +\infty.$$

We will use the **Bernoulli inequality**

$$\boxed{\forall \alpha > -1, \ \forall n \in \mathbb{N}: \ (1+\alpha)^n \geq 1 + n\,\alpha.}$$

Since $a > 1$, then there exists $x > 0$ such that $a = 1 + x$. Then

$$v_n = a^n = (1+x)^n \geq 1 + nx = u_n, \quad \text{where } \lim_{n\to\infty} u_n = +\infty.$$

Thus $v_n \geq u_n$ and $\lim_{n\to\infty} u_n = +\infty$ imply that $\lim_{n\to\infty} a^n = \lim_{n\to\infty} v_n = +\infty$. ■

Definition 6.4.3 A given sequence (u_n) is said to converge to $l \in \mathbb{R}$ if and only if $\lim\limits_{n\to\infty} u_n = l$.

(R) [Reference Sequences] The following sequences are defined via reference functions:

$$\boxed{\lim_{n\to\infty} \frac{1}{n} = \lim_{n\to\infty} \frac{1}{n^2} = \lim_{n\to\infty} \frac{1}{\sqrt{n}} = 0},$$

$$\boxed{\lim_{n\to\infty} \frac{1}{\ln(n)} = 0, \qquad \lim_{n\to\infty} e^{-n} = 0},$$

$$\boxed{If \quad t > 0,\ \lim_{n\to\infty} \frac{1}{n^t} = 0, \qquad If \quad |a| < 1,\ \lim_{n\to\infty} a^n = 0}.$$

■ **Example 6.14** If $0 < a < 1$, then

$$\lim_{n\to\infty} a^n = 0.$$

For $a \in (0,\ 1)$, there exists $b > 1$ such that $a = \dfrac{1}{b}$, and since $\lim\limits_{n\to\infty} b^n = +\infty$ (see Example 6.13), we have $\lim\limits_{n\to\infty} a^n = \lim\limits_{n\to\infty} \dfrac{1}{b^n} = 0$. ■

Theorem 6.4.3 Let (u_n) and (v_n) be convergent sequences of real numbers. Then we have

$$\lim_{n\to\infty} a u_n + b v_n = a \lim_{n\to\infty} u_n + b \lim_{n\to\infty} v_n.$$

Theorem 6.4.4 Let (u_n) and (v_n) be convergent sequences of real numbers. Then we have

$$\forall\, n \geq 0,\quad u_n \leq v_n \Longrightarrow \lim_{n\to\infty} u_n \leq \lim_{n\to\infty} v_n.$$

Proof. By contradiction, we assume that $\lim\limits_{n\to\infty} u_n - \lim\limits_{n\to\infty} v_n > 0$, and then

$$\lim_{n\to\infty} u_n - v_n = l > 0 \Longrightarrow \forall\, \varepsilon > 0,\ \exists\, n_0 \in \mathbb{N}:\ \forall\, n \geq n_0 \Longrightarrow |u_n - v_n - l| < \varepsilon.$$

Thus for all $n \geq n_0$, we have:

$$-\varepsilon < u_n - v_n - l < \varepsilon \Longleftrightarrow l - \varepsilon < u_n - v_n < l + \varepsilon,$$

and if we take $\varepsilon = \dfrac{l}{2}$, we obtain

$$0 < \frac{l}{2} < u_n - v_n < \frac{3l}{2} \Longrightarrow u_n - v_n > 0 \Longrightarrow \exists\, n_0 \in \mathbb{N}:\ \forall\, n \geq n_0,\ u_n > v_n,$$

which contradicts $\forall\, n \in \mathbb{N},\ u_n \leq v_n$. Therefore, what we assumed is wrong and then $\lim\limits_{n\to\infty} u_n \leq \lim\limits_{n\to\infty} v_n$. ■

(R)

$$\forall\, n \in \mathbb{N},\ u_n < v_n, \text{ and then } \lim_{n\to\infty} u_n \leq \lim_{n\to\infty} v_n.$$

Strict inequalities are not preserved when we apply limits. For instance, $x_n = \dfrac{1}{n+1} > 0$ for any $n \geq 0$ but $\lim\limits_{n\to\infty} x_n = 0$.

Theorem 6.4.5 Let (u_n), (v_n), and (w_n) be sequences of real numbers such that
- $u_n \leq v_n \leq w_n, \quad \forall\, n \geq n_0$, and
- (u_n) and (w_n) are convergent to the same value l.

Then the sequence (v_n) also converges to l.

Proof. Direct applications of Theorem 6.4.4. ■

Proposition 6.4.6 Let (a_n) be a sequence that converges to zero, and let (b_n) be a bounded sequence. Then the sequence (c_n) with the nth term $c_n = a_n\, b_n$ also converges to zero.

Proof.

(i) *Idea 1.* Observe that $\lim_{n\to\infty} a_n = 0 \Longrightarrow \lim_{n\to\infty} |a_n| = 0$ and using the fact that (b_n) is bounded,

$$\exists\, M > 0,\ \forall\, n \in \mathbb{N}:\ |b_n| \leq M.$$

We obtain

$$|a_n\, b_n| \leq M\,|a_n| \implies -M|a_n| \leq c_n \leq M|a_n|.$$

Now, since $\lim_{n\to\infty} |a_n| = 0$, and using the squeeze theorem, we obtain

$$\lim_{n\to\infty} c_n = \lim_{n\to\infty} a_n\, b_n = 0.$$

(ii) *Idea 2.* Since (b_n) is bounded, then

$$\exists\, M > 0,\ \forall\, n \in \mathbb{N}:\ |b_n| \leq M,$$

and using the definition of $\lim_{n\to\infty} a_n = 0$, we have:

$$\forall\, \varepsilon' > 0,\ \ \exists\, N > 0,\ \ n \geq N \implies |a_n| < \varepsilon'.$$

Now, for any arbitrary $\varepsilon > 0$, if we consider $\varepsilon' = \dfrac{\varepsilon}{M}$, then

$$\exists\, N > 0,\ \forall\, n \geq N,\ \ |c_n| = |a_n\, b_n| = |a_n||b_n| < \varepsilon' M = \frac{\varepsilon}{M} M = \varepsilon.$$

Thus the (c_n) converges to 0.

■

Theorem 6.4.7 Let (u_n) be a sequence:
1. (u_n) is convergent if (u_n) is increasing and bounded above.
2. (u_n) is convergent if (u_n) is decreasing and bounded below.

Proof. See [8]. ■

■ **Example 6.15** The sequence $x_n = \dfrac{1}{n^2+1}$, $n \geq 0$ is convergent.
It is easy to see that (x_n) is bounded above and below: $0 < x_n \leq 1$. We need to study its variations. Since $(n+1)^2 + 1 > n^2 + 1$ for any non-negative integer n. Thus

$$x_{n+1} = \frac{1}{(n+1)^2+1} < \frac{1}{n^2+1} = x_n,$$

and (x_n) is decreasing. Finally, (x_n) is decreasing and bounded below by 0, then it is converging. ■

■ **Example 6.16** Let

$$z_{n+1} = \frac{z_n^2}{16} + 4, \; z_0 = 4.$$

Study the the convergence of (z_n).
It is shown (see Examples 6.6 and 6.8) that (z_n) is increasing, and bounded from above by 8, therefore, (z_n) is convergent.
■

We finish this section with an interesting remark and a challenging question (Example 6.18).

(R) Let us consider $a_n = (-1)^n, \; n \geq 0$, and the sequence

$$1, \; -1, \; 1, \; -1, \; 1, \; -1, \; 1, \ldots$$

- When $n \to \infty$, n may take huge even values (take $n = 2k$), and then $a_n \to +1$.
- When $n \to \infty$, n may take huge odd values (take $n = 2k+1$), and then $a_n \to -1$.

Thus we conclude the nonexistence of $\lim_{n\to\infty} (-1)^n$, since we have two different possible limits.
In real analysis courses, students will study such situations with a proof based on the notion of `sub-sequences` (see [8]).

$$\boxed{\lim_{n\to\infty} (-1)^n \quad \text{does not exist}}$$

Proposition 6.4.8 Any convergent sequence is bounded.

Proof. Let (a_n) be a sequence converging to l:

$$\forall \, \varepsilon > 0, \, \exists \, N > 0, \;\; n \geq N \implies |a_n - l| < \varepsilon.$$

If we take $\varepsilon = 1$, we will have for an $n \geq N$: $|a_n - l| < 1$, and then

$$|a_n| = |a_n - l + l| \leq |a_n - l| + |l| < 1 + |l|.$$

Thus $\forall \; i \in \{1, 2, ..., N-1\}$; we have: $|a_i| \leq \max_{1 \leq i \leq N-1} \{|a_i|\}$, and for all $i \geq N$, $|a_i| < 1 + |l|$.

Therefore, there exists $M = Max\left\{1 + |l|, \; \max_{1 \leq i \leq N-1} \{|a_i|\}\right\}$, such that $|a_n| \leq M$. for all n. ■

(R) The converse of the preceding result is not true. Indeed, there exists a bounded sequence that is diverging. For instance, $x_n = (-1)^n$ is bounded $|x_n| = 1$, but (x_n) is not converging.

■ **Example 6.17** Let $x_n = \cos(n\pi), \;\; n \geq 0$.
It is well-known that

$$\cos(0) = 1, \quad \cos(\pi) = -1, \quad \cos(2\pi) = 1, \quad \text{and} \quad \cos(x + 2\pi k) = cos(x), \; k \in \mathbb{Z}.$$

For example

$$\cos(2022\pi) = \cos(0 + 2\pi(1011)) = \cos(0) = 1, \;\; \text{and} \; \cos(2021) = \cos(\pi + 2\pi(1010)) = \cos(\pi) = -1.$$

Thus it is easy to conclude that we can re-express $\cos(n\pi)$ as follows:

$$\boxed{a_n = \cos(n\pi) \;\; = \;\; \begin{cases} +1 & \text{if} \quad n \text{ is even}, \\ -1 & \text{if} \quad n \text{ is odd}. \end{cases}}$$

It is useful to keep in mind that $\boxed{\cos(2k\pi) = 1}$ and $\boxed{\cos[(2k+1)\pi] = -1}$ for integer k as well as (x_n) has no limit.
■

■ **Example 6.18** Let $w_n = \sin\left(\pi\sqrt{4n^2-n}\right)$. Determine $\lim_{n\to\infty} w_n$ (if any).
It is clear that we are dealing with a $\sin(N)$, where $N \to \infty$, but the presence of the π may change the whole picture, as we will see. It is known that

$$\sin(x-2n\pi) = \sin(x) \Longrightarrow w_n = \sin\left(\pi\sqrt{4n^2-n}\right) = \sin\left(\pi\sqrt{4n^2-n}-2n\pi\right).$$

We suggest letting $v_n = \sqrt{4n^2-n}-2n$, and then

$$\begin{aligned} v_n &= \sqrt{4n^2-n}-2n, \\ &= \frac{-n}{\sqrt{4n^2-n}+2n}, \\ &= \frac{-n}{\sqrt{4n^2\left(1-\frac{n}{4n^2}\right)}+2n}, \\ &= \frac{-n}{|2n|\sqrt{1-\frac{1}{4n}}+2n}, \quad n>0, \\ &= \frac{-n}{2n\left(\sqrt{1-\frac{1}{4n}}+1\right)}, \\ &= \frac{-1}{2\left(\sqrt{1-\frac{1}{4n}}+1\right)}. \end{aligned}$$

Thus we obtain

$$w_n = \sin(\pi\, v_n) = \sin\left[\frac{-\pi/2}{\left(\sqrt{1-\frac{1}{4n}}+1\right)}\right],$$

and by using the continuity of the function $t \mapsto \sin(t)$, we obtain

$$\lim_{n\to\infty} w_n = \sin\left[\lim_{n\to\infty}\frac{-\pi/2}{\left(\sqrt{1-\frac{1}{4n}}+1\right)}\right] = \sin\left[\frac{-\pi/2}{2}\right] = -\sin\left(\frac{\pi}{4}\right)$$

and, finally, $\lim_{n\to\infty} w_n = \frac{-1}{\sqrt{2}}$. ■

6.5 Arithmetic Progression

Definition 6.5.1 A sequence (u_n) is called **Arithmetic Sequence**, which is also called **Arithmetic Progression (AP)** if there exists a real number r such that

$$\boxed{u_{n+1} = u_n + r, \quad \forall n \in \mathbb{N}}$$

The number r is called the **common difference**.

We can write the sequence (u_n) as follows

$$\boxed{(u_n) = (u_0,\ u_1,\ u_2,\ u_3, \cdots, u_{(n-1)},\ u_n,\ \ldots)}.$$

■ **Example 6.19** Let $(a_n) = (1,2,3,4,5,\cdots,100,\cdots)$ which defines an arithmetic sequence with a common difference $r = 1$. Moreover, the sequence of even integers

$$(e_n) = (0,2,4,6,\cdots,100,\cdots,) \quad \text{or} \quad e_n = 2n,\ n \geq 0.$$

It is clear, that the common difference is $r = e_{n+1} - e_n = 2(n+1) - 2n = 2$. Thus (e_n) is an arithmetic progression, with, a common difference $r = 2$.
Similarly, let (O_n) be the sequence of odd integers, $(O_n) = (1,3,5,7,\cdots,101,\cdots)$ or $O_n = 2n+1,\ n \geq 0,$ and the common difference $r = 2$. ■

6.5.1 Properties of Arithmetic Progressions

It is clear that for a given arithmetic sequence (u_n) with common difference r, the difference $u_{n+1} - u_n$ is independent of n. Indeed, the monotonicity of an Arithmetic Progression (AP) depends on the sign of r.

$$r = u_{n+1} - u_n \Longrightarrow \begin{cases} (u_n) \text{ is increasing,} & \text{if } r > 0, \\ (u_n) \text{ is constant,} & \text{if } r = 0, \\ (u_n) \text{ is decreasing,} & \text{if } r < 0. \end{cases}$$

Ⓡ

Let a, b, and c be three consecutive terms of an AP. Then $b = \dfrac{a+c}{2}$. In general,

$$u_n = \frac{u_{n-1} + u_{n+1}}{2}, \qquad \forall n \geq 1.$$

We observe that the three consecutive terms a, b, and c of an AP can be expressed as

$$a,\ b = a+r,\ c = b+r = a+2r \Longrightarrow \frac{a+c}{2} = \frac{a+a+2r}{2} = a+r = b.$$

Conversely,

$$b = \frac{a+c}{2} \Longrightarrow 2b = a+c \Longrightarrow b+b = a+c \Longrightarrow b = a+c-b \Longrightarrow b-a = c-b.$$

Thus the difference between the second and first term is equal to the difference between the third and second term. This proves that a,b,c are consecutive terms of an AP.

Proposition 6.5.1 Let (u_n) be an AP with a common difference r. Then $u_n = u_0 + n \times r,\quad \forall n \geq 0$.

Proof. Using the definition of an AP, we have: $u_k = u_{k-1} + r$ for $k \geq 1$., then we get:

$$\begin{array}{rcl} u_n & = & u_{n-1} + r, \\ u_{n-1} & = & u_{n-2} + r, \\ u_{n-2} & = & u_{n-3} + r, \\ \vdots & \vdots & \vdots \\ u_2 & = & u_1 + r, \\ u_1 & = & u_0 + r. \end{array}$$

We wish to express the general term u_n in terms of u_0 and r. By adding the listed terms, we obtain[3]:

$$u_n = u_0 + \underbrace{r + r + r + \ldots + r}_{n \text{ times}} = u_0 + nr.$$

■

$$u_n = u_0 + n \times r, \qquad \forall n \geq 0, \tag{6.1}$$

moreover,

$$u_n = u_1 + (n-1) \times r, \qquad \forall n \geq 1. \tag{6.2}$$

The following terms are obtained by applying (6.1) twice:

$$\begin{cases} u_p & = u_0 + p \times r \\ u_q & = u_0 + q \times r \end{cases}$$

Consequently, we have

$$u_p - u_q = (p-q) \times r, \quad \forall p,\, q \geq 0 \quad \Longrightarrow \quad u_p = u_q + (p-q) \times r. \tag{6.3}$$

Thus the sum of u_0 and u_n can be rewritten as

$$u_0 + u_n = u_0 + u_0 + n \times r = u_0 + r + u_0 + (n-1)r = u_1 + u_{n-1} = u_2 + u_{n-2} = \cdots.$$

Conclusively, we obtain

$$u_0 + u_n = u_k + u_{n-k}, \quad k \geq 1 \quad \text{or} \quad u_1 + u_n = u_k + u_{n+1-k}, \quad k \geq 2. \tag{6.4}$$

In the upcoming example, we illustrate the practicality of employing Arithmetic Progressions to represent a given iterative scheme in terms of n.

■ **Example 6.20** Let $(u_n) = \begin{cases} u_0 = 2, \\ u_{n+1} = \dfrac{3u_n - 2}{2u_n - 1}. \end{cases}$

1. Show that $u_n > 1$ for $n \geq 0$.
2. If $v_n = \dfrac{1}{u_n - 1}$, express u_n in terms of n.

1. Let $P(n):\ u_n > 1, n \geq 0$.
   We employ induction on $P(n)$:
   Base step. For $n = 0$, $u_0 = 2 > 1 \Longrightarrow P(0)$ is true.
   Inductive step. Assume now that $P(k)$ is true, and let us prove that $P(k+1)$ also remains true. Then,

$$u_{k+1} - 1 = \frac{3u_k - 2}{2u_k - 1} - 1 = \frac{u_k - 1}{2u_k - 1}.$$

[3]From 1 to N there are N numbers, where the common difference is one. But from p to q, there are $q - p + 1$ numbers.

Since $P(k)$ is true,

$$u_k > 1 \implies 2u_k > 2 \implies 2u_k - 1 > 0 \implies u_{k+1} - 1 > 0.$$

Thus $P(k+1)$ is true; and hence, we conclude that for any $n \geq 0,\ u_n > 1$.

2. Let us study the nature of the sequence (v_n). For any $n \geq 0,\ u_n > 1$ and $u_{n+1} > 1$, which implies that the sequence (v_n) is well defined. Consider

$$v_{n+1} - v_n = \frac{1}{u_{n+1}-1} - \frac{1}{u_n - 1} = \frac{2u_n - 1}{u_n - 1} - \frac{1}{u_n - 1} = \frac{2u_n - 1 - 1}{u_n - 1} = 2\frac{u_n - 1}{u_n - 1} = 2$$

Therefore (v_n) is an AP with a common difference $d = 2$ and $v_0 = \dfrac{1}{u_0 - 1} = 1$.
Thus $v_n = v_0 + nd = 1 + 2n$, so

$$u_n = \frac{1}{v_n} + 1 = \frac{1}{1+2n} + 1.$$

Finally, we have

$$u_n = \frac{2+2n}{1+2n}, \quad n \geq 0.$$

■

6.5.2 Sum of an Arithmetic Progression

Let (u_n) be an AP with the common difference r and first term u_0. We wish to evaluate the sum S_n of the first $(n+1)$ terms of (u_n). To this end, let us recall first the well-known Gauss Sum:

$$\boxed{GS = \sum_{k=1}^{n} k = 1 + 2 + 3 + \ldots + n = \frac{n(1+n)}{2} \hookrightarrow \text{Gauss sum}}$$

(R) Johann Carl Friedrich Gauss a German mathematician (1777–1855) had a bright idea to write the original sum from n to 1 and add the two equal sums as follows:

$$\begin{array}{rcl} GS & = & \left[1 + \ 2 \ + \ 3 \ + \ldots + (n-2) + (n-1) + n\right] \\ GS & = & \left[n + (n-1) + (n-2) + \ldots + \ 3 \ + \ 2 \ + \ 1\right] \\ \hline 2GS & = & (1+n) + (2+n-1) + (3+n-2) + \ldots + (n-2+3) + (n-1+2) + (n+1) \\ 2GS & = & n \times (n+1) \\ GS & = & \dfrac{n(1+n)}{2}. \end{array}$$

This formula is also proved by mathematical induction in Example 4.77 in the previous chapter.

■ **Example 6.21** Determine the following sum:

$$S = 2022^2 - 2021^2 + 2020^2 - 2019^2 + 2018^2 - 2017^2 + \ldots + 2^2 - 1^2.$$

The sum S can be rewritten as follows:
$$(2022^2 - 2021^2) + (2020^2 - 2019^2) + (2018^2 - 2017^2) + \ldots + (2^2 - 1^2) = \sum_{k=1}^{1011} \left((2k)^2 - (2k-1)^2\right),$$

which is equal to

$$S = \sum_{k=1}^{1011}\left((2k)^2-(2k-1)^2\right) = \sum_{k=1}^{1011} 4k^2-4k^2+4k-1 = \sum_{k=1}^{1011}(4k-1) = 4\sum_{k=1}^{1011} k - \sum_{k=1}^{1011} 1.$$

The first sum is the well-known Gauss sum: $\sum_{k=1}^{1011} k = \frac{1011\times 1012}{2} = 511,566$, and $\sum_{k=1}^{1011} 1 = 1011$.

Finally $S = 2,045,253$. ■

Proposition 6.5.2 The sum S_n of an AP (u_n) is given by

$$S_n = u_0+u_1+u_2+\ldots+u_{n-1}+u_n = (n+1)\left(\frac{u_0 + u_n}{2}\right).$$

Proof. According the Gauss Sum, we can write:

$$\begin{aligned}
S_n &= u_0+u_1+u_2+\ldots+u_{n-1}+u_n,\\
&= u_0+(u_0+r)+(u_0+2r)+\ldots+(u_0+nr),\\
&= \underbrace{u_0+u_0+\ldots+u_0}_{(n+1)\ terms}+(r+2r+3r+\ldots+nr),\\
&= (n+1)u_0+r(1+2+3+\ldots+n)_{GS},\\
&= (n+1)u_0+r\,\frac{n(n+1)}{2},\\
&= \frac{(n+1)}{2}\left(2u_0+rn\right) = \frac{(n+1)}{2}\left(u_0+(u_0+rn)\right),\\
&= \frac{(n+1)}{2}\left(u_0+u_n\right).
\end{aligned}$$

■

As a conclusion, we write the formula for the sum S_n of an AP (u_n) as the following:

$$S_n = u_0+u_1+u_2+\ldots+u_{n-1}+u_n = (n+1)\left(\frac{u_0 + u_n}{2}\right).$$

The sum formula can also be expressed

$$S_n = (\text{number of terms in the sum})\left(\frac{first\ term + last\ term}{2}\right)$$

Another alternate of the sum formula is stated below in terms of the average [4] of two numbers:

$$\sum_{k=0}^{n} u_k = (\text{number of terms in the sum}) \times average\ (first\ term\ ,\ last\ term\ )$$

[4] $Average(a,b) = \frac{a+b}{2}$.

(R) We often need to choose consecutive numbers from a sequence. As an example, three consecutive arbitrary numbers from an AP could be $x,\ x+r$, and $x+2r$, where r represents the common difference. A set of four consecutive numbers could be $x,\ x+1,\ x+2,\ x+3$ with their corresponding sum and product:

$$S_4 = x+(x+1)+(x+2)+(x+3) = 4x+6 \text{ and } P = x(x+1)(x+2)(x+3).$$

The following methods are more convenient for selecting such numbers:

- odd number of consecutive terms in AP with common difference r

  $$\boxed{x-r,\ x,\ x+r} \qquad \boxed{x-2r,\ x-r,\ x,\ x+r,\ x+2r},$$

  or more generally:

  $$\boxed{x-(2k+1)r, \cdot, x-r,\ x,\ x+r,\ x+2r, \cdot, x+(2k+1)r}$$

  One can see that

  $$x-2r,\ x-r,\ x,\ x+r,\ x+2r \rightarrow S=5x,\ \ P = x(x^2-4r^2)(x^2-r^2).$$

- even number of consecutive terms in AP with the common difference $r=2a$:

  $$\boxed{x-3a,\ x-a,\ x+a,\ x+3a}, \text{ and } \boxed{x-5a,\ x-3a,\ x-a,\ x+a,\ x+3a,\ x+5a,}$$

  or more generally,

  $$\boxed{x-(2k+1)a,\ x-(2k-1)a, \cdots,\ x-a,\ x+a,\ x+3a, \cdot, x+(2k+1)a.}$$

  For instance, let us find a general expression for the sum and product of 6 consecutive numbers in AP with a common difference $r=2a$:

  $$x-5a,\ x-3a,\ x-a,\ x+a,\ x+3a,\ x+5a \rightarrow S=6x, \text{ and } P = (x^2-25a^2)(x^2-9a^2)(x^2-a^2).$$

■ **Example 6.22** Find an expression for four numbers of an AP that has a common difference of 6.
Let $2a=6$, so $a=3$. For $x \in \mathbb{R}$, the numbers are

$$x-9,\ x-3,\ x+3, \text{ and } x+9.$$

■

■ **Example 6.23** Find an expression for six consecutive numbers of an AP that has a common difference of 7.
Let $2a=7$, so $a=\frac{7}{2}$. For any $x \in \mathbb{R}$, the numbers are:

$$x-\frac{35}{2},\ x-\frac{21}{2},\ x-\frac{7}{2},\ x+\frac{7}{2},\ x+\frac{21}{2},\ x+\frac{35}{2}.$$

■

■ **Example 6.24** Determine four consecutive numbers of an AP such that their sum is 12 and the sum of their squares is 116.

The general form for the numbers is $x-3a,\ x-a,\ x+a$, and $x+3a$, where the common difference $r=2a$. Then, $4x=12$, so $x=3$, and $(x-3a)^2+(x-a)^2+(x+a)^2+(x+3a)^2=4x^2+20a^2=116$, so $a^2=4$, and then $a=\pm 2$. Thus

- the numbers are: 9, 5, 1, and -3, when $a=-2$.
- the numbers are: -3 , 1, 5, and 9, when $a=2$.

■

6.6 Geometric Progression

Definition 6.6.1 A **geometric progression** (v_n), also called geometric sequence, is a sequence of non-zero numbers, where each term after the first is found by multiplying the previous one by a fixed, non-zero number called the common ratio q such that

$$\boxed{v_n = v_{n-1} \times q, \quad \forall n \in \mathbb{N}}$$

In other words, a geometric progression (GP) is a sequence of consecutive terms with a constant ratio q.

Example 6.25

- The sequence $(s_n) = (1, 2, 4, 8, 16, \cdots, 1024, \cdots)$ describes a GP with the common ratio $q = 2$, so $s_n = 2^n$, $n \geq 0$.
- $(v_n) = (1, 3, 9, 27, 81, \cdots, 59049, \cdots)$ also describes a GP whose elements can be listed as powers of 3; that is, $v_n = 3^n$, $n \geq 0$.

■

6.6.1 Properties of Geometric Progression

Let a, b, and c be nonzero numbers that represent three consecutive terms of a (GP). Then $b^2 = a \times c$. We observe that

$$a, \; b = a \times q, \quad c = b \times q \Longrightarrow \quad b \times b = a \times q \times b = a \times c.$$

On the other hand,

$$b^2 = ac \Longrightarrow \quad b = \frac{ac}{b} \Longrightarrow \quad \frac{b}{a} = \frac{c}{b}.$$

In other words, the ratio between the second and the first equals the ratio between the third and the second. Thus a, b, c are three consecutive terms of a GP.
If (v_n) is a geometric sequence, the aforementioned result can be expressed as follows:

$$\boxed{v_n^2 = v_{n-1} \times v_{n+1}, \quad \forall \, n \geq 1.}$$

Proposition 6.6.1 The general term of a GP (v_n) with initial term v_0 is given by:

$$\boxed{v_n = v_0 \times q^n, \text{ where the common ratio } q = \frac{v_{n+1}}{v_n}}$$

Proof. For a geometric sequence (v_n) with common ratio q, we wish to recursively express the general term v_n in terms of its initial term v_0:

$$\begin{array}{lcl} v_n & = & v_{n-1} \times q, \\ v_{n-1} & = & v_{n-2} \times q, \\ v_{n-2} & = & v_{n-3} \times q, \\ \vdots & \vdots & \vdots \\ v_2 & = & v_1 \quad \times q, \\ v_1 & = & v_0 \quad \times q. \end{array}$$

Thus $v_n = v_0 \times \underbrace{q \times q \times q \times \ldots \times q}_{n \text{ times}} = v_0 \times q^n, \; \forall \, n \geq 0.$ ■

The nth term of a geometric sequence (v_n) with the common ratio q and first term v_0 is

$$v_n = v_0 \times q^n, \qquad \forall\, n \geq 0. \tag{6.5}$$

Moreover,

$$v_n = v_m \times q^{n-m}, \qquad \forall\, m,\, n \geq 0. \tag{6.6}$$

Using (6.5), the product of the first and nth terms, $v_0 \times v_n$ is

$$v_0\, v_n = v_0 \times v_0\, q^n = v_0\, q \times v_0\, q^{n-1} = v_1\, v_{n-1} = v_2\, v_{n-2} \;\cdots = .$$

Thus in general, we have

$$v_0\, v_n = v_k\, v_{n-k}, \qquad \text{or} \qquad v_1\, v_n = v_k\, v_{(n+1)-k}, \qquad k \geq 0. \tag{6.7}$$

R The nth term of a GP is expressed as

$$v_n = v_1\, q^{n-1} \quad \text{or} \quad v_n = v_0\, q^n.$$

Example 6.26 Show that the following sequences are geometric, and then find their common ratio:

1. $v_n = (-3)^{2n-1}$.
2. $w_n = \begin{cases} 11, & \text{if } n \text{ is even,} \\ -11, & \text{if } n \text{ is odd.} \end{cases}$
3. $t_n = \dfrac{(-1)^n}{4^{5n-1}}$.
4. $r_n = 5^{n-1} \times 3^{2-n}$.

We evaluate the common ratio and the nth term of each sequence to determine if they are geometric:

1. $v_n = (-3)^{2n-1}$.

$$\frac{v_{n+1}}{v_n} = \frac{(-3)^{2(n+1)-1}}{(-3)^{2n-1}} = (-3)^{2n+1-2n+1} = (-3)^2.$$

(v_n) has a common ratio $q = 9$, thus it is a GP. Moreover, the first term is $v_0 = (-3)^{-1}$ and $v_n = (-3)^{-1} \times (-3)^{2n} = (-3)^{2n-1}$.
Alternatively, we observe that the exponent $2n-1$ of the given sequence is odd:

$$v_n = -\,3^{2n-1} = \left(\frac{-1}{3}\right) 3^{2n} = \left(\frac{-1}{3}\right) 9^n = -\frac{9^n}{3}.$$

2. If n is odd, $n+1$ is even. The common ratio of w_n is

$$\frac{w_{n+1}}{w_n} = \frac{11}{-11} = -1.$$

On the other hand, if n is even, then $n+1$ is odd, and the common ratio is

$$\frac{w_{n+1}}{w_n} = \frac{-11}{+11} = -1.$$

Thus in each case $q=-1$. Moreover, the first term is $w_0 = 11$; hence, by Equation (6.5), (w_n) is a GP, and its general term, $w_n = (11)\times(-1)^n,\ \ n \geq 0$.

3. Observe the the common ratio for sequence (t_n):

$$\frac{t_{n+1}}{t_n} = \frac{\frac{(-1)^{n+1}}{4^{5(n+1)-1}}}{\frac{(-1)^n}{4^{5n-1}}} = \frac{(-1)^{n+1}}{4^{5(n+1)-1}} \times \frac{4^{5n-1}}{(-1)^n} = \frac{(-1)^n(-1)^1}{4^{5n-1}\times 4^5} \times \frac{4^{5n-1}}{(-1)^n} = \frac{-1}{4^5}.$$

Thus (t_n) is a GP with the common ratio $q=-4^{-5}$ and first term $t_0 = 4$.
Let us find its nth term:

$$t_n = \frac{(-1)^n}{4^{5n-1}} = \frac{(-1)^n}{4^{5n}\times 4^{-1}} = 4\left(\frac{(-1)^n}{(4^5)^n}\right) = 4\left(\frac{-1}{4^5}\right)^n.$$

4. The nth term of (r_n) can be expressed as

$$5^{n-1}\times 3^{2-n} = \frac{5^{n-1}}{3^{n-2}} = \frac{5^{-1}}{3^{-2}}\times\frac{5^n}{3^n} = \frac{3^2}{5}\times\left(\frac{5}{3}\right)^n = \frac{9}{5}\times\left(\frac{5}{3}\right)^n.$$

Thus (r_n) is a GP with the common ratio $q=\frac{5}{3}$ and first term $r_0 = \frac{9}{5}$.
Alternatively, one can find the common ratio q as follows:

$$q = \frac{r_{n+1}}{r_n} = \frac{\frac{5^n}{3^{n-1}}}{\frac{5^{n-1}}{3^{n-2}}} = \frac{5^n}{3^{n-1}} \times \frac{3^{n-2}}{5^{n-1}} = \frac{5^{n-n+1}}{3^{n-1-n+2}} = \frac{5}{3}.$$

■

■ **Example 6.27** Show that for any geometric progression (g_n) we have:

$$(g_{n+2}-g_{n+3})^2 + (g_{n+3}-g_{n+1})^2 + (g_{n+4}-g_{n+2})^2 = (g_{n+1} - g_{n+4})^2.$$

One can use any four consecutive terms[5] (a,b,c,d) of a GP with ratio q and show

$$(b-c)^2+(c-a)^2+(d-b)^2 = (a-d)^2 \iff (aq-aq^2)^2+(aq^2-a)^2+(aq^3-aq)^2 = (a-aq^3)^2.$$

Using the fact that $b=aq,\ c=aq^2$, and $d=aq^3$, and after expanding and factoring, we get

$$a^2\left(q^2+q^4-2q^3+q^4+1-2q^2+q^6+q^2-2q^4\right) = a^2\left(q^6-2q^3+1\right) = a^2(1-q^3)^2 = (a-aq^3)^2.$$

■

In the upcoming example, we illustrate the practicality of employing Geometric Progressions to represent a given iterative scheme in terms of n.

■ **Example 6.28** Let $(a_n) = \begin{cases} a_0 = 1, \\ a_{n+1} = \dfrac{a_n}{2+a_n}. \end{cases}$

[5] same if we set $g_{n+1}=a,\ g_{n+2}=b,\ g_{n+3}=c,\ g_{n+4}=d$.

1. Show that $a_n > 0$ for any $n \geq 0$.
2. If $b_n = \dfrac{a_n}{1+a_n}$, express a_n in terms of n.

Solution.

1. Let $P(n)$: $a_n > 0$.
   Base step. For $n = 0$, $a_0 = 1 > 0$, which implies that $P(0)$ is true.
   Inductive step. Let us assume $P(k)$ is true; we wish to prove that $P(k+1)$ is also true. Consider
   $$a_k > 0 \Longrightarrow a_k + 2 > 0 \Longrightarrow \frac{a_k}{a_k+2} > 0 \Longrightarrow a_{k+1} > 0.$$
   Thus $P(k+1)$ is true. Hence, we conclude that $a_n > 0$ for all $n \geq 0$.

2. Let $b_n = \dfrac{a_n}{1+a_n}$. We encourage the reader to verify that (b_n) does not follow an Arithmetic Progression (AP) and let us now investigate whether it conforms to a Geometric Progression (GP).
   $$b_{n+1} = \frac{a_{n+1}}{1+a_{n+1}} = \frac{\dfrac{a_n}{2+a_n}}{1+\dfrac{a_n}{a_n+2}} = \frac{\dfrac{a_n}{2+a_n}}{\dfrac{2+2a_n}{a_n+2}} = \frac{a_n}{2+2a_n} = \frac{1}{2}\left(\frac{a_n}{1+a_n}\right) = \frac{1}{2}b_n.$$
   Thus (b_n) is a geometric sequence with a common ratio $q = 1/2$. Therefore, $b_0 = \dfrac{a_0}{a_0+1} = \dfrac{1}{2}$.
   Now
   $$b_n = \frac{a_n}{a_n+1} \Longrightarrow b_n(a_n+1) = a_n \Longrightarrow a_n b_n + b_n = a_n \Longrightarrow b_n = (1-b_n)a_n.$$
   Since $a_n > 0$, $b_n > 0$, which implies that $1 - b_n > 0$.
   Thus $a_n = \dfrac{b_n}{1-b_n}$.
   Moreover,
   $$b_n = \frac{1}{2}\left(\frac{1}{2^n}\right) = \frac{1}{2^{n+1}},$$
   and
   $$a_n = \frac{b_n}{1-b_n} = \frac{\dfrac{1}{2^{n+1}}}{1-\dfrac{1}{2^{n+1}}} = \frac{\dfrac{1}{2^{n+1}}}{\dfrac{2^{n+1}-1}{2^{n+1}}}.$$
   Hence,
   $$a_n = \frac{1}{2^{n+1}-1}, \qquad \forall n \geq 0.$$

■

6.6.2 Sum of a Geometric Progression

Let (v_n) be a geometric progression with the common ratio q and first term v_0. We now evaluate the sum S_n of the first $(n+1)$ in terms of the geometric progression (v_n):

$$\begin{aligned} S_n &= v_0 + v_1 + v_2 + \ldots + v_{n-1} + v_n, \\ &= v_0 + v_0 q + v_0 q^2 + \ldots + v_0 q^n, \\ &= v_0\left(1 + q^2 + q^3 + \ldots + q^n\right). \end{aligned}$$

Let $T_n = 1 + q^2 + q^3 + \ldots + q^n$. We consider the following two cases:

- For $q=1$, T_n becomes:

$$T_n=\sum_{i=0}^{n} 1 = (n-0+1)(1) = n+1.$$

- For $q \neq 1$, let us multiply the sum T_n by q:

$$q \times T_n = q+q^2+q^3+...+q^n+q^{n+1},$$

then,

$$T_n - q\times T_n = \left(1+q+q^2+q^3+...+q^n\right)-\left(q+q^2+q^3+...+q^n+q^{n+1}\right).$$

That is,

$$(1-q)T_n = 1-q^{n+1}, \text{ so } T_n=\frac{1-q^{n+1}}{1-q}, \quad q\neq 1.$$

Therefore,

$$T_n=\sum_{k=0}^{n} q^k = \frac{1-q^{n+1}}{1-q} = \frac{q^{n+1}-1}{q-1}, \qquad q\neq 1.$$

Moreover,

$$S_n = \sum_{k=0}^{n} v_k = v_0\left(\frac{1-q^{n+1}}{1-q}\right) = v_0\left(\frac{q^{n+1}-1}{q-1}\right), \quad q\neq 1.$$

■ **Example 6.29** Determine

- $S=1+2+4+8+...+1024.$
- $T=\frac{1}{8}+\frac{1}{16}+...+\frac{1}{2^{n-5}}, \ n\geq 8.$

- Observe that S denotes the sum of a sequence (s_n), which is geometric with the first term, $s_0=1$, and $q=2$. In addition, observe that the last term in the sum, 1024, corresponds to $s_{10}=2^{10}$, and therefore,

$$S=\sum_{i=0}^{10} s_i = \sum_{i=0}^{10} 2^i = 1+2+4+8+...+1024 = 1\times\left(\frac{1-2^{11}}{1-2}\right) = 2^{11}-1.$$

- T is the sum of the GP $(t_n)_{n\geq 3}$, with $q=1/2$ and the first term $t_3=\frac{1}{8}=\frac{1}{2^3}$, and we have

$$t_4=\frac{1}{16}=\frac{1}{2^4}, \ ..., \qquad t_{n-5}=\frac{1}{2^{n-5}}.$$

Thus the sum T is

$$\begin{aligned}
T &= \sum_{k=3}^{n-5}\frac{1}{2^k},\\
&= \frac{1}{8}+\frac{1}{16}+...+\frac{1}{2^{n-5}},\\
&= \frac{1}{8}\left(\frac{1-\left(\frac{1}{2}\right)^{n-7}}{1-\frac{1}{2}}\right),\\
&= \frac{1}{4}\left(1-\left(\tfrac{1}{2}\right)^{n-7}\right) = \frac{1}{4} - \frac{1}{2^{n-5}}.
\end{aligned}$$

Alternatively, one may proceed as follows:

$$\begin{aligned}
T &= \frac{1}{8}+\frac{1}{16}+\ldots+\frac{1}{2^{n-5}} = \frac{1}{8}\left(1+\frac{1}{2}+\ldots+\frac{1}{2^{n-8}}\right), \\
&= \frac{1}{8}\left(\frac{1}{2^0}+\frac{1}{2^1}+\ldots+\frac{1}{2^{n-8}}\right) = \frac{1}{8}\sum_{k=0}^{n-8}\frac{1}{2^k}, \\
&= \frac{1}{8}\left(\frac{1-\left(\frac{1}{2}\right)^{n-7}}{1-\frac{1}{2}}\right), \\
T &= \frac{1}{4}\left(1-\frac{1}{2^{n-7}}\right).
\end{aligned}$$

■

■ **Example 6.30** Determine the following sums:

- $U = 8-16+32-64+\ldots-1048576.$
- $V_n = \frac{1}{3}+\frac{1}{3^3}+\frac{1}{3^5}+\ldots+\frac{1}{3^{2n+1}}.$

(i) Observe that U denotes the sum of a geometric sequence (u_n) with the common ratio $q=-2$. Then,

$$U = 8(1-2+4-8+\ldots-131072) = 2^3(2^0-2^1+2^2-2^3+\ldots-2^{17}).$$

Thus

$$U = 8-16+32-64+\ldots-1048576 = \frac{8(1-(-2)^{18})}{1-(-2)} = \frac{8(1-2^{18})}{3}.$$

(ii) Let V_n denotes the sum of a geometric sequence (v_n), with the common ratio $q=\frac{1}{3^2}$, and initial term $v_1 = 1/3$. Then,

$$v_k = v_1 q^{n-1} = \frac{1}{3}\left(\frac{1}{3^2}\right)^{k-1} = \frac{1}{3}\left(\frac{1}{3}\right)^{2k-2} = \left(\frac{1}{3}\right)^{2k+1}.$$

Thus

$$V_n = \sum_{k=0}^{n+1} v_k = \sum_{k=0}^{n+1}\left(\frac{1}{3}\right)^{2k+1} = \frac{1}{3}\sum_{k=0}^{n+1}\left(\frac{1}{3}\right)^{2k} = \frac{1}{3}\sum_{k=0}^{n+1}\left(\frac{1}{9}\right)^{k}.$$

$$V_n = \frac{1}{3}\left(\frac{1-(1/9)^{n+1}}{1-(1/9)}\right).$$

Thus

$$\boxed{V = \frac{1}{3}+\frac{1}{3^3}+\frac{1}{3^5}+\ldots+\frac{1}{3^{2n+1}} = \frac{3}{8}\left(1-\frac{1}{9^{n+1}}\right)}$$

■

■ **Example 6.31** Determine the following sum:

$$C = \frac{4}{3} + \frac{2\sqrt{3}}{3} + 1 + \ldots + \frac{27\sqrt{3}}{128} + \frac{81}{256}.$$

Let us determine if the sum is of a geometric sequence by calculating the common ratio of the consecutive ratios. We find

$$\frac{2\sqrt{3}}{3} \div \frac{4}{3} = \frac{\sqrt{3}}{2} = 1 \div \frac{2\sqrt{3}}{3}.$$

Thus C is a sum of the G.P (c_n) with the common ratio $q = \frac{\sqrt{3}}{2}$, and $c_1 = \frac{4}{3}$. Therefore, the sum

$$C = \frac{4}{3} \times \frac{\left(\frac{\sqrt{3}}{2}\right)^n - 1}{\frac{\sqrt{3}}{2} - 1}.$$

We still need to find the value of n for which is the number of terms in the required sum:

$$c_n = \frac{4}{3}\left(\frac{\sqrt{3}}{2}\right)^{n-1} \Longrightarrow \frac{4}{3}\left(\frac{\sqrt{3}}{2}\right)^{n-1} = \frac{81}{256}.$$

That is,

$$\frac{4}{3} \times \frac{3^{\frac{n-1}{2}}}{4^{\frac{n-1}{2}}} = \frac{3^4}{4^4} \Longrightarrow \left(\frac{3}{4}\right)^{\frac{n-1}{2}} = \left(\frac{3}{4}\right)^5.$$

So, $\frac{n-1}{2} = 5 \Longrightarrow n = 11$. Thus

$$C = \frac{4}{3} \times \frac{\left(\frac{\sqrt{3}}{2}\right)^{11} - 1}{\frac{\sqrt{3}}{2} - 1} = \frac{8}{3} \times \frac{\left(\frac{\sqrt{3}}{2}\right)^{11} - 1}{\sqrt{3} - 2} = \frac{8}{3} \times \frac{(\sqrt{3}+2)\left[\left(\frac{\sqrt{3}}{2}\right)^{11} - 1\right]}{(\sqrt{3}-2)(\sqrt{3}+2)}.$$

Therefore,

$$\boxed{C = \frac{4}{3} + \frac{2\sqrt{3}}{3} + 1 + \ldots + \frac{27\sqrt{3}}{128} + \frac{81}{256} = \frac{8(2+\sqrt{3})}{3}\left[1 - \left(\frac{\sqrt{3}}{2}\right)^{11}\right]}$$

■

■ **Example 6.32** Determine the following sum.

$$D = \sqrt{2} - 2 + 2\sqrt{2} - 4 + 4\sqrt{2} \ldots - 64 + 64\sqrt{2}.$$

Clearly, D is sum of the terms of a geometric sequence (d_n) with the first term $d_0 = \sqrt{2}$ and common ratio $q = -\sqrt{2}$. We wish to determine the number of the terms in the given sum D, where $d_n = \sqrt{2}(-\sqrt{2})^n, \ n \geq 0$ is the nth term.
To this end, let us solve for n : $\sqrt{2}(-\sqrt{2})^n = 64\sqrt{2}$. Directly, we observe that n is even, so

$$(\sqrt{2})^n = 2^6 \Longleftrightarrow 2^{n/2} = 2^6 \Longrightarrow n = 12.$$

Thus the required sum is $D = \sum_{k=0}^{12} d_k$:

$$D = \sqrt{2} - 2 + 2\sqrt{2} - 4 + 4\sqrt{2} \ldots - 64 + 64\sqrt{2} = \sqrt{2}\left(\frac{1-(-\sqrt{2})^{13}}{\sqrt{2}+1}\right) = \sqrt{2}\left(\frac{1+(\sqrt{2})^{13}}{1+\sqrt{2}}\right).$$

Hence, using the conjugate of $\sqrt{2}+1$, we have

$$\boxed{D = \sqrt{2} - 2 + 2\sqrt{2} - 4 + 4\sqrt{2} \ldots - 64 + 64\sqrt{2} = 127\sqrt{2} - 126}$$

■

■ **Example 6.33** The main objective of this example is to determine the closed form for $\sum_{k=1}^{n} u_k$ for the following sequence

$$(u_n) = \frac{1}{3}, \frac{2}{3}, \frac{2}{3^2}, \frac{2^2}{3^2}, \frac{2^2}{3^3}, \frac{2^3}{3^3}, \frac{2^3}{3^4}, \frac{2^4}{3^4}, \ldots.$$

1. Express u_{2n} in terms of n and conclude $\sum_{i=1}^{n} u_{2i}$.
2. Express u_{2n-1} in terms of n and conclude $\sum_{i=1}^{n} u_{2i-1}$.
3. Conclude $\sum_{i=1}^{n} u_i$.

$$u_1 = \frac{1}{3}, \ u_2 = \frac{2}{3}, \ u_3 = \frac{2}{3^2}, \ u_4 = \frac{2^2}{3^2}, \ldots.$$

1. Observe that when n is even:

$$u_2 = u_{2\times 1} = \frac{2^1}{3^1}, \ u_4 = u_{2\times 2} = \frac{2^2}{3^2},$$

thus $u_{2n} = \frac{2^n}{3^n} = \left(\frac{2}{3}\right)^n$, and then

$$\sum_{i=1}^{n} u_{2i} = \sum_{i=1}^{n}\left(\frac{2}{3}\right)^i = \frac{2}{3} \times \frac{1-\left(\frac{2}{3}\right)^n}{1-\frac{2}{3}} = 2\left[1-\left(\frac{2}{3}\right)^n\right].$$

2. Similarly, when n is odd,

$$u_1 = u_{2\times 0+1} = \frac{2^0}{3^{0+1}}, \quad u_3 = u_{2\times 1+1} = \frac{2^1}{3^{1+1}}, \ \ldots,$$

$$u_{2n+1} = \frac{2^n}{3^{n+1}} = \frac{1}{3}\left(\frac{2}{3}\right)^n,$$

which implies that $u_{2n-1} = \frac{1}{3}\left(\frac{2}{3}\right)^{n-1}$.

$$\sum_{i=1}^{n} u_{2i-1} = \frac{1}{3}\sum_{i=1}^{n}\left(\frac{2}{3}\right)^{i-1} = \frac{1}{3}\frac{1-\left(\frac{2}{3}\right)^n}{1-\frac{2}{3}} = 1-\left(\frac{2}{3}\right)^n.$$

3. If n is even, $n = 2p,\ p \geq 1$. Then

$$\sum_{i=1}^{n} u_i = \sum_{i=1}^{2p} u_i = \sum_{i=1}^{p} u_{2i-1} + \sum_{i=1}^{p} u_{2i} = 1 - \left(\frac{2}{3}\right)^p + 2\left[1 - \left(\frac{2}{3}\right)^p\right].$$

If n is odd, $n = 2p+1,\ p \in \mathbb{N}$. Then

$$\sum_{i=1}^{n} u_i = \sum_{i=1}^{2p+1} u_i = \sum_{i=1}^{p} u_{2i} + \sum_{i=1}^{p+1} u_{2i-1} = 2\left[1 - \left(\frac{2}{3}\right)^p\right] + 1 - \left(\frac{2}{3}\right)^{p+1}.$$

Hence,

$$\boxed{\sum_{i=1}^{n} u_i = \begin{cases} 3 - 3\left(\dfrac{2}{3}\right)^{\frac{n}{2}}, & \text{if } n \text{ is even} \\ 3 - \dfrac{8}{3}\left(\dfrac{2}{3}\right)^{\frac{n-1}{2}} & \text{if } n \text{ is odd} \end{cases}}.$$

■

■ **Example 6.34** Determine the following sum

$$\sum_{k=1}^{n} \frac{1}{u_k - 1}, \quad \text{where } u_n = \frac{3^n - 4^n}{3^n + 4^n}.$$

Since for any $k \geq 1$, consider

$$u_k - 1 = \frac{3^k - 4^k}{3^k + 4^k} - 1 = \frac{3^k - 4^k - 3^k - 4^k}{3^k + 4^k} = -2\frac{4^k}{3^k + 4^k}.$$

Then,

$$\frac{1}{u_k - 1} = \frac{-1}{2}\left(\frac{3}{4}\right)^k - \frac{1}{2};$$

therefore,

$$\begin{aligned} \sum_{k=1}^{n} \frac{1}{u_k - 1} &= \sum_{k=1}^{n} \frac{-1}{2}\left(\frac{3}{4}\right)^k - \sum_{k=1}^{n} \frac{1}{2}, \\ &= \frac{-1}{2}\sum_{k=1}^{n}\left(\frac{3}{4}\right)^k - \frac{1}{2}\sum_{k=1}^{n} 1, \\ &= \frac{-1}{2} \times \frac{3}{4}\left(\frac{1 - (3/4)^n}{1 - 3/4}\right) - \frac{1}{2}n, \\ &= \frac{-3}{2}\left(1 - (3/4)^n\right) - \frac{n}{2}. \end{aligned}$$

Hence, we have the following sum

$$\boxed{\sum_{k=1}^{n} \frac{1}{u_k - 1} = \frac{1}{2}\left[3\left(\frac{3}{4}\right)^n - (n+3)\right]}.$$

■

■ **Example 6.35** There are a certain number of candidates in the first round of the presidential election in a country C. They have each received exactly half the number of votes of the candidate immediately above them. Should a second round be held?

Let n be the number of candidates and x be the total number of votes for the first candidate. In round one, the second candidate will receive $\frac{x}{2}$ votes, the third candidate will have $\frac{1}{2} \times \frac{x}{2} = \frac{x}{2^2}$ votes, and so on. Thus the total number of votes is given by the following sum:

$$x + \frac{x}{2} + \frac{x}{2^2} + \ldots + \frac{x}{2^{n-1}} = x\left[1 + \frac{1}{2} + \frac{1}{2^2} + \ldots + \frac{1}{2^{n-1}}\right].$$

It is evident that this is the sum of terms in a geometric progression (a_n), with, nth term given by $a_n = \frac{1}{2^n}$, $n \geq 0$ and the common ratio $q = \frac{1}{2}$.
Thus the sum

$$x + \frac{x}{2} + \frac{x}{2^2} + \ldots + \frac{x}{2^{n-1}} = x\left[\frac{1 - \frac{1}{2^n}}{1 - \frac{1}{2}}\right] = x\left(2 - \frac{1}{2^{n-1}}\right).$$

Finally, to determine whether a second round is necessary, it is sufficient to evaluate the percentage of votes obtained by the first candidate, which is the ratio between the number of votes for the preferred candidate and the total number of votes. Thus

$$\frac{x}{x\left(2 - \frac{1}{2^{n-1}}\right)} = \frac{1}{\left(2 - \frac{1}{2^{n-1}}\right)} > \frac{1}{2}, \quad \text{because the denominator is less than } 2.$$

The favorite candidate has received more than 50% of the votes, so a second round is unnecessary. ■

(R) When calculating the sum of an AP or GP, it is very important to count the terms. Occasionally, it is possible to rewrite a sequence in this manner:

$$a_{p+1},\ a_{p+2}, \cdots,\ a_{p+n}.$$

For example, the following sum has 46 terms:

$$q^{13} + q^{14} + q^{15} + \ldots q^{58} = q^{12+1} + q^{12+2} + q^{12+3} + \ldots q^{12+46}$$

We note that the number of terms can also be obtained by $58 - 13 + 1 = 46$.

6.7 Fibonacci Sequence

The Fibonacci Sequence is one of the most famous sequences and was introduced by Leonardo of Pisano (1175-1250) in his book *Liber Abaci* in 1202. The Fibonacci sequence was investigated by an Indian mathematician known as Pingala, who was born 200 BCE and is considered the father of the binary system (see [7, 9] for more details). Beyond the few facts given in his mathematical writings, little is known about Fibonacci's life. Leonardo studied rabbit populations for over a year and came up with this fascinating list of numbers called the Fibonacci sequence;

$$\begin{cases} F_0 = 0, \quad F_1 = 1, \\ F_{n+2} = F_{n+1} + F_n, \quad n \geq 0. \end{cases}$$

That is,

$$0,\ 1,\ 1,\ 2,\ 3,\ 5,\ 8,\ 13,\ 21,\ 34,\ 55,\ 89,\ 144,\ 233,\ 377,\ 610,\ 987,\ 1597,\ 2584,\ldots$$

Leonardo of Pisano studied mathematics in the North African port of Bugia (now Bejaïa, Algeria), a town on the Mediterranean coast. Then he traveled to Egypt, Syria, and Greece with an Arab master and studied different numerical systems and mathematical methods.

R Here are some interesting facts about the Fibonacci sequence:

- The terms of the Fibonacci sequence are easy to calculate, but their values increase rapidly. For example, F_{2000} has 400 digits while $F_{20,000}$ has 10 times more digits.
- The **Tribonacci,** or **3-bonacci,** sequence defined as follows:

$$\begin{cases} F_0 = 0, \quad F_1 = 1, \quad F_2 = 1, \\ F_{n+3} = F_{n+2} + F_{n+1} + F_n, \ n \geq 0. \end{cases}$$

- In addition to that, we find another sequence called Lukas sequence defined as follows:

$$\begin{cases} F_0 = 2, \quad F_1 = 1, \\ F_{n+2} = F_{n+1} + F_n, \ n \geq 0. \end{cases}$$

The French mathematician Edouard Lukas (1842-1891) specialized in Fibonacci sequences and their properties. Lukas called the terms of the Fibonacci sequence as the Fibonacci numbers.

6.7.1 Properties and Results

The following sections present various properties and results of Fibonacci sequences and numbers. Here we present the proofs that do not require a mathematical induction, and the remaining proofs as part of the practice problems of Chapter 7.
We start by giving an idea for deriving a closed formula for the Fibonacci sequence (F_n). This unusual rule is widely known as **Binet**'s formula, was developed by J. Binet (1786 – 1856) in 1843. To this end, let us consider a sequence $x_n = a r^n$, $n \geq 0$ that may satisfy the Fibonacci iterative formula for some particular non zero values of the constants a and r. In other words,

$$x_{n+2} = x_{n+1} + x_n \iff a r^{n+2} = a r^{n+1} + a r^n \ = \ a r^n (r^2 - r - 1) = 0.$$

Thus $r_{1,2} = \dfrac{1 \pm \sqrt{5}}{2}$, which suggests we consider our Fibonacci sequence as follows: $F_n = a r_1^n + b r_2^n$. To finalize our reasoning, we need to identify the corresponding values of a and b.

$$\begin{aligned} &0 = F_0 = a + b \Longrightarrow b = -a, \\ &1 = F_1 = a r_1 + b r_2 = a(r_1 - r_2) = a\frac{-2\sqrt{5}}{2} = -a\sqrt{5} \Longrightarrow a = \frac{-1}{\sqrt{5}}. \end{aligned}$$

Finally, we obtain: $F_n = \dfrac{-1}{\sqrt{5}}\left(\left[\dfrac{1-\sqrt{5}}{2}\right]^n - \left[\dfrac{1+\sqrt{5}}{2}\right]^n\right)$, which can be summarized in the theorem below.

Theorem 6.7.1 Let (F_n) be the Fibonacci sequence defined by

$$\begin{cases} F_0 = 0, \quad F_1 = 1, \\ F_{n+2} = F_{n+1} + F_n, \quad n \geq 0. \end{cases}$$

Then, its closed formula is given by

$$F_n = \frac{\left(1+\sqrt{5}\right)^n - \left(1-\sqrt{5}\right)^n}{2^n\sqrt{5}}, \quad \forall\, n \geq 0.$$

Proof. Let us proceed by strong induction. Let $P(n): \ F_n = \dfrac{r_2^n - r_1^n}{\sqrt{5}}$, $r_{1,2} = \dfrac{1 \pm \sqrt{5}}{2}$, $r_1 < r_2$ are the roots of the quadratic equation: $x^2 - x - 1 = 0$.

Base step. $n = 0,\ F_0 = 0 = \dfrac{r_2^0 - r_1^0}{\sqrt{5}} = \dfrac{0}{2\sqrt{5}} = 0$, then $P(0)$ is true.

$n = 1,\ F_1 = 1 = \dfrac{r_2^1 - r_1^1}{\sqrt{5}} = \dfrac{2\sqrt{5}}{2\sqrt{5}} = 1$, then $P(1)$ is true.

Inductive step. Assume that $P(k)$ and $P(k+1)$ are true, and let us prove that $P(k+2)$ is also true.

Since

$$\begin{aligned}
F_{k+2} &= F_{k+1} + F_k \\
&= \frac{r_2^{k+1} - r_1^{k+1}}{\sqrt{5}} + \frac{r_2^k - r_1^k}{\sqrt{5}} \\
&= \frac{1}{\sqrt{5}}\left[r_2^{k+1} - r_1^{k+1} + r_2^k - r_1^k\right] \\
&= \frac{1}{\sqrt{5}}\left[r_2^k(r_2+1) - r_1^k(r_1+1)\right], \text{ the roots satisfy } x^2 = x+1, \\
&= \frac{1}{\sqrt{5}}\left[r_2^k(r_2^2) - r_1^k(r_1^2)\right] \\
&= \frac{1}{\sqrt{5}}\left[r_2^{k+2} - r_1^{k+2}\right] \implies P(k+2) \text{ is true.}
\end{aligned}$$

Thus $P(n)$ is true for any $n \geq 0$.

(R) The positive root of the equation $x^2 - x - 1 = 0$ denoted $r_2 = \dfrac{1+\sqrt{5}}{2} = \phi$ is known as the `Golden Number` satisfying:

1. $\phi^2 = \phi + 1$,
2. $\phi^3 = \phi^2 + \phi$,
3. $\phi^3 = 2\phi + 1$,
4. $\phi = \dfrac{1+\phi}{\phi} = 1 + \dfrac{1}{\phi}$,
5. $\phi^{-1} = \dfrac{1}{\phi} = \phi - 1$.

■

Theorem 6.7.2 Let (F_n) represent the Fibonacci sequence, defined by

$$(F_n) = \begin{cases} F_0 = 0, \ F_1 = 1, \\ F_{n+2} = F_{n+1} + F_n. \end{cases}$$

Show that

$$F_{n+1}^2 - F_{n+1}F_n - F_n^2 = (-1)^n, \quad \forall\, n \geq 0.$$

Proof. Since (F_n) represents the Fibonacci sequence, let us evaluate some of its terms as follows:

$$F_2 = F_1 + F_0 = 1 + 0 = 1, \ F_3 = F_2 + F_1 = 1 + 1 = 2, \text{ and } F_4 = F_3 + F_2 = 2 + 1 = 3.$$

Let $P(n)$: $F_{n+1}^2 - F_{n+1}F_n - F_n^2 = (-1)^n, \ n \geq 0$. We employ induction on n:

Base step. For $n = 0$, $F_1^2 - F_1F_0 - F_0^2 = 1 = (-1)^0$, then $P(0)$ is true.

Inductive step. Assume that $P(k)$ is true for some arbitrary positive integer k. We show that $P(k+1)$ is also true; that is

$$P(k+1): F_{k+2}^2 - F_{k+2}F_{k+1} - F_{k+1}^2 = (-1)^{k+1}.$$

Using the definition of the Fibonacci sequence, we have

$$\begin{aligned} F_{k+2}^2 - F_{k+2}F_{k+1} - F_{k+1}^2 &= (F_{k+1} + F_k)^2 - (F_{k+1} + F_k) \times F_{k+1} - F_{k+1}^2, \\ &= F_{k+1}^2 + 2F_{k+1}F_k + F_k^2 - F_{k+1}^2 - F_kF_{k+1} - F_{k+1}^2, \\ &= F_{k+1}F_k + F_k^2 - F_{k+1}^2, \\ &= -\underbrace{\left(F_{k+1}^2 - F_{k+1}F_k - F_k^2\right)}_{P(k)}, \\ &= -(-1)^k = (-1)^{k+1}. \end{aligned}$$

Thus $P(k+1)$ is true. Hence, $F_{n+1}^2 - F_{n+1}F_n - F_n^2 = (-1)^n, \ \forall\, n \geq 0$. ■

Theorem 6.7.3 Let (F_n) be the Fibonacci sequence defined by

$$\begin{cases} F_0 = 0, \quad F_1 = 1, \\ F_{n+2} = F_{n+1} + F_n, \quad n \geq 0. \end{cases}$$

Then F_{2n} is divisible by F_n, $\forall\, n \geq 1$.

Proof. See Question 50. ■

The following theorem is presented without a proof and is included among the unresolved practice questions (see Question 45).

Theorem 6.7.4 Let (F_n) be the Fibonacci sequence defined by

$$\begin{cases} F_0 = 0, \quad F_1 = 1, \\ F_{n+2} = F_{n+1} + F_n, \quad n \geq 0. \end{cases}$$

Then,

$$\forall\, n \geq 1, \quad F_{n+1} \;<\; \left(\frac{7}{4}\right)^n.$$

Theorem 6.7.5 — Sums. Let (F_n) be the Fibonacci sequence defined by

$$\begin{cases} F_0 = 0, \quad F_1 = 1, \\ F_{n+2} \;=\; F_{n+1} + F_n, \quad n \geq 0. \end{cases}$$

Then, the Fibonacci sums are as follows:

(a) For all $n \geq 1$,

$$\sum_{k=1}^{n} F_k \;=\; F_{n+2} - 1.$$

(b) For all $n \geq 1$,

$$\sum_{k=1}^{n} F_{2k-1} \;=\; F_{2n}.$$

(c) For all $n \geq 1$,

$$\sum_{k=1}^{n} F_{2k} \;=\; F_{2n+1} \;-\; 1.$$

Proof.

(a) Let us give a direct proof.

$$\begin{aligned} F_{n+2} &= F_{n+1} + F_n, \\ F_{n+1} &= F_n + F_{n-1}, \\ F_n &= F_{n-1} + F_{n-2}, \\ &\vdots \\ &\vdots \\ F_3 &= F_2 + F_1, \\ F_2 &= F_1 + F_0. \end{aligned}$$

By summing and simplifying, we end up with:

$$F_{n+2} = \sum_{k=0}^{n} F_k \;+\; F_1 \;=\; \sum_{k=1}^{n} F_k \;+\; 1,$$

which completes the proof.

(b) Let us give a direct proof

$$\begin{aligned} F_{2n-1} &= F_{2n-2} + F_{2n-3}, \\ F_{2n-3} &= F_{2n-4} + F_{2n-5}, \\ F_{2n-5} &= F_{2n-6} + F_{2n-7}, \\ &\vdots \\ &\vdots \\ F_5 &= F_4 + F_3, \\ F_3 &= F_2 + F_1. \end{aligned}$$

By summing and simplifying, we end up with

$$\sum_{k=2}^{n} F_{2k-1} \;=\; \sum_{k=1}^{2n-2} F_k \implies \sum_{k=1}^{n} F_{2k-1} \;=\; F_1 + \sum_{k=1}^{2n-2} F_k.$$

Finally, by applying the result obtained in (a), we obtain:

$$\sum_{k=1}^{n} F_{2k-1} = F_1 + \sum_{k=1}^{2n-2} F_k = 1 + (F_{2n-2+2} - 1) = F_{2n}.$$

(c) Let us give a direct proof:

$$\begin{aligned}
F_{2n} &= F_{2n-1} + F_{2n-2}, \\
F_{2n-2} &= F_{2n-3} + F_{2n-4}, \\
F_{2n-4} &= F_{2n-5} + F_{2n-6}, \\
\vdots \quad & \vdots \quad \vdots \\
\vdots \quad & \vdots \quad \vdots \\
F_6 &= F_5 + F_4, \\
F_4 &= F_3 + F_2, \\
F_2 &= F_1 + F_0.
\end{aligned}$$

By summing, simplifying, and applying the result obtained in (a), we obtain

$$\sum_{k=1}^{n} F_{2k} = \sum_{k=0}^{2n-1} F_k = F_{2n-1+2} - 1 = F_{2n+1} - 1.$$

■

■ **Example 6.36** Let $(F_n)_n$ be the Fibonacci sequence defined by $F_0 = 0,\ F_1 = 1$, and for any $n \geq 0,\ F_{n+2} = F_{n+1} + F_n$. The objective is to explore the parity of F_{3k} for any $k \geq 0$. To this end, we suggest to follow these steps:

1. Look at the parity of the first 11 terms.
2. Use conjecture to determine the parity of all subsequent terms.
3. Use induction to prove the conjectured result.

We will go through the suggestions:

1. Let us list the first 11 terms of F_n:

n	0	1	2	3	4	5	6	7	8	9	10
F_n	0	1	1	2	3	5	8	13	21	34	55

2. Direct observation gives the following:

n	0	1	2	3	4	5	6	7	8	9	10
F_n	0	1	1	2	3	5	8	13	21	34	55
parity	even	odd	odd	even	odd	odd	even	odd	odd	even	odd

We have a pattern regarding to the parity of the following terms: even, odd, odd and from this, we can conjecture that

$$F_{3k} \text{ is even, } \ F_{3k+1} \text{ is odd, and } F_{3k+2} \text{ is odd.}$$

3. Let us use induction on $k \geq 0$ to prove the following:

$$P(k)\colon F_{3k} \text{ is even, } \ F_{3k+1} \text{ is odd, and } F_{3k+2} \text{ is odd.}$$

`Base Step`: For $k = 0,\ F_0 = 0$ is even, and $F_1 = F_2 = 1$ are odd, so $P(0)$ is true.
Inductive step. Assume now that $P(n)$ is true, and let us show that

$P(n+1)$: F_{3n+3} is even, and F_{3n+4} is odd, F_{3n+5} is odd is also true.
We note that

$$\begin{aligned} F_{3n+3} &= F_{3n+2}+F_{3n+1} && \text{is even as a sum of two odd terms,} \\ F_{3n+4} &= F_{3n+3}+F_{3n+2} && \text{is even as a sum of even and odd terms,} \\ F_{3n+5} &= F_{3n+4}+F_{3n+3} && \text{is even as a sum of even and odd terms.} \end{aligned}$$

Thus $P(n+1)$ is true.
Hence, we conclude that F_{3k} is even for any $k \geq 0$.

■

6.8 Adjacent Sequences

Definition 6.8.1 Any two sequences (u_n) and (v_n) are called **adjacent sequences** if and only if

- (u_n) is increasing,
- (v_n) is decreasing,
- $\lim\limits_{n\to\infty} (u_n - v_n) = 0.$

Theorem 6.8.1 Let (u_n) and (v_n) be adjacent sequences; then both sequences converge to the same limit l such that

$$u_n \leq l \leq v_n, \quad \forall\, n \in \mathbb{N}.$$

Proof. Let us assume that (u_n) is the increasing sequence and (v_n) the decreasing one.

- For all $n \in \mathbb{N}, u_n \leq v_n$?
  Consider $t_n = v_n - u_n$ and we want to prove that (t_n) is a decreasing sequence.

$$\begin{aligned} t_{n+1} - t_n &= (v_{n+1} - u_{n+1}) - (v_n - u_n), \\ &= (v_{n+1} - v_n) - (u_{n+1} - u_n). \end{aligned}$$

  Since (u_n) is an increasing sequence, $(u_{n+1} - u_n)$ is a positive number, and since (v_n) is decreasing sequence, $(v_{n+1} - v_n)$ is negative. This implies that $t_{n+1} - t_n \leq 0$.
  Thus (t_n) is a decreasing sequence.
  Moreover, $\lim\limits_{n\to\infty} t_n = \lim\limits_{n\to\infty} (v_n - u_n) = -\lim\limits_{n\to\infty} (u_n - v_n) = 0$, which implies that (t_n) decreases to 0, thus $t_n = v_n - u_n \geq 0$, and then for all $n \in \mathbb{N}, \ v_n \geq u_n$.

- Convergence of (u_n) and (v_n)
  Since, $u_0 \leq u_n \leq v_n \leq v_0$, we conclude that the sequence (u_n) is bounded from above by v_0, and the sequence (v_n) is bounded from below by v_0. Finally, using Theorem 6.4.7, we have the convergence of both sequences.

- Limits of (u_n) and (v_n)
  Since (u_n) and (v_n) converge,

$$\lim_{n\to\infty} (u_n - v_n) = \lim_{n\to\infty} u_n - \lim_{n\to\infty} v_n = 0 \Longrightarrow \lim_{n\to\infty} u_n = \lim_{n\to\infty} v_n = l.$$

  Thus using the fact that (u_n) increases to l and (v_n) decreases to l, then we conclude that

$$\forall\, n \in \mathbb{N}: \ u_n \leq l \leq v_n.$$

■

6.9 Chapter 6 Practice Problems

Question 1 Let (u_n) be a sequence defined by $u_0 = 4, \quad u_{n+1} = \frac{u_n^2}{16} + 4$. Determine if (u_n) is monotone and bounded.

Question 2 Study the convergence of $s_n = \sin(n\pi), \quad n \geq 0$.

Question 3 Study the convergence of $s_n = \sin\left[\frac{(2n+1)\pi}{2}\right], \quad n \geq 0$.

Question 4 Consider a function f defined by $f(x) = \frac{x+4}{x+1}$, $x \in \mathbb{R} \setminus \{-4; -1\}$.

1. Study the variations of f, and sketch its graph.
2. Let (u_n) be the sequence defined by: $\begin{cases} u_0 = 0, \\ u_{n+1} = f(u_n). \end{cases}$
   Place points $A_i(u_i, 0)$ for $i = 0, 1, 2, 3, 4$ on the graph of f.
3. Determine the nature of (v_n) if $v_n = \frac{u_n - 2}{u_n + 2}$.
4. Use (v_n) to express u_n in terms of n.
5. Prove that $\lim_{n\to\infty} u_n = 2$.

Question 5 Let (x_n) and (y_n) be two sequences such that

$$\begin{cases} x_{n+1} = \dfrac{x_n}{1 - x_n}, \quad x_1 = -2, \quad n \geq 1. \\ y_n = \dfrac{1 + x_n}{x_n}. \end{cases}$$

Evaluate $\sum_{k=1}^{2022} y_k$, and then express x_n in terms of n.

Question 6 Let (z_n) be defined as follows: $\begin{cases} z_0 = 2, \\ z_{n+1} = \dfrac{3 + z_n}{5 - z_n}. \end{cases}$

1. Express z_n in terms of n using the sequence $t_n = \frac{3 - z_n}{z_n - 1}$.
2. Determine $\lim_{n\to\infty} z_n$.

Question 7 Let (a_n) be the sequence defined by:

$$\begin{cases} a_0 = 1, \quad a_1 = 2, \\ a_{n+2} = a_{n+1} - a_n. \end{cases}$$

Show that $a_{n+6} = a_n$ for any natural number n.

Question 8 Let (b_n) be the sequence defined by: $\begin{cases} b_0 = -1, \\ b_{n+1} = 1 - \dfrac{1}{b_n}. \end{cases}$

Show that $b_{n+3} = b_n$ for any natural number n.

Question 9 Let (c_n) be the sequence defined by: $\begin{cases} c_0 = 2, \\ c_{n+1} = \dfrac{c_n - 5}{c_n - 1}. \end{cases}$

1. Show that $c_{n+2} = c_n$ for any natural number n.

2. Find the nature of the sequences (r_n) and (s_n) given by $\begin{cases} r_n = c_{2n}, \\ s_n = c_{2n+1}. \end{cases}$

Question 10 Let (x_n) and (y_n) be the sequences defined by: $\begin{cases} x_1 = 1, \quad y_1 = 12, \\ 3x_{n+1} = x_n + 2y_n, \\ 4y_{n+1} = x_n + 3y_n. \end{cases}$

1. Use the sequence defined by $t_n = y_n - x_n$ to study the variations of (x_n) and (y_n).
2. Study the boundedness of (x_n) and (y_n).
3. Let $w_n = 3x_n + 8y_n$ express x_n and y_n in terms of n.
4. Can we say that (x_n) and (y_n) are adjacent? (Justify.)

Question 11 Express v_n and u_n in terms of n for

$$\begin{cases} u_{n+1} = \dfrac{u_n}{1-u_n}, \quad v_n = \dfrac{1+u_n}{u_n} \\ u_1 = -2, \quad \forall n \geq 1. \end{cases}$$

Question 12 Let (u_n) be the sequence given by $u_0 = 1$, $u_{n+1} = \dfrac{3+2u_n}{2+u_n}$, $\forall n \geq 0$.

(i) Show that (u_n) is bounded from above by $\sqrt{3}$.
(ii) Show that (u_n) converges and find its limit.
(iii) Express u_n in terms of n. (Hint consider the auxiliary sequence $w_n = \dfrac{u_n - \sqrt{3}}{u_n + \sqrt{3}}$, $n \geq 0$.)

Question 13 Let (t_n) be the sequence given by $t_n = \sqrt{n^2+n} - n$, $\forall n \geq 1$.
(i) Show that (t_n) is bounded.
(ii) Show that (t_n) converges, and find its limit.

Question 14 Assume that the angles A, B, and C of triangle (ABC) are in AP. Calculate the expression $a^2 - b^2 + c^2 - ac$, where a, b, and c represent the respective lengths of the triangle sides.

Question 15 Evaluate the following sums:

$$\sum_{k=1}^{2022} (k-3)(k-1), \text{ and } \sum_{m=10}^{n+3} (m-2)^3.$$

Question 16 Determine $\sum_{k=1}^{n} k^4$.

Question 17 Determine $\sum_{i=2}^{n} \dfrac{1}{i^2 - 7i + 12}$.

Question 18

(i) Show that $1 + 4k^4 = [1 + 2k(k-1)][1 + 2k(k+1)]$.
(ii) Express x_{k+1} if $x_k = 1 + 2k(k-1)$.
(iii) Express k in terms of x_k and x_{k+1}.
(iv) Determine the sum $\sum_{j=1}^{p} \dfrac{4j}{4j^4+1}$.

Question 19 Determine the sum

$$\sum_{n=1}^{N} \frac{1}{\sqrt{n + \sqrt{n^2 - 1}}}.$$

Question 20 Determine the sum

$$\sum_{n=1}^{N} n!(1+n+n^2).$$

Question 21 Determine the sum

$$\sum_{a=1}^{p} \frac{a}{a^4+a^2+1}.$$

Question 22 Check whether 1968 is a term of the list of numbers $102, 113, 124, 135, \cdots$

Question 23 Given a sequence (z_n) satisfying $z_1+z_2+z_3+...+z_n = 2n^2-3n$ for any $n \geq 1$, show that (z_n) is an AP

Question 24 Find the 2022th term of the AP, $\sqrt{12}, \sqrt{27}, \sqrt{48}, \cdots$.

Question 25 Is there any k so that $u_k = 0$ for the AP, $2012, 2008, 2004, 2000, \cdots$?

Question 26 How many two-digit numbers are multiples of 7?

Question 27 Two AP have the same common difference r. The difference between their 100th terms is 100, what is the difference of their 1000th terms?

Question 28 Let (a_n) be an AP. Show that

$$(2-n^{-1})\sum_{k=1}^{2n}(-1)^{k-1}a_k^2 = a_1^2 - a_{2n}^2.$$

Question 29 Let $s_n = 3n^2+5n$ be the nth term of an AP. Find the value of k, if $s_k = 164$.

Question 30 Let (v_n) be a sequence defined by $v_n = u_{n+1}^2 - u_n^2$, where (u_n) is an AP. Prove that (v_n) is an AP.

Question 31 Give an explicit expression of the sequence (t_n) defined by $\begin{cases} t_1 &= 1, \\ t_{n+1} &= t_n + n. \end{cases}$

Question 32 Determine 4 terms of an AP such that their product equals 385 and the common difference is 6.

Question 33 For a, b, c, d, and p distinct real numbers satisfying

$$(a^2+b^2+c^2)p^2 - 2p(ab+bc+cd) + (b^2+c^2+d^2) \leq 0.$$

Determine if a, b, c, d, are in GP or AP as well as if they satisfy $ab = cd$.

Question 34 If the roots of the cubic equation $ax^3+bx^2+cx+d = 0$ are in G.P, show that $\left(\frac{b}{c}\right)^3 = \frac{a}{d}$.

Question 35 Evaluate $81a_n + 10 + 9n$ for

$$a_n = 1 + 11 + 111 + ... + \underbrace{111...1}_{n \text{ ones}}.$$

Question 36 Consider a sequence (u_n) defined by $\begin{cases} u_0 &= 10, \\ u_{n+1} &= \frac{u_n}{2} + n^2 + n. \end{cases}$

1. Determine a quadratic polynomial P such that the sequence $\alpha_n = P(n)$ satisfies the same iterative scheme as (u_n).
2. Show that the sequence $v_n = u_n - \alpha_n$ is a GP
3. Express v_n and then u_n in terms of n.

Question 37 Express the nth term of the sequence (z_n) in two ways, if

$$z_{n+1} = 2z_n - z_{n-1} + 2, \quad z_1 = 1, \; z_2 = 4.$$

Question 38 Assume that you receive 1 euro the first day of February, 2 euros the second day, 4 euros the third day and so on. If you get more than one billion of euros at the end of the month; explain why this transaction could not happen in 2022.

Question 39 Let (x_n) and (y_n) be two sequences defined by:

$$x_{n+1} = x_n + 4, \quad 5y_n = x_n + 1, \text{ and } x_1 = 1, \; n \geq 1.$$

(i) Determine x_n as a function in n.

(ii) Evaluate

$$T = \sum_{k=1}^{n} \frac{\sqrt{y_k}}{\sqrt{y_{k+1}} - \sqrt{y_k}}.$$

Question 40 Determine whether the following two sequences are adjacent:

$$u_n = \sum_{k=1}^{n} \frac{1}{k^2}, \qquad v_n = u_n + \frac{1}{n}.$$

Question 41 Evaluate $\sum_{k=15}^{30} x_k$, where (x_n) is an AP and $x_3 + x_9 + 2x_{15} + x_{21} + x_{27} = 300$.

Question 42 Determine whether the sequences (a_n) and (b_n) are adjacent, and then find their common limit, given that

$$a_{n+1} = \frac{1}{2(a_n + b_n)}, \quad \text{and} \quad b_{n+1} = \frac{2a_n b_n}{a_n + b_n}.$$

Question 43 Study the following sequences:

$$a_n = -2\sqrt{n+1} + \sum_{k=1}^{n} k^{-1/2} \text{ and } b_n = -2\sqrt{n} + \sum_{k=1}^{n} k^{-1/2}, \; \forall n \geq 1.$$

(Hint think about adjacent sequences.)

Question 44 Study the following sequences:

$$x_n = \prod_{k=1}^{n} \left(1 + \frac{1}{k^2}\right), \quad \text{and} \quad y_n = x_n \left(1 + \frac{1}{n}\right), \; \forall n \geq 1.$$

(Hint think about adjacent sequences.)

Question 45 For all $n \geq 0$, $F_{n+2} = F_{n+1} + F_n$, with $F_0 = 0$, $F_1 = 1$. Prove that for all $n \geq 1$ one has

$$F_{n+1} < \left(\frac{7}{4}\right)^n.$$

Question 46 For all $n \geq 0$, $F_{n+2} = F_{n+1} + F_n$, with $F_0 = 0$, $F_1 = 1$. Prove that for all $n \geq 1$ one has

$$\sum_{k=1}^{n} F_{2k-1} = F_{2n}.$$

Question 47 For all $n \geq 0$, $F_{n+2} = F_{n+1} + F_n$, with $F_0 = 0$, $F_1 = 1$. Prove that For all $n \geq 1$,

$$\sum_{k=1}^{2n} F_k F_{k+1} = F_{2n}^2.$$

Question 48 For all $n \geq 0$, $F_{n+2} = F_{n+1} + F_n$, with $F_0 = 0$, and $F_1 = 1$. Prove that for all $n \geq 1$ one has

$$\sum_{k=0}^{n} (-1)^k F_{k+1} = (-1)^n F_n + 1.$$

Question 49 For all $n \geq 0$, $F_{n+2} = F_{n+1} + F_n$, with $F_0 = 0$, and $F_1 = 1$. Prove that for all $n \geq 1$ one has

$$\sum_{k=1}^{n} F_k^2 = F_n F_{n+1}.$$

Question 50 For all $n \geq 0$, $F_{n+2} = F_{n+1} + F_n$, with $F_0 = 0$, and $F_1 = 1$. Prove that for all $n \geq 1$ one has

$$\sum_{k=1}^{2n} F_k F_{k+1} = F_{2n}^2.$$

6.10 Chapter 6 Additional Practice Problems with Solutions

Question 1 Determine the domain of each sequence:

$$a_n = \frac{2\sqrt{n}}{2n^2-1},\ b_n = \frac{1}{\sqrt{5n-11}},\ c_n = \sqrt{n^2-2n-15},\ d_n = \sqrt{3^n-29},\ \text{and } e_n = \frac{1}{2022^{2n}-2022}.$$

Solution.

1. $a_n \hookrightarrow n \geq 0.$
2. $b_n \hookrightarrow 5n-11 > 0 \Longrightarrow n \geq 3.$
3. $c_n \hookrightarrow n^2-2n-15 > 0 \Longrightarrow (n-5)(n+3) > 0 \Longrightarrow n > 5.$
4. $d_n \hookrightarrow 3^n - 29 \geq 0 \Longrightarrow n \geq 4.$
5. $e_n \hookrightarrow 2n \neq 1 \Longrightarrow n \geq 0.$

Question 2 Give the first five terms of each sequence:

$$x_n = \frac{1}{4n},\ y_n = \frac{2n-1}{\sqrt{n-2}},\ z_n = (-1)^n\sqrt{9^n},\ \text{and } t_n = 2^{3n+1}.$$

Solution.

1. $n \geq 1:\ x_1 = \frac{1}{4},\ x_2 = \frac{1}{8},\ x_3 = \frac{1}{12},\ x_4 = \frac{1}{16},\ x_4 = \frac{1}{20}.$
2. $n \geq 3:\ y_3 = 5,\ y_4 = \frac{7}{\sqrt{2}},\ y_5 = \frac{9}{\sqrt{3}},\ y_6 = \frac{11}{2},$ and $y_7 = \frac{13}{\sqrt{5}}.$
3. $n \geq 0:\ z_0 = 1,\ z_1 = -3,\ z_2 = 9,\ z_3 = -27,$ and $z_4 = 81.$
4. $n \geq 0:\ t_0 = 2,\ t_1 = 2^4 = 16,\ t_2 = 2^7,\ t_3 = 2^{10} = 1024,$ and $t_4 = 2^{13}.$

Question 3 Give the first four terms of each sequence

$$\begin{cases} x_0 = 0, \\ x_{n+1} = \dfrac{3}{4+x_n} \end{cases},\quad \begin{cases} y_0 = 2, \\ y_{n+1} = \dfrac{y_n - 1}{y_n + 1} \end{cases}.$$

Solution.

1. $x_1 = \frac{3}{4},\ x_2 = \frac{3}{4+\frac{3}{4}} = \frac{12}{19},\ x_3 = \frac{3}{4+\frac{12}{19}} = \frac{57}{88},$ and $x_4 = \frac{3}{4+\frac{57}{88}} = \frac{264}{409}.$
2. $y_1 = \frac{1}{3},\ y_2 = \frac{\frac{1}{3}-1}{\frac{1}{3}+1} = \frac{-1}{2},\ y_3 = \frac{\frac{-1}{2}-1}{\frac{-1}{2}+1} = -3,$ and $y_4 = \frac{-3-1}{-3+1} = 2.$

Question 4 Guess the general term of the following sequence:

$$(0,-1,2,-3,4,-5,6,-7,\cdots),$$

Solution. Since the sign alternates, it implies the presence of $(-1)^n$, and in this list of numbers it is easy to see that the general term is $(-1)^n \times n, \ \ n \geq 0$.

Question 5 For $u_n = 3n - 2n^2 - 1$ express in terms of n:

$$u_{n+1}, \ u_{n-1}, \ u_{2n-1}, \ u_{2n}, \text{ and } u_{2n+1}.$$

Solution. It suffices to replace by the following in the expression of u_n:

1. $u_{n+1} = 3(n+1) - 2(n+1)^2 - 1 = 3n+3-2n^2-4n-2-1 = -2n^2-n.$
2. $u_{n-1} = 3(n-1) - 2(n-1)^2 - 1 = 3n-3-2n^2+4n-2-1 = -2n^2+7n-6.$
3. $u_{2n-1} = 3(2n-1) - 2(2n-1)^2 - 1 = 6n-3-8n^2+8n-2-1 = -8n^2+14n-6.$
4. $u_{2n} = 3(2n) - 2(2n)^2 - 1 = 6n - 8n^2 - 1 = -8n^2 + 6n - 1.$
5. $u_{2n+1} = 3(2n+1) - 2(2n+1)^2 - 1 = 6n+3-8n^2-8n-2-1 = -8n^2-2n.$

Question 6 We have same question with $u_n = n(3-n)$.

Solution. It suffices to replace by the following in the expression of u_n:

1. $u_{n+1} = (n+1)(3-n-1) = (n+1)(2-n) = -n^2+n+2.$
2. $u_{n-1} = (n-1)(3-n+1) = (n-1)(4-n) = -n^2+5n-4.$
3. $u_{2n-1} = (2n-1)(3-2n+1) = (2n-1)(4-2n) = -4n^2+10n-4.$
4. $u_{2n} = 2n(3-2n) = -4n^2+6n.$
5. $u_{2n+1} = (2n+1)(3-2n-1) = (2n+1)(2-2n) = -4n^2+2n+2.$

Question 7 Express v_{n-4} in terms of n for $n \geq 4$, where $v_n = \dfrac{n^2+5n+10}{3n+8}$.

Solution. We replace n with $(n-4)$ in the given expression:

$$v_{n-4} = \frac{(n-4)^2+5(n-4)+10}{3(n-4)+8} = \frac{n^2-8n+16+5n-20+10}{3n-12+8} = \frac{n^2-3n+6}{3n-4}.$$

Question 8 Express w_{3n}, w_{3n+1}, and w_{3n+2} in terms of n, where $w_n = \cos\left[\dfrac{(4n-1)\pi}{6}\right]$.

Solution.

1. $u_{3n} = \cos\left[\dfrac{(4(3n)-1)\pi}{6}\right] = \cos\left(2\pi n - \dfrac{\pi}{6}\right) = \cos\left(-\dfrac{\pi}{6}\right) = \cos(\dfrac{\pi}{6}) = \dfrac{\sqrt{3}}{2}.$
2. $u_{3n+1} = \cos\left[\dfrac{(4(3n+1)-1)\pi}{6}\right] = \cos\left(2\pi n + \dfrac{\pi}{2}\right) = \cos\left(\dfrac{\pi}{2}\right) = 0.$
3. $u_{3n+2} = \cos\left[\dfrac{(4(3n+2)-1)\pi}{6}\right] = \cos\left(2\pi n + \dfrac{7\pi}{6}\right) = \cos\left(\dfrac{7\pi}{6}\right) = \cos\left(\pi + \dfrac{\pi}{6}\right) = -\cos\left(\dfrac{\pi}{6}\right),$

   Thus $u_{3n+2} = -\dfrac{\sqrt{3}}{2}.$

Question 9 Express w_{3n}, w_{3n+1}, and w_{3n+2} in terms of n, where $w_n = \sin\left[\dfrac{(4n-1)\pi}{6}\right]$.

Solution.

1. $u_{3n} = \sin\left[\dfrac{(4(3n)-1)\pi}{6}\right] = \sin\left(2\pi n - \dfrac{\pi}{6}\right) = \sin\left(-\dfrac{\pi}{6}\right) = -\sin(\pi/6) = \dfrac{-1}{2}.$
2. $u_{3n+1} = \sin\left[\dfrac{(4(3n+1)-1)\pi}{6}\right] = \sin\left(2\pi n + \dfrac{\pi}{2}\right) = \sin\left(\dfrac{\pi}{2}\right) = 1.$

3. $u_{3n+2} = \sin\left[\frac{(4(3n+2)-1)\pi}{6}\right] = \sin\left(2\pi n + \frac{7\pi}{6}\right) = \sin\left(\frac{7\pi}{6}\right) = \sin\left(\pi + \frac{\pi}{6}\right) = -\sin\left(\frac{\pi}{6}\right)$,

   Thus $w_{3n+2} = \frac{-1}{2}$.

Question 10 Study the monotony of each sequence:

$$a_n = 3\sqrt{n+2},\ b_n = n - n^2,\ c_n = 3n + \frac{1}{2^n},\ d_n = (0.99)^n + \frac{n}{100},\ e_n = 2^n - n, \text{ and } f_n = \frac{2^n\sqrt{n}}{5^n},\ n \geq 1.$$

Solution.

1. $a_{n+1} - a_n = 3(\sqrt{n+3} - \sqrt{n+2}) = 3\frac{(n+3)-(n+2)}{(\sqrt{n+3}+\sqrt{n+2})} = \frac{3}{(\sqrt{n+3}+\sqrt{n+2})} > 0$

   $\Longrightarrow (a_n)$ is strictly increasing.

2. $b_{n+1} - b_n = [(n+1)-(n+1)^2] - (n-n^2) = n+1-n^2-2n-1-n+n^2 = -2n \leq 0$
   $\Longrightarrow (a_n)$ is decreasing.

3. $c_{n+1} - c_n = \left(3(n+1) + \frac{1}{2^{n+1}}\right) - \left(3n + \frac{1}{2^n}\right) = 3 + \frac{1}{2^{n+1}} - \frac{1}{2^n} = 3 - \frac{1}{2^{n+1}} = \frac{3 \times 2^{n+1} - 1}{2^{n+1}}$, and since for any $n \geq 0$, $2^{n+1} > 1 \Longrightarrow 3 \times 2^{n+1} - 1 > 0 \Longrightarrow c_{n+1} - c_n > 0$, and then (c_n) is strictly increasing.

4. $d_{n+1} - d_n = (0.99)^{n+1} + \frac{n+1}{100} - (0.99)^n - \frac{n}{100} = (0.99)^n(0.99-1) + \frac{1}{100}$,

   $$d_{n+1} - d_n = (0.99)^n(-0.01) + \frac{1}{100} = (0.99)^n(-10^{-2}) + 10^{-2} = 10^{-2}[1-(0.99)^n],$$

   and since $0.99 < 1 \Longrightarrow 0.99^n < 1 \Longrightarrow d_{n+1} - d_n > 0$, so (d_n) is strictly increasing.

5. $e_{n+1} - e_n = 2^{n+1} - (n+1) - (2^n - n) = 2^n(2-1) - 1 = 2^n - 1 \geq 0$, $\forall n \geq 0$. Thus (e_n) is increasing.

6. Since the terms of the sequence (f_n) are positive for $n \geq 1$, and contains powers, it is advised to compare the ratio $\frac{f_{n+1}}{f_n}$ to 1.

   $$\forall n \geq 1,\ \frac{f_{n+1}}{f_n} = \frac{\frac{2^{n+1}\sqrt{n+1}}{5^{n+1}}}{\frac{2^n\sqrt{n}}{5^n}} = \frac{2}{5}\sqrt{\frac{n+1}{n}} = \frac{2}{5}\sqrt{1+\frac{1}{n}}.$$

   Now, since for any $n \geq 1$: $\frac{1}{n} \leq 1 \Longrightarrow 1 + \frac{1}{n} \leq 2 \Longrightarrow \sqrt{1+\frac{1}{n}} \leq \sqrt{2}$, and then we get:

   $$\frac{2}{5}\sqrt{1+\frac{1}{n}} \leq \frac{2\sqrt{2}}{5} = \frac{\sqrt{8}}{5} < \frac{\sqrt{25}}{5} = 1.$$

   Thus we have: $\frac{f_{n+1}}{f_n} < 1$, and then (f_n) is strictly decreasing.

Question 11 Study the monotony of each sequence:

$$a_n = n^3 - 2n^2 - n, \quad b_n = 2^{n^2}, \text{ ad } c_n = \frac{3^n}{2^n \times n^3}.$$

Solution.

1. $a_{n+1} - a_n = (n+1)^3 - 2(n+1)^2 - (n+1) - n^3 + 2n^2 + n = 3n^2 - n - 3, \quad n \geq 0.$ We need to study the sign of the quadratic polynomial $3x^2 - x - 3$ for $x \geq 0$. The positive root is $x_2 = \dfrac{1+\sqrt{37}}{6}$, thus $3x^2 - x - 3 \geq 0$ for $x \geq x_2$, and since $2 > x_2 > \dfrac{1+\sqrt{36}}{6} = \dfrac{7}{6} > 1$, we can conclude that for $n \geq 2$, we have $a_{n+1} - a_n > 0$, so (a_n) is strictly increasing for $n \geq 2$.[6]

2. Since it is clear that all terms of (b_n) are strictly positive,
$$\frac{b_{n+1}}{b_n} = \frac{2^{(n+1)^2}}{2^{n^2}} = 2^{(n+1)^2 - n^2} = 2^{2n+1} = 2 \times 4^n > 1, \ \forall n \geq 0.$$
Thus (b_n) is strictly increasing.

3. This question can be considered as an exploration situation, where students may use computational tools (Excel is enough).
For $n \geq 1$ all terms of (b_n) are strictly positive; then
$$\frac{c_{n+1}}{c_n} = \frac{\dfrac{3^{n+1}}{2^{n+1} \times (n+1)^3}}{\dfrac{3^n}{2^n \times n^3}} = \frac{3}{2}\left(\frac{n}{n+1}\right)^3.$$
Now, let us compare this ratio to one:
$$\frac{3}{2}\left(\frac{n}{n+1}\right)^3 \geq 1 \Longleftrightarrow \left(\frac{n}{n+1}\right)^3 \geq \frac{2}{3} \Longleftrightarrow \frac{n}{n+1} \geq \sqrt[3]{\frac{2}{3}} \approx 0.87,$$
$$n(1-0.87) \geq 0.87 \Longleftrightarrow n \geq \frac{0.87}{0.13} = \frac{87}{13} \approx 6.7 \Longleftrightarrow n \geq 7.$$
Thus (c_n) is decreasing for the first 6 terms, and then it is increasing for all $n \geq 7$. Finally, (c_n) is not monotone[7] for any $n \geq 1$. Also, in such questions the student is invited to use Excel to have an idea about the behavior of the sequence. For instance, the first following 12 terms are given clearly showing the result.

n	1	2	3	4	5	6	7
c_n	1.5	0.281	0.125	0.079	0.061	0.053	0.049

n	8	9	10	11	12
c_n	0.050	0.053	0.058	0.065	0.075

Question 12 Study whether each sequence is bounded or not.

$$u_n = 2 - \frac{1}{3n}, \quad v_n = \frac{4}{3} - \cos(n), \quad w_n = \frac{5n-1}{3+5n}, \text{ and } x_n = (-1)^n \times n.$$

[6] Student should observe that for $n = 1$, we have: $3n^2 - n - 3 = -1$, this is why we compared x_2 to 1.
[7] We can also conclude that (c_n) is monotone for $n \geq 7$.

Solution.

1. $n > 0 \Longleftrightarrow -\frac{1}{3n} < 0 \Longleftrightarrow 2 - \frac{1}{3n} < 2 \Longleftrightarrow u_n < 2$, and for all $n \geq 1$ we have:

$$3n \geq 3 \Longleftrightarrow \frac{1}{3n} \leq \frac{1}{3} \Longleftrightarrow \frac{-1}{3n} \geq \frac{-1}{3} \Longleftrightarrow 2 - \frac{1}{3n} \geq 2 - \frac{1}{3} = \frac{5}{3}.$$

   Thus $\frac{5}{3} \leq u_n < 2$ for all $n \geq 1$. Then (u_n) is bounded.

2. Since for any n we have:

$$-1 \leq \cos(n) \leq 1 \Longleftrightarrow -1 \leq -\cos(n) \leq 1 \Longleftrightarrow \frac{4}{3} - 1 \leq \frac{4}{3} - \cos(n) \leq 1 + \frac{4}{3} \Longleftrightarrow \frac{1}{3} \leq v_n \leq \frac{7}{3},$$

   then (v_n) is bounded.

3. For $n \geq 0$,

$$w_n = \frac{5n-1}{3+5n} = \frac{5n-1}{5n+3} = \frac{5n+3-4}{5n+3} = 1 - \frac{4}{5n+3},$$

   thus we have

$$n \geq 0 \Longrightarrow 5n+3 > 0 \Longrightarrow -\frac{4}{3+5n} < 0 \Longrightarrow w_n < 1.$$

   Also,

$$5n+3 \geq 3 \Longrightarrow \frac{1}{3+5n} \leq \frac{1}{3} \Longrightarrow \frac{4}{3+5n} \leq \frac{4}{3} \Longrightarrow -\frac{4}{3+5n} \geq -\frac{4}{3},$$

   and then
$$1 - \frac{4}{3+5n} \geq 1 - \frac{4}{3} \Longleftrightarrow w_n \geq \frac{-1}{3}.$$

   Thus we have: $\frac{-1}{3} \leq w_n < 1$, and (w_n) is bounded.

4. $|x_n| = |(-1)^n n| = n$, which is not bounded, so (x_n) is not bounded. [8]

Question 13 Find an upper bound for the sequence $a_n = \frac{2n - n^2}{3992n}$, $n \geq 1$, among the list of numbers $1, \frac{1}{2}, \frac{1}{3}, \ldots$.

Solution. For $n \geq 1$, $a_n = \frac{2n - n^2}{3992n} = \frac{2-n}{3992}$, and using the fact that $n \geq 1$, we obtain

$$\frac{n}{3992} \geq \frac{1}{3992} \Longrightarrow -\frac{n}{3992} \leq -\frac{1}{3992} \Longrightarrow \frac{2}{3992} - \frac{n}{3992} \leq \frac{2}{3992} - \frac{1}{3992} \Longrightarrow a_n \leq \frac{1}{3992}.$$

Thus (a_n) is bounded from above with $\frac{1}{3992}$.

Question 14 From which rank (value of n), are the following sequences well defined?

$$u_n = \frac{1}{8 \times 2^n - 2048}, \quad v_n = (n-3)^{-n}.$$

[8] Direct application of the definition of the boundedness of sequences.

Solution.

1. $u_n = \dfrac{1}{8\times 2^n - 2048}$ is defined if the denominator is not zero. [9] i.e.,

$$8\times 2^n - 2048 \neq 0 \Longleftrightarrow 2^{n+3} \neq 2\times 1024 \Longleftrightarrow 2^{n+3} \neq 2^{11} \Longleftrightarrow n+3 \neq 11 \Longleftrightarrow n \neq 8.$$

Thus (u_n) is defined for all $n \geq 9$.

2. $v_n = (n-3)^{-n} = \dfrac{1}{(n-3)^n}$ is defined if $n-3 \neq 0$. Thus for $n \geq 4$, (v_n) is well defined.

Question 15 Express λ_{n-2}, λ_{n+1}, λ_{2n} in terms of n, where $\lambda_n = 2^{2^n}$.

Solution.

$$\lambda_{n-2} = 2^{2^{n-2}}, \quad \lambda_{n+1} = 2^{2^{n+1}}, \quad \lambda_{2n} = 2^{2^{2n}}.$$

Question 16 Let $u_0 = 1$, $u_1 = \dfrac{1}{2+1}, \ldots$:

$$u_n = \cfrac{1}{2+\cfrac{1}{2+\cfrac{1}{2+\cfrac{1}{2+\frac{1}{2+\ldots}}}}} \quad \text{with n fractions.}$$

Determine the function f such that $u_n = f(u_{n-1})$, $n \geq 1$.

Solution. It is clear that the terms of the sequence are all positive, so the function is

$$f(x) = \frac{1}{2+x}, \quad x > 0.$$

Question 17 Let a be a positive real number, where $a \neq 1$. Show that the sequence of general term $x_n = a^n + \dfrac{1}{a^n}$ is increasing.

Solution. Let us consider the following difference:

$$x_{n+1} - x_n = (a^{n+1} + \frac{1}{a^{n+1}}) - (a^n + \frac{1}{a^n}) = a^n(a-1) + \frac{1}{a^{n+1}}(1-a) = (a-1)\left[a^n - \frac{1}{a^{n+1}}\right],$$

i.e.,

$$x_{n+1} - x_n = (a-1)\left[\frac{a^{2n+1}-1}{a^{n+1}}\right].$$

It is worth remarking that $(a^{2n+1}-1)$ has the same sign as $(a-1)$, and since $a > 0$ and $a \neq 1$, we conclude that the difference is positive and (x_n) is strictly increasing. (See the following remark.)

R

- Using the classical identity, we have

$$a^{2n+1} - 1 = a^{2n+1} - 1^{2n+1} = (a-1)\left[a^{2n} + a^{2n-1} + a^{2n-2} + \ldots a^3 + a^2 + a + 1\right],$$

  and since $a > 0$, the second factor $\left[a^{2n} + a^{2n-1} + a^{2n-2} + \ldots a^3 + a^2 + a + 1\right] > 0$, and thus the sign of $a^{2n+1} - 1$, is the sign of $(a-1)$.

[9] The number 1024 is a well-known power equal to 2^{10}.

- In general, for $a, b \in [0, \infty)$ and $n \in \mathbb{N}$, we can conclude that:
$$(a^n - b^n)(a-b) \geq 0.$$
  - If $a > b > 0$, it is trivial that we get: $a^n > b^n$, which proves our above result.
  - If $0 < a < b$, directly we have: $a^n < b^n$, which proves our result
- Also, it is useful to know that: for $a, b \in [0, \infty)$:
$$(a^n - b^n)(b-a) \leq 0.$$

Question 18 Let $z_n = \dfrac{(-1)^n \sin(n) + \cos(n)}{4+3n}$. Prove that $\forall n \geq 1, \ \ |z_n| \leq \dfrac{2}{3n}$.

Solution. Using the triangular inequality,

$$|(-1)^n \sin(n) + \cos(n)| \leq |(-1)^n||\sin(n)| + |\cos(n)| = 1 \times |\sin(n)| + |\cos(n)| < 1+1 = 2,$$

we obtain the following for $n \geq 1$:

$$|z_n| = \frac{|(-1)^n \sin(n) + \cos(n)|}{|4+3n|} < \frac{2}{4+3n}.$$

Since $4+3n > 3n \Longrightarrow \dfrac{1}{4+3n} < \dfrac{1}{3n}$, we finally get

$$|z_n| = \frac{|(-1)^n \sin(n) + \cos(n)|}{|4+3n|} < \frac{2}{3n}, \quad \forall n \geq 1.$$

Question 19 Express v_n and u_n in terms of n for

$$\begin{cases} u_{n+1} = \dfrac{u_n}{1-u_n}, \text{ and } v_n = \dfrac{1+u_n}{u_n}, \\ u_1 = -2, \ \ \forall n \geq 1, \end{cases}, \quad \text{assuming } (1-u_n)(1-v_n)u_n \neq 0.$$

Solution. The main idea here is to prove that one of the two sequences is AP or GP To this end, let us focus first on any relation, gathering both sequences and assuming:[10] $(1-u_n)(1-v_n)u_n \neq 0$.

$$v_{n+1} = \frac{1+u_{n+1}}{u_{n+1}} = \frac{1+\dfrac{u_n}{1-u_n}}{\dfrac{u_n}{1-u_n}} = \frac{1}{u_n}.$$

Also from $v_n = \dfrac{1+u_n}{u_n}$, we get $u_n = \dfrac{1}{v_n - 1}$, and then we obtain $v_{n+1} = v_n - 1$, which implies the sequence (v_n) is an AP with the common difference $d = -1$. Thus

$$v_n = v_1 + (n-1)d \ = v_1 + (n-1) \times (-1) = \frac{3}{2} - n, \qquad u_n = \frac{2}{1-2n}, \ \ n \geq 1.$$

Question 20 Determine if the sequence (b_n) is increasing or decreasing, where $b_n = \sqrt{n+1} - \sqrt{n}$.

Solution. Let us consider the difference $b_{n+1} - b_n$. Using the two conjugates, we get the following for any $n \geq 0$:

$$b_{n+1} - b_n = (\sqrt{n+2} - \sqrt{n+1}) - (\sqrt{n+1} - \sqrt{n}) = \frac{1}{\sqrt{n+2} + \sqrt{n+1}} - \frac{1}{\sqrt{n+1} + \sqrt{n}}.$$

[10] The condition $(1-u_n)(1-v_n)u_n \neq 0$ is assumed to be true in this question, but normally it should be proved (see Chapter 6, Question 11).

Now,

$$n+2>n \Longrightarrow \sqrt{n+2}>\sqrt{n} \Longrightarrow \sqrt{n+2}+\sqrt{n+1}>\sqrt{n+1}+\sqrt{n},$$

which implies

$$\frac{1}{\sqrt{n+2}+\sqrt{n+1}}<\frac{1}{\sqrt{n+1}+\sqrt{n}},$$

and then $b_{n+1}-b_n<0$, for any non-negative value of n and finally (b_n) is strictly decreasing.

Question 21 Show that the sequence whose general term is $u_n=1+2^n(n-1)$ satisfies $u_{n+1}=u_n+2^n\,(n+1),\ u_1\ =\ 1$, and then evaluate

$$1+2\times 2+3\times 2^2+4\times 2^3+5\times 2^4+\ldots+11\times 2^{10}.$$

Solution. For $n=1$, we get $u_1=1+2^1(1-1)=1$, and if we write the difference

$$u_{n+1}-u_n=1+2^{n+1}(n+1-1)-1-2^n(n-1)=n\times 2^n(2-1)+2^n=2^n(n+1),$$

we get: $u_{n+1}\ =\ u_n\ +(n+1)\,2^n$ for any $n\geq 1$. To complete the answer, we need to evaluate the sum. To this end, let us iterate the following proved expression:

$$\begin{array}{lcl}
u_{n+1} & = & u_n+(n+1)\times 2^n, \\
u_n & = & u_{n-1}+n\times 2^{n-1}, \\
u_{n-1} & = & u_{n-2}+(n-1)\times 2^{n-2}, \\
\vdots & & \vdots, \\
u_3 & = & u_2+3\times 2^2, \\
u_2 & = & u_1+2\times 2^1.
\end{array}$$

By adding all above equations, we obtain the following after simplification:

$$u_{n+1}\ =\ (n+1)\times 2^n+n\times 2^{n-1}+\ldots+3\times 2^2+2\times 2^1+\ 1.$$

Thus our sum corresponds to the case with $n=10$; in other words, our sum is equal to $1+2^{11}\times(11-1)=1+5\times 2^{12}\ =\ u_{11}=\ 20,481.$

Question 22 Show that

$$0<u_n<\frac{3}{2},\quad \text{with}\ \ u_n=\frac{2n+(-1)^n}{3n+(-1)^{n+1}},\quad \forall n\geq 1.$$

Solution. Let us show first that $u_n>0$. To this end, let us start by stating some trivial useful inequalities:

$$-1\leq(-1)^n<1,\qquad n\geq 1 \Longrightarrow 2n-1\geq 1$$

and

$$-1\leq(-1)^{n+1}<1,\qquad n\geq 1 \Longrightarrow 3n-1\geq 2.$$

Since $3n-1\leq 3n+(-1)^{n+1}\leq 3n+1$ and $3n-1\geq 2$, we conclude that $2\leq 3n-1\leq 3n+(-1)^{n+1}\leq 3n+1$, Thus the numerator is positive, and we have

$$\frac{1}{3n-1}\geq\frac{1}{3n+(-1)^{n+1}}.$$

Now, using the fact that $2n+(-1)^n\geq 2n-1\geq 1>0$, we have proven that the numerator $2n+(-1)^n$ is positive, and then $u_n>0$.

Since the numerator is positive, we can write

$$0<u_n\leq\frac{2n+(-1)^n}{3n-1}\leq\frac{2n+1}{3n-1}=\frac{\frac{2n}{3n}+\frac{1}{3n}}{\frac{3n}{3n}-\frac{1}{3n}}=\frac{\frac{2}{3}+\frac{1}{3n}}{1-\frac{1}{3n}}.$$

Now, using the fact that $\frac{1}{3n} \leq \frac{1}{3}$ and $\frac{1}{2n} \leq \frac{1}{2}$, it is easy to see that

$$1 - \frac{1}{3n} \geq \frac{2}{3} \Longrightarrow \frac{\frac{2}{3} + \frac{1}{3n}}{1 - \frac{1}{3n}} \leq \frac{3}{2} \times \left(\frac{2}{3} + \frac{1}{3n}\right) = 1 + \frac{1}{2n}.$$

Finally, we have:

$$0 < u_n \leq \frac{\frac{2}{3} + \frac{1}{3n}}{1 - \frac{1}{3n}} \leq 1 + \frac{1}{2} = \frac{3}{2}.$$

R In the sequence, many students may give a direct answer and think that the upper bound is $\frac{2}{3}$ but if one checks via Excel, will get: u_4, u_6, u_{10}, u_{12} which are greater than $\frac{2}{3}$. Thus the preceding upper bound is correct.

Question 23 Express explicitly in terms of n the sequence with general term u_n that represents the unit digit of $(1789)^n - (1515)^n$.

Solution. For $n = 0$, $(1789)^0 - (1515)^0 = 1 - 1 = 0 \Longrightarrow u_0 = 0$, and now for $n \geq 1$, it is clear that 5^n always has 5 as a unit digit but the unit digit corresponding to 9^n alternates between 1 and 9 upon the parity of n. Thus we have the following If n is even ($n \geq 2$), the unit digit of 9^n equals 1, and for n odd ($n \geq 1$), the unit digit equals 9. Finally, the sequence (u_n) can be expressed as follows:

$$u_n = \begin{cases} 0 & (\text{if } n = 0), \\ 9 - 5 = 4 & (\text{if } n \text{ is odd}), \\ 6 & (\text{if } n \text{ is even}). \end{cases}$$

Here the case where n is even, $u_n = 1 - 5$, which is, in fact $11 - 5 = 6$.

Question 24 Let (t_n) be the sequence defined by $\begin{cases} t_0 = 1, \\ t_{n+1} = -1.5t_n^2 + 2.5t_n + 1. \end{cases}$

Calculate the first six terms of (t_n) and determine what you can conclude.

Solution. The first six terms are:

$$t_0 = 1,\ t_1 = 2,\ t_2 = 0,\ t_3 = 1,\ t_4 = 2,\ t_5 = 0, \text{ and } t_6 = 1.$$

It is a kind of periodic function : $t_{n+3} = t_n$. Thus our sequence is periodic with period 3.

Question 25 Find an explicit formula for the sequence with general term

$$\begin{cases} b_1 = 1, \\ b_{n+1} = \dfrac{1}{1 + \frac{1}{b_n}}, \quad n \geq 1. \end{cases}$$

Solution. It is easy to calculate $b_2 = \frac{1}{2}$, $b_3 = \frac{1}{3}$, ..., thus we can expect that $b_n = \frac{1}{n}$, $\forall n \geq 1$. To prove this formula, we set another sequence $c_1 = 1$ and $c_n = \frac{1}{n}$, $n \geq 1$. The next term is

$$c_{n+1} = \frac{1}{n+1} = \frac{1}{1+n} = \frac{1}{1 + \frac{1}{\frac{1}{n}}} = \frac{1}{1 + \frac{1}{c_n}}.$$

Thus we have two sequences with the same recursive formula and the same initial values. Then $b_n = c_n$, for any $n \geq 1$, and finally, $b_n = \dfrac{1}{n}$, $n \geq 1$.

Question 26 Let $u_n = \dfrac{n+5}{n+3}$, $n \geq 0$. Study the monotony and the boundedness of (u_n).

Solution. For any $n \geq 0$,

$$u_{n+1} - u_n = \frac{(n+1)+5}{(n+1)+3} - \frac{n+5}{n+3} = \frac{n+6}{n+4} - \frac{n+5}{n+3} = \frac{-2}{(n+3)(n+4)} < 0.$$

Thus (u_n) is strictly decreasing, and then it is monotone.
Since (u_n) is decreasing, we have: $u_n \leq u_0 = \dfrac{5}{3}$, thus (u_n) is bounded above by $5/3$. Using $n+5>0$ and $n+3>0$, we conclude that $u_n > 0$ for any n, and then (u_n) is bounded from below. Finally, the sequence (u_n) is bounded.

Question 27 Let $u_{n+1} = 2u_n - 3$, $u_0 = 2$. Study the monotony and the boundedness of (u_n).

Solution. Let us find some terms of the sequence (u_n).

$$u_1 = 1,\ u_2 = -1, \text{ and } u_3 = -5;$$

these values suggest the sequence is decreasing and bounded from above by $u_0 = 2$.
Let us set a sequence (v_n) defined by $v_n = u_n - t$, where t will be selected to make (v_n) a GP (if possible). In other words,

$$\begin{aligned} v_{n+1} &= u_{n+1} - t, \\ &= 2u_n - 3\ - t \quad \text{(by subtracting and adding } 2t) \\ &= 2(u_n - t) + 2t - 3 - t, \\ &= 2v_n\ +\ (t-3) \Longrightarrow (v_n) \text{ is a G.P if } t = 3. \end{aligned}$$

Thus $v_n = v_0 \times 2^n$, and $v_0 = u_0 - 3 = -1$, and then $u_n - 3 = v_n = -(2^n) \Longrightarrow u_n = 3 - 2^n$ for any $n \geq 0$. Thus $u_{n+1} - u_n = (3 - 2^{n+1}) - (3 - 2^n) = 2^n - 2^{n+1} = 2^n\,(1-2) < 0$, then (u_n) is strictly decreasing.
In addition, $\forall\, n \geq 0,\ u_0 = 2 \geq u_n$; thus (u_n) is bounded from above. But since

$$\lim_{n\to\infty} u_n = \lim_{n\to\infty} 3 - 2^n = -\infty,$$

the sequence cannot be bounded from below. Finally, (u_n) is not bounded.

Question 28 Determine whether the sequence (u_n) is increasing or decreasing:

$$\begin{cases} u_{n+1} = \sqrt{2+u_n} \\ u_0 = 0. \end{cases}$$

Solution. Let us consider the difference:

$$u_{n+1} - u_n = \sqrt{2+u_n} - u_n = \frac{(\sqrt{2+u_n} - u_n)(\sqrt{2+u_n} + u_n)}{\sqrt{2+u_n} + u_n} = \frac{2 + u_n - u_n^2}{\sqrt{2+u_n} + u_n}.$$

At this stage we need to study the sign of the last ratio. The numerator corresponds to a quadratic expression $-x^2 + x + 2 = -(x-2)(x+1)$ which is positive on $(-1,\ 2)$ and negative elsewhere. It is clear that $u_n \geq 0$, $\forall n \geq 0$, as an output of a square root. It can be shown directly by induction. Thus we need to know if u_n is in the interval $[0,\ 2)$. Since $u_1 = \sqrt{2} < 2$, we will try to prove by induction the $u_n < 2$, $\forall n \geq 0$, which is $P(n)$.

$P(0): u_0 = 0 < 2$ and $P(1): u_1 = \sqrt{2} < 2$ are true. Let us assume $P(k)$ is true, and let us show that $P(k+1)$ remains true:

$$u_k < 2 \Longrightarrow 2 + u_k < 4 \Longrightarrow u_{k+1} = \sqrt{2+u_k} < 2.$$

Thus for any $n \geq 0$, $u_n \in [0, 2)$, which implies that $u_{n+1} - u_n = \dfrac{2 + u_n - u_n^2}{\sqrt{2+u_n} + u_n} > 0$, so the sequence (u_n) is increasing.

Question 29 Let $u_{n+1} = \dfrac{u_n^2}{16} + 4$ and $u_0 = 4$. Study the monotony and the boundedness of (u_n).

Solution. It is clear that $u_n > 0$ for any non-negative integer n, and if we consider the function f generating the given sequence, $f(x) = \dfrac{x^2}{16} + 4$, $x > 0$, we can prove easily that for any real x,

$$f(x) - x = \frac{(x-8)^2}{16} \geq 0 \Longrightarrow f(x) \geq x \Longrightarrow u_{n+1} = f(u_n) \geq u_n,$$

then (u_n) is increasing.
For the boundedness of (u_n), we need to calculate some terms to have a more appropriate idea.

$$u_1 = 5,\ u_2 = 5.5625,\ u_3 = 5.933, \ldots, u_{12} = 7.03.$$

We show that $u_n < c$, $c > 7$. If we use induction in our proof, we obtain the following implication:

$$u_n \leq c \Longrightarrow \frac{u_n^2}{16} + 4 \leq \frac{c^2}{16} + 4 \leq c \Longleftrightarrow \frac{(c-8)^2}{16} \leq 0 \Longrightarrow c = 8.$$

This is why, we will prove that $u_n < 8$ for any $n \geq 0$ by induction.
$P(n):\ u_n < 8$.

Base step. For $n = 0$ and $n = 1$, $P(n)$ is true.
Inductive step. Let us assume $P(k)$ is true, and we will show that $P(k+1)$ remains true:

$$u_k < 8 \Longrightarrow u_{k+1} = \frac{u_k^2}{16} + 4 < \frac{8^2}{16} + 4 = 8.$$

Thus $P(k+1)$ is true, and then $u_n < 8$ for any $n \geq 0$. Finally, (u_n) is increasing, and bounded from below[11] by 4 and from above by 8.

Question 30 Check whether 1968 is a term of the list of numbers $102, 113, 124, 135, \ldots$

Solution. It is easy to check that the list of numbers are in AP with a common difference $d = 11$, and then we have

$$102 + 11(n-1) = 1968 \Longleftrightarrow 11n = 1968 - 102 + 11 = 1877 \Longrightarrow n = \frac{1877}{11} \notin \mathbb{N}.$$

Thus 1968 is not a term of the given list.

Question 31 How many two-digit numbers are multiples of 3?

Solution. Let us write some terms of the multiples of 3 with two digits:

$$12, 15, 18, 21, 24, \ldots \Longrightarrow \text{it forms an AP } (x_n) \text{ with } x_0 = 12,\ d = 3.$$

Thus our question is equivalent to determining n so that $x_n = 99$ (the last multiple of 3 with two digits), i.e.,

$$x_n = 99 \Longrightarrow x_0 + 3n = 99 \Longrightarrow 3n = 99 - 12 = 87 \Longrightarrow n = \frac{87}{3} = 29.$$

Thus there are $29 + 1 = 30$ multiples of 3. We have added 1 because the list starts from x_0.

[11] Since (u_n) is an increasing sequence, $u_n \geq u_0 = 4$

Question 32 An AP consists of 75 terms, of which the fifth term is 52, and the last term is 112. Find the 31st term.

Solution. Let (a_n) denote the given AP such that:

$$\begin{cases} a_{75} = 112, \\ a_5 = 52, \end{cases} \Longrightarrow \begin{cases} a_5 + (75-5)d = 112, \\ a_5 = 52, \end{cases} \Longrightarrow d = \frac{6}{7} \Longrightarrow a_{31} = a_5 + (31-5) \times \frac{6}{7} = \frac{520}{7}.$$

Question 33 Which term of the AP: $-1, 19, 39, 59, 79, \dots$ will be 1040 more than its 1970th term?

Solution. Let a_n denote the given AP with common difference $d = 20$ and $a_1 = -1$.[12] We want to find n such that:

$$a_n = 1040 + a_{1970} \Longrightarrow -1 + 20(n-1) = 1040 - 1 + 1969 \times 20 \Longrightarrow n = 2022.$$

Question 34 Two APs have the same common difference d. The difference between their 100th terms is 100. What is the difference of their 1000th terms?

Solution. Let us denote by (x_n) and (y_n) the two APs with the common difference d. Since

$$x_n - y_n = (x_0 + (n-1)d) - (y_0 + (n-1)d) = x_0 - y_0,$$

which implies that the sequence of the general term $x_n - y_n$ is constant and $x_n - y_n = x_{100} - y_{100} = 100$ for any n, we have $x_{1000} - y_{1000} = 100$.

Question 35 How many three-digit numbers are multiples of 7?

Solution. The three-digit multiples of 7 [13] are $105, 112, 119, \dots, 994$. So, now it is easy to find their number. It is an AP with the common difference $d = 7$ and first term $a_1 = 105$, where we have $a_n = 105 + 7(n-1) = 994$ then $n = 128$. Finally, there are 128 multiples of 7 with three digits.

Question 36 How many multiples of 5 lie between 1968 and 2022?

Solution. Following the same method as the Exercise 35 with $d = 5$, and the first term is 1970. Since 2020 is the last multiple of 5 less than or equal to 2022 Then, we have:

$$2020 = 1970 + 5(n-1) \Longrightarrow n = 11.$$

Finally, there are 11 multiples of 5 between 1968 and 2022.

Question 37 Determine 5 consecutive terms in AP so that the sum of the first three terms is -24 and the sum of the last three terms is 18.

Solution. Let d be the common difference, and let the 5 consecutive numbers be denoted by

$$x-2d,\ x-d,\ x,\ x+d,\ x+2d \Longrightarrow \begin{cases} 3x - 3d &= -24, \\ 3x + 3d &= 18 \end{cases} \Longrightarrow x = -1,\ d = 7 \Longrightarrow -15, -8, -1, 6, 13.$$

Question 38 If the sum of the first 27 terms of an AP is 1971 and its second term is 13, find the 31st term.

Solution. Let (b_n) be the required AP with first term b_1. Since the $b_2 = 13 \Longrightarrow b_1 = 13 - r$ and the sum of the 27 is given by:

$$1971 = 27\left(\frac{b_1 + b_{27}}{2}\right) = 27\left(\frac{b_1 + b_1 + 26r}{2}\right) = b_1 + 13r = 27\,(13 + 12r),$$

thus $r = 5$ and then $b_{31} = b_2 + (31-2)r = 13 + 5 \times 29 = 158$.

[12] Easy to see that $d = 19 - (-1) = 39 - 19 = 20$.

[13] To determine the first and last three-digits multiples of 7, divide 100 by 7 and take the integer part, (i.e., 14 and the next integer 15 will be the first multiple of 7 and in the same way, divide 1000 by 7 to obtain 142.8; and thus the last multiple of 7 is 142).

Question 39 How many terms of the AP: $118, 113, 108, \ldots$ must be added to obtain 1288?

Solution. Let (c_n) be the required AP with $r = -5$ and $c_1 = 118$. The sum of n terms is given by:

$$1288 = n \times \frac{c_1 + c_n}{2} = \frac{n[2c_1 - 5(n-1)]}{2} \Longrightarrow 2576 = n(241 - 5n),$$

which is a quadratic equation $5n^2 - 241n + 2576 = 0$ in n, where the roots are $161/5$ and 16. Only the value $n = 16$ is accepted, since the second one is not integer. Finally, we need to add 16 terms of the given AP to obtain 1288.

Question 40 Evaluate $\sum_{k=1}^{36} x_k$, where (x_n) is an AP and

$$x_2 + x_6 + x_{12} + x_{25} + x_{31} + x_{35} = 337.$$

Solution. By writing the sum
$S_{36} = \sum_{k=1}^{36} x_k = \frac{36(x_1 + x_{36})}{2} = 18(x_1 + x_{36})$ it remains to evaluate $x_1 + x_{36}$. But here, we directly use the property, affirming that $x_1 + x_{36} = x_2 + x_{35} = x_6 + x_{31} = x_{12} + x_{25}$, and since $337 = x_2 + x_6 + x_{12} + x_{25} + x_{31} + x_{35} = (x_2 + x_{35}) + (x_6 + x_{31}) + (x_{12} + x_{25}) = 3(x_2 + x_{35})$. Thus $x_2 + x_{35} = \frac{337}{3}$, and finally, we obtain:

$$S_{36} = 18(x_1 + x_{36}) = 18(x_2 + x_{35}) = 18 \times \frac{337}{3} = 2022.$$

Question 41 Let (u_n) be a sequence defined by $u_{n+1} = a u_n + b$ for any $n \geq 0$, where a and b are real numbers with $a \neq 1$. Prove that

$$u_n = a^n u_0 + b\left(\frac{a^n - 1}{a - 1}\right), \; n \geq 1.$$

Solution. We will solve it with and without induction.

(i) *Idea 1.* By induction, let $P(n): \; u_n = a^n u_0 + b\left(\frac{a^n - 1}{a - 1}\right)$.

Base step. $n = 1$, $u_1 = a u_0 + b(1)$, so $P(1)$ is true.

Inductive step. Assume $P(k)$ is true, and let us show that $P(k+1)$ is also true.

$P(k+1): \; u_{k+1} = a^{k+1} u_0 + b\left(\frac{a^{k+1} - 1}{a - 1}\right)$.

Since,

$$\begin{aligned} u_{k+1} = a u_k + b &= a\left[a^k u_0 + b\left(\frac{a^k - 1}{a - 1}\right)\right] + b, \\ &= a^{k+1} u_0 + b\left(\frac{a(a^k - 1)}{a - 1} + 1\right), \\ &= a^{k+1} u_0 + b\left(\frac{a^{k+1} - a + a - 1}{a - 1}\right), \\ &= a^{k+1} u_0 + b\left(\frac{a^{k+1} - 1}{a - 1}\right). \end{aligned}$$

Thus $P(k+1)$ is true, and then $P(n)$ is true for all $n \geq 1$.

(ii) *Idea 2.* Let us consider $a \neq 0$ and $a \neq 1$. We will use an auxiliary sequence $v_n = u_n - \alpha$, where α will be chosen in such a way that (v_n) is a geometric progression.

$$v_{n+1} = u_{n+1} - \alpha = a u_n + b - \alpha = a(u_n - \alpha) + b - \alpha + a\alpha = a v_n + b - \alpha + a\alpha,$$

Thus (v_n) is a GP if $b - \alpha + a\alpha = 0 \Longrightarrow \alpha = \dfrac{-b}{a-1}$, and then we obtain

$$v_{n+1} = a v_n \Longrightarrow v_n = v_0 \times a^n, \quad v_0 = u_0 - \alpha = u_0 + \frac{b}{a-1}.$$

Since $u_n = v_n + \alpha = (u_0 + \dfrac{b}{a-1}) \times a^n - \dfrac{b}{a-1}$, we get:

$$u_n = u_0 \times a^n + b\left(\frac{a^n - 1}{a-1}\right), \; a \neq 1, \; n \geq 0.$$

Question 42 Given an AP (x_n) with nonzero terms, find the following sum:

$$\sum_{i=1}^{n} \frac{1}{x_i x_{i+2}}.$$

Solution. The following nice idea that students are invited to learn and adopt :

$$\frac{1}{ab} = \left(\frac{1}{b} - \frac{1}{a}\right) \div (a-b), \quad ab \neq 0, \; a \neq b.$$

Thus our ratio becomes:

$$\frac{1}{x_i x_{i+2}} = \left(\frac{1}{x_i} - \frac{1}{x_{i+2}}\right) \div (x_{i+2} - x_i),$$

and since $x_{i+2} - x_i = x_i + 2r - x_i = 2r$, we obtain

$$\frac{1}{x_i x_{i+2}} = \frac{1}{2r}\left(\frac{1}{x_i} - \frac{1}{x_{i+2}}\right).$$

We will use the sigma notation and a change of sum's index to complete the answer:

$$2r \sum_{i=1}^{n} \frac{1}{x_i x_{i+2}} = \sum_{i=1}^{n}\left(\frac{1}{x_i} - \frac{1}{x_{i+2}}\right) = \sum_{i=1}^{n} \frac{1}{x_i} - \sum_{i=1}^{n} \frac{1}{x_{i+2}},$$

i.e.,

$$2r \sum_{i=1}^{n} \frac{1}{x_i x_{i+2}} = \frac{1}{x_1} + \frac{1}{x_2} + \sum_{i=3}^{n} \frac{1}{x_i} - \sum_{i=3}^{n+2} \frac{1}{x_i},$$

and thus

$$2r \sum_{i=1}^{n} \frac{1}{x_i x_{i+2}} = \frac{1}{x_1} + \frac{1}{x_2} + \sum_{i=3}^{n} \frac{1}{x_i} - \sum_{i=3}^{n} \frac{1}{x_i} - \frac{1}{x_{n+1}} - \frac{1}{x_{n+2}},$$

i.e.,

$$2r \sum_{i=1}^{n} \frac{1}{x_i x_{i+2}} = \frac{1}{x_1} + \frac{1}{x_2} - \frac{1}{x_{n+1}} - \frac{1}{x_{n+2}},$$

and finally,

$$\sum_{i=1}^{n} \frac{1}{x_i x_{i+2}} = \frac{1}{2r}\left[\frac{1}{x_1} + \frac{1}{x_1 + r} - \frac{1}{x_1 + nr} - \frac{1}{x_1 + (n+1)r}\right].$$

Question 43 Place 3 missed numbers between 96 and 1520 to obtain a GP

Solution. Having 5 terms of G.P, we can select them as follows: $\frac{x}{q^2}$, $\frac{x}{q}$, x, xq, and xq^2, where q is the common ratio. We have $(x > 0)$,

$$\begin{cases} \frac{x}{q^2} = 96, \\ xq^2 = 1520 \end{cases} \Longrightarrow x^2 = 1520 \times 96 \Longrightarrow x = \sqrt{1520 \times 96},\ q^2 = \frac{1520}{\sqrt{1520 \times 96}} = \frac{\sqrt{1520}}{\sqrt{96}}.$$

Thus we obtain:

$$x = \sqrt{1520 \times 96},\ q = \pm\sqrt{\sqrt{\frac{1520}{96}}} = \pm\sqrt[4]{\frac{1520}{96}}$$

and the missed numbers when $q = -\sqrt[4]{\frac{1520}{96}}$,

$$\frac{\sqrt{1520 \times 96}}{-\sqrt[4]{\frac{1520}{96}}},\quad \sqrt{1520 \times 96},\ \text{and}\ -\sqrt[4]{\frac{1520}{96}}\sqrt{1520 \times 96},$$

i.e.,

$$-\sqrt[4]{1520 \times 96^3},\quad \sqrt{1520 \times 96},\quad -\sqrt[4]{1520^3 \times 96},$$

and when $q = \sqrt[4]{\frac{1520}{96}}$,

$$\sqrt[4]{1520 \times 96^3},\quad \sqrt{1520 \times 96},\ \text{and}\ \sqrt[4]{1520^3 \times 96}.$$

Question 44 Place 3 missed numbers between 96 and 1520 to obtain an AP

Solution. Having 5 terms of an AP, we can select them as follows: $x-2r$, $x-r$, x, $x+r$, and $x+2r$, where r is the common difference. We have

$$\begin{cases} x - 2r = 96, \\ x + 2r = 1520 \end{cases} \Longrightarrow x = 808,\ \text{and}\ r = 356.$$

Thus , the missed numbers are: 452, 808, 1164.

Question 45 If x, y and z are in G.P of ratio q, show that

$$a = \frac{1}{3}(x+y+z),\quad b = \sqrt{\frac{1}{3}(xy+yz+zx)},\quad \text{and}\ c = \sqrt[3]{xyz}$$

are in GP

Solution. Let our three numbers selected be $x = \frac{y}{q}$, y, and $z = yq$. Directly, we have:

$$a = \frac{y}{3}\left(1+q+\frac{1}{q}\right),\quad b = \sqrt{\frac{1}{3}\left(\frac{y^2}{q}+y^2q+y^2\right)} = |y|\sqrt{\frac{1}{3}\left(\frac{1}{q}+q+1\right)},\ \text{and}$$

$c = \sqrt[3]{xyz} = \sqrt[3]{y^3} = y$, and then

$$b^2 = \frac{y^2}{3}\left(\frac{1}{q}+q+1\right) = \frac{y}{3}\left(1+q+\frac{1}{q}\right) \times y = a \times c.$$

Thus a, b, and c are in GP.

Question 46 Express the following sum in terms of n: $\sum_{k=0}^{n}(\ 2^k - 5k + 6\ )$.

Solution.

- *Idea 1.* Using the sigma properties, the Gauss sum, and the sum of the GP $v_n = 2^n$, we obtain.

$$\sum_{k=0}^{n}(2^k - 5k + 6) = \sum_{k=0}^{n} 2^k - 5\sum_{k=0}^{n} k + \sum_{k=0}^{n} 6 = 2^0 \times \frac{2^{n+1}-1}{2-1} - 5\frac{n(1+n)}{2} + 6(n+1),$$

  i.e.,

$$\sum_{k=0}^{n}(\ 2^k - 5k + 6\ ) = \frac{1}{2}(2^{n+2} - 5n^2 + 7n + 10).$$

- *Idea 2.* We can also consider the two sequences of general terms: $v_n = 2^n$ and $u_n = 5n - 6$. It is easy to see that our sum is the difference of the sum of terms in a G.P with ratio 2, and a sum of terms in an AP with the common difference of 5. Thus we have

$$\sum_{k=0}^{n}(2^k - 5k + 6) = \sum_{k=0}^{n} v_k - \sum_{k=0}^{n} u_k = v_0 \times \frac{2^{n+1}-1}{2-1} - (n+1)\frac{u_0 + u_n}{2},$$

  i.e.,

$$\sum_{k=0}^{n}(2^k - 5k + 6) = 2^{n+1} - 1 - \frac{(n+1)(-6+5n-6)}{2} = \frac{1}{2}(2^{n+2} - 5n^2 + 7n + 10).$$

Question 47 Find the square of the product of the n terms of a G.P in terms of the first term, last term and n.

Solution. Let us denote the first term as a and the last one as c and let P be the product of the n terms. It is clear that

$$P = a \times aq \times aq^2 \times \ldots \times aq^{n-1} = a^n q^{1+2+3+\ldots+(n-1)} = a^n q^{\frac{(n-1)n}{2}},$$

and since $c = aq^{n-1}$, we conclude that

$$P^2 = a^{2n} q^{n(n-1)} = \left(a^2 q^{n-1}\right)^n = \left(a \times aq^{n-1}\right)^n = (a\, c)^n.$$

Question 48 A writer, Al-Sephadi, says that the mathematician Sesa, who invented the game of chess, was presented to the king of Persia (now Iran). As a reward, he was promised to get all what he may wish. Sesa, modestly, asked the king to place 1 grain of wheat for the first box of the game, 2 for the second, 4 for the third, 8 for the fourth, and so on, until the last box (the chess game contains 64 boxes). The king considered the wish unworthy of his generosity, but he was surprised to know that it was impossible to satisfy. Answer the following questions related to this story.

1. How many grains of wheat should be put on the last box?
2. How many grains of wheat does the king need to fulfill Sesa's wish?
3. Let us imagine that Sesa got his reward and wanted to go with the whole wheat. Knowing that a grain of wheat weights, on average 0.02 grams, and a ship of the time could carry 10,000 tons, what size of armada of ships was necessary to transport the whole wheat?

Solution.

1. The last box should contain $2^{64-1} = 2^{63}$ grains of wheat.
2. The total number of grains is the sum of the G.P with ratio 2 and is equal to

$$\frac{2^{64}-1}{2-1} = 2^{64} - 1.$$

3. We start by calculating the weight of the total grains, which corresponds to $0.02 \times (2^{64} - 1) = (2^{65} - 2) \times 10^{-2}$ grams, which can be converted to tonnes,

$$(2^{65} - 2) \times 10^{-2} \times 10^{-6} = (2^{65} - 2) \times 10^{-8} \approx 3.6 \times 10^{19} \times 10^{-8} \approx 3.6 \times 10^{11} \text{ tonnes.}$$

In other words, the total number of grains is about 3.6×10^{11} tonnes, and since one ship could carry 10^4 tonnes of grains, Sesa needed about 3.6×10^7 ships. In other words, Sesa needed about 36 millions of ships[14].

R According to Wikipedia, the world production of wheat in 2020 was about 650,000,000 tonnes which is less than 7×10^8 tonnes and a direct calculation shows that

$$\frac{3.5 \times 10^{11}}{7 \times 10^8} \approx 500,$$

which means that Sesa will get 500 times the world annual production of wheat; this production is only possible with all the technology of the century.

Question 59 Let (x_n) be a sequence defined by $x_{n+1} = \frac{1}{2}\left(x_n + \frac{a}{x_n}\right)$, with $a > 0$ and $x_0 > 0$. Our objective is to show that (x_n) converges to $\sqrt{a}$:

1. Show that $x_{n+1}^2 - a = \dfrac{(x_n^2 - a)^2}{4x_n^2}$.
2. Show that if $n \geq 1$, then $x_n \geq \sqrt{a}$.
3. Show that (x_n) is decreasing, and conclude that (x_n) is convergent.
4. Show that $x_{n+1} - \sqrt{a} \leq \dfrac{(x_n - \sqrt{a})^2}{2\sqrt{a}}$.
5. Show that if $x_1 - \sqrt{a} \leq k$, then

$$x_n - \sqrt{a} \leq 2\sqrt{a}\left(\frac{k}{2\sqrt{a}}\right)^{2^{n-1}}, \quad \forall n \geq 1.$$

Solution.

1. Using the definition, we have:

$$x_{n+1}^2 - a = \frac{1}{4}\left(x_n + \frac{a}{x_n}\right)^2 - a = \frac{x_n^4 + a^2 - 2ax_n^2}{4x_n^2} = \frac{(x_n^2 - a)^2}{4x_n^2}.$$

2. According to Step 1 we have

$$x_n^2 - a = \frac{(x_{n-1}^2 - a)^2}{4x_{n-1}^2} \geq 0 \Longrightarrow x_n^2 - a \geq 0 \Longrightarrow x_n \geq \sqrt{a} \text{ (assuming } x_n > 0).$$

Let us prove $x_n > 0$ for all $n \geq 0$ by induction. To this end, let $P(n): \; x_n > 0$.
Base step. $n = 0$, $x_0 > 0$ (given), then $P(0)$ is true.
Inductive step. Assume that $P(k)$ is true, and let us prove that $P(k+1)$ is also true:

$$x_{k+1} = \frac{1}{2}\left(x_k + \frac{a}{x_k}\right) > 0, \text{ since } x_k > 0 \text{ and } a > 0.$$

Thus $x_n > 0$ for any $n \geq 0$.

[14] It is unlikely, that, all over the world, we could gather this number of ships.

3. To discuss the monotony of (x_n), we need to determine the sign of

$$x_{n+1}-x_n=\frac{1}{2}\left(x_n+\frac{a}{x_n}\right)-x_n=\frac{x_n^2+a-2x_n^2}{2x_n}=-\frac{(x_n^2-a)}{2x_n}\leq 0.$$

Thus (x_n) is decreasing, and since it is also bounded from below by $\sqrt{a}$, we conclude that (x_n) is convergent.

4. We have

$$\begin{aligned} x_{n+1}-\sqrt{a} &= \frac{1}{2}\left(x_n+\frac{a}{x_n}\right)-\sqrt{a}, \\ &= \frac{x_n^2+a-2x_n\sqrt{a}}{2x_n}, \\ &= \frac{(x_n-\sqrt{a})^2}{2x_n}. \end{aligned}$$

Since $x_n \geq \sqrt{a} \Longrightarrow 2x_n \geq 2\sqrt{a} \Longrightarrow \frac{1}{2x_n} \leq \frac{1}{2\sqrt{a}}$, and then

$$x_{n+1}-\sqrt{a} \leq \frac{(x_n-\sqrt{a})^2}{2\sqrt{a}}.$$

5. Let us proceed by induction. $P(n):\ x_n-\sqrt{a}\ \leq\ 2\sqrt{a}\left(\frac{k}{2\sqrt{a}}\right)^{2^{n-1}}.$

Base step. $n=1$, $x_1-\sqrt{a}\leq k=2\sqrt{a}\left(\frac{k}{2\sqrt{a}}\right)^{2^{1-1}}$, so $P(1)$ is true.

Inductive step. Assume that $P(m)$ is true, and let us prove that $P(m+1)$ is also true.

$$x_{m+1}-\sqrt{a}\ \leq \frac{(x_m-\sqrt{a})^2}{2\sqrt{a}}\ \leq\ \frac{1}{2\sqrt{a}}\left[2\sqrt{a}\left(\frac{k}{2\sqrt{a}}\right)^{2^{m-1}}\right]^2\ =2\sqrt{a}\left(\frac{k}{2\sqrt{a}}\right)^{2^m}.$$

Thus $P(m+1)$ is true, and then $\forall\, n\geq 1,\ \ x_n-\sqrt{a}\ \leq\ 2\sqrt{a}\left(\frac{k}{2\sqrt{a}}\right)^{2^{n-1}}.$

Let l be the limit of the sequence (x_n). If we take the limit in the result in (1), we obtain:

$$\lim_{n\to\infty} x_{n+1}^2-a\ =\ \lim_{n\to\infty}\frac{(x_n^2-a)^2}{4x_n^2}\Longleftrightarrow l^2-a\ =\ \frac{(l^2-a)^2}{4l^2}\Longleftrightarrow 4l^4-4al^2=l^4+a^2-2al^2$$

that is equivalent to $3l^4-2a\,l^2-a^2=0$, and if we set $l^2=t>0$, the quadratic equation has one positive root $t=l^2=a$. Thus since $l>0$, we conclude finally that $l\ =\ \lim\limits_{n\to\infty} x_n\ =\ \sqrt{a}.$

Question 49 For all $n\geq 0:\ F_{n+2}=F_{n+1}+F_n$ with $F_0\ =\ F_1\ =\ 1.$

(a) Prove that for all $n\geq 1$ one has:

$$\begin{pmatrix} F_{n+1} & F_n \\ F_n & F_{n-1}\end{pmatrix}=\begin{pmatrix} 1 & 1 \\ 1 & 0\end{pmatrix}^{n+1}.$$

(b) Deduce that for all $n\geq 1$:

$$F_{n+1}F_{n-1}=(-1)^{n+1}+F_n^2.$$

Solution.

(a) We proceed by induction. Let $P(n): \begin{pmatrix} F_{n+1} & F_n \\ F_n & F_{n-1} \end{pmatrix} = \begin{pmatrix} 1 & 1 \\ 1 & 0 \end{pmatrix}^{n+1}$.

Base step. $n = 1$, $\begin{pmatrix} F_2 & F_1 \\ F_1 & F_0 \end{pmatrix} = \begin{pmatrix} 2 & 1 \\ 1 & 1 \end{pmatrix} = \begin{pmatrix} 1 & 1 \\ 1 & 0 \end{pmatrix}^2$ (can be checked easily)

Inductive step. Assume that $P(k)$ is true, and let us prove that $P(k+1)$ is also true.

$$\begin{pmatrix} F_{k+2} & F_{k+1} \\ F_{k+1} & F_k \end{pmatrix} = \begin{pmatrix} F_{k+1}+F_k & F_{k+1} \\ F_k+F_{k-1} & F_k \end{pmatrix} = \begin{pmatrix} F_{k+1} & F_k \\ F_k & F_{k-1} \end{pmatrix}\begin{pmatrix} 1 & 1 \\ 1 & 0 \end{pmatrix}.$$

Thus $\begin{pmatrix} F_{k+2} & F_{k+1} \\ F_{k+1} & F_k \end{pmatrix} = \begin{pmatrix} 1 & 1 \\ 1 & 0 \end{pmatrix}^{k+1}\begin{pmatrix} 1 & 1 \\ 1 & 0 \end{pmatrix} = \begin{pmatrix} 1 & 1 \\ 1 & 0 \end{pmatrix}^{k+2}$, and then $P(k+1)$ is true.

Hence, $P(n)$ is true for any $n \geq 0$.

(b) Using the result in (a), and the fact that $F_{n+1}F_{n-1} - F_n^2 = det\begin{pmatrix} F_{n+1} & F_n \\ F_n & F_{n-1} \end{pmatrix}$, we obtain:

$$det\begin{pmatrix} F_{n+1} & F_n \\ F_n & F_{n-1} \end{pmatrix} = det\left[\begin{pmatrix} 1 & 1 \\ 1 & 0 \end{pmatrix}^{n+1}\right] = \left[det\begin{pmatrix} 1 & 1 \\ 1 & 0 \end{pmatrix}\right]^{n+1} = (-1)^{n+1}.$$

Question 50 For all $n \geq 0: F_{n+2} = F_{n+1} + F_n$, with $F_0 = 0$, $F_1 = 1$. Prove that

$$F_{2n} \text{ is divisible by } F_n, \quad \forall\, n \geq 1.$$

Solution. It is easy to prove by induction the following result relating to Fibonacci sequence, as was done in Question 49. To this end, let us define our Fibonacci sequence as follows:

$$\forall\, n \geq 0: F_{n+2} = F_{n+1} + F_n, \text{ with } F_0 = 0, \ F_1 = 1,$$

we have:

$$\begin{pmatrix} F_{n+1} & F_n \\ F_n & F_{n-1} \end{pmatrix} = \begin{pmatrix} 1 & 1 \\ 1 & 0 \end{pmatrix}^n, \quad n \geq 1.$$

A nice idea to prove the requested result consists in equating A^{2n} to $(A^n)^2$, where

$$A = \begin{pmatrix} 1 & 1 \\ 1 & 0 \end{pmatrix}.$$

In other words,

$$\begin{aligned} \begin{pmatrix} F_{2n+1} & F_{2n} \\ F_{2n} & F_{2n-1} \end{pmatrix} &= \begin{pmatrix} 1 & 1 \\ 1 & 0 \end{pmatrix}^{2n} = \begin{pmatrix} 1 & 1 \\ 1 & 0 \end{pmatrix}^n \begin{pmatrix} 1 & 1 \\ 1 & 0 \end{pmatrix}^n, \\ &= \begin{pmatrix} F_{n+1} & F_n \\ F_n & F_{n-1} \end{pmatrix}\begin{pmatrix} F_{n+1} & F_n \\ F_n & F_{n-1} \end{pmatrix}, \\ &= \begin{pmatrix} F_{n+1}^2 + F_n^2 & F_{n+1}F_n + F_nF_{n-1} \\ F_{n+1}F_n + F_nF_{n-1} & F_n^2 + F_{n-1}^2 \end{pmatrix}. \end{aligned}$$

Thus by considering the first row, second column element, we obtain

$$F_{2n} = F_{n+1}F_n + F_nF_{n-1} = F_n\left(F_{n+1} + F_{n-1}\right) \Longrightarrow F_n \mid F_{2n}.$$

Question 51 If ϕ denotes the golden number (the positive root of $x^2 - x - 1 = 0$), show that

$$\phi^{2n} = \sum_{k=0}^{n} \binom{n}{k} \phi^k.$$

Solution. Using the binomial formula

$$\sum_{k=0}^{n} \binom{n}{k} \phi^k = \sum_{k=0}^{n} \binom{n}{k} \phi^k \, 1^{n-k} = (1+\phi)^n,$$

and using the fact that $\phi^2 - \phi - 1 = 0 \Longrightarrow 1+\phi = \phi^2$, we have

$$\phi^{2n} = (\phi^2)^n = (1+\phi)^n = \sum_{k=0}^{n} \binom{n}{k} \phi^k.$$

7. Relations

Chapter Key Terms

(a) *Relations*
(b) *Equivalence Relation*
(c) *Equivalence Classes*
(d) *Congruence Modulo n*
(e) *The Integer Modulo n*

Just as the theory of sets serves as a common platform for all branches of mathematics, so too does that of relations add another unifying layer to that foundation. In Section 1, we introduce the basic notation associated with relations. Results for functions and their properties are presented in Section 2. In Section 3, we discuss the cardinality and compare the size of sets using it. Additionally, we outline some special functions that can be used to develop formal classification of functions, and we indicate some of the axioms involved in the process.

7.1 Relations

There are many ways to associate two things in a mathematical relationship. It is true that if all relations in mathematics were taken away, the world would be a very different one. Therefore, it is important to understand not only relations but also the symbols that bind them. We express their relationship with well-defined symbols, most of which you know already. For instance, in $7 < 9$, the symbol $<$ expresses the relation between 7 and 9. In the same way, symbols shown in the following statements are used to express a relationship between two quantities:

$$5 > 8, \quad 3 \equiv 9, \quad 5 = \frac{45}{9}, \quad a \notin \mathbb{R}, \quad 3 \equiv 5(mod 2), \quad A \subseteq B, \quad \text{and } \sqrt{3} \notin \mathbb{Z}.$$

There are infinitely many types of relations, making it impossible to study them all individually.

We develop a general, yet comprehensive framework that encompasses all relations. By gaining an understanding of a general notion, we can create the conceptual framework and foundation necessary to discuss any specific relation. Let us begin with a motivational example that will naturally lead us to our definition of a relation. Consider sets $A = \{2,6,8\}$ and $B = \{1,3/2,5,7,9\}$. If we are to define relations between elements in A and those in B, then a relation `less than` or $<$ would be a logical choice. For instance $2 < 5,\ 2 < 7,\ 2 < 9,\ 6 < 7,\ 6 < 9$, In other words, we can describe the "less than" relation as the following set:

$$R = \{(2,5),(2,7),(2,9),(6,7),(6,9),(8,9)\}.$$

Definition 7.1.1 A relation from a set A to a set B is a subset $R \subseteq A \times B$. Whenever the ordered pair $(a,b) \in R$, we say that a is related to b by the relation defined in R and write $a\,R\,b$. If the ordered pair $(a,b) \notin R$, then a is not related to b by R, and we write $a \not R\ b$.

Let R be a relation from set A to a set B. The **domain** of R is a subset of A, which we denote $dom(R)$, given by

$$dom(R) = \{a \in A : (a,b) \in R \text{ for some } b \in B\}.$$

Likewise, the **range** of R is a subset of B denoted by $range(R)$ and is defined as

$$range(R) = \{b \in B : (a,b) \in R \text{ for some } a \in A\}.$$

In other words, the domain of a relation R comprises the elements of A that appear as the first coordinates in the ordered pairs within R, while $range(R)$ consists of the elements of B that appear as the second coordinates within the relation R.

Definition 7.1.2 A relation on A is a subset of $A \times A$.

As an example, let us take a set $A = \{2,4,8,10\}$. When defining relationships among the elements within set A, a suitable choice could be the "divides" relation denoted as "$|$." For instance, 2 divides 4, 2 divides 8, and 2 divides 10. In other words, we can describe the "divides" relation as the following set:

$$R = \{(2,2),(2,4),(2,8),(2,10),(4,4),(4,8),(8,8),(10,10)\}.$$

The set R encodes the meaning of the divisibility or $|$ relation for elements in A. The ordered pair (a,b) is in the set R if and only if $a\,|\,b$. We notice that the order pairs $(4,10)$ and $(8,10)$ are not included in R. The set R, which is a subset of $A \times A$, completely describes the relation divides for A.

■ **Example 7.1** Let $A = \{1,2,3\}$ and $S = \{(a,b) : a,b \in A \text{ and } a+b \text{ is odd}\}$ be a relation on A. Then the domain and range of set S is the set A. ■

■ **Example 7.2** Let $E = \{2,3,4,5,6,7,8,9,10\}$ and $T = \{(a,b) \in E^2,\ a \neq b,\ a \text{ is a multiple of } b\}$ be a relation on E. It is clear that T is given by

$$T = \{(4,2);(6,2);(6,3);(8,2);(8,4);(9,3);(10,2);(10,5)\}, \text{ where}$$

$$dom(T) = \{10,9,8,6,4\} \text{ and } range(T) = \{2,3,4,5\}.$$

■

R If R is a relation from a set A to a set B, the R^{-1} is the **inverse relation** from set B to set A and is defined by

$$R^{-1} = \{(b,a) : (a,b) \in R\}.$$

For instance the inverse relation for T is given by

$$T^{-1} = \{(2,4);(2,6);(3,6);(2,8);(4,8);(3,9);(2,10);(5,10)\}.$$

7.2 Properties of Relations

Some relations have special properties that other relations defined on a single set. For example, the relation $\geqslant$ on $\mathbb{N}$ has the property that $n \geqslant n$ for every $n \in \mathbb{N}$, but the relation $<$ on $\mathbb{N}$ does not have this property, since $n < n$ is never true. There are three significant special properties of a that will be of special interest to us.

Definition 7.2.1 — Reflexive, Symmetric and Transitive Relations. Let R be a relation defined on a set A.

1. Relation R is called **reflexive** if $a\ R\ a$ for every $a \in A$.
   That is, $(a,a) \in R, \quad \forall a \in A$.
2. Relation R is called **symmetric** $a\ R\ b$, then $b\ R\ a$ for every $a,b \in A$.
   That is, for all $a,b \in A$ if $(a,b) \in R$, then $(b,a) \in R$.
3. Relation R is called **transitive** if whenever $a\ R\ b$ and $b\ R\ c$, then $a\ R\ c$ for every $a,b,c \in A$.
   That is, for all a, b and $c \in A$, if $(a,b) \in R$, and $(b,c) \in R$, then $(a,c) \in R$.

Definition 7.2.2 — Equivalence Relation. A relation R defined on a single set A is called an **equivalence relation** if R is reflexive, symmetric, and transitive.

For example, the *equal* relation R defined on $\mathbb{R}$ by $a\ R\ b$ if $a = b$ is an equivalence relation on $\mathbb{R}$.

■ **Example 7.3** Let $A = \{1,2,3,4,5\}$ and R be the relation defined on A that consists of the following ordered pairs:
$R = \{(1,1),(1,3),(1,5),(2,2),(2,4),(3,1),(3,3),(3,5),(4,2),(4,4),(5,1),(5,3),(5,5)\}$. We determine whether R is an equivalence relation.

1. Relation R is reflexive. For all $a \in A$, we note that $(a,a) \in R$. That is, $a\ R\ a$ for every $a \in R$. Note that $1\ R\ 1$, $2\ R\ 2$, $3\ R\ 3$, $4\ R\ 4$, and $5\ R\ 5$.
2. The relation R is symmetric, because whenever we have $a\ R\ b$, it follows that $b\ R\ a$. Observe that $1\ R\ 3$ and $3\ R\ 1$, $1\ R\ 5$ and $5\ R\ 1$, $2\ R\ 4$ and $4\ R\ 2$, $1\ R\ 5$ and $5\ R\ 1$.
3. The relation R is transitive, because whenever we have $a\ R\ b$ and $b\ R\ c$ it follows that $a\ R\ c$. Observe that $1\ R\ 1$ and $1\ R\ 3$ implies that $1\ R\ 3$, $1\ R\ 3$ and $3\ R\ 1$ implies that $1\ R\ 1$, $1\ R\ 5$ and $5\ R\ 3$ implies that $1\ R\ 3$, and so on.

Thus R is an equivalence relation on A. ■

■ **Example 7.4** Let R be a symmetric and transitive relation on a set A, and every element a of A be related to something in A. Then R is also a reflexive relation. Since every element $a \in A$ is related to some element, let $b \in A$ such that $a\ R\ b$. Then, $b\ R\ a$ because R is symmetric. By the transitivity property of R, we have $a\ R\ b$, and $b\ R\ a$ implies that $a\ R\ a$. Thus R is reflexive relation. ■

R Let R be a relation on A, and let $a,b \in A$.

- R is **irreflexive** if $a\ R\ b$, then $a \neq b$.
- R is **anti-symmetric** if $a\ R\ b$ and $b\ R\ a$, then $a = b$.

■ **Example 7.5** Using the relations $=$, $<$, and $\leqslant$ on $\mathbb{Z}$, we have

- $=$ is reflexive ($3 = 3$).
- $=$ is symmetric (if $x = 4$, then $4 = x$).
- $<$ is transitive ($1 < 2$ and $2 < 5$ implies $1 < 5$).
- $<$ is irreflexive ($1 < 2$ implies $1 \neq 2$).
- $\leqslant$ is anti-symmetric ($a \leqslant b$ and $b \leqslant a$ implies $a = b$).

■

■ **Example 7.6** Let R be a relation on $A = \{1,2,3,4,5\}$ defined as $x\,R\,y$ if and only if $|x-y|$ is even. Is R an equivalence relation?

(i) Since $0 = |x-x|$ is even, R is Reflexive.
(ii) Since $|x-y| = |y-x|$, R is Symmetric.
(iii) We will check all of the elements of A. It is easy to see from

$$R = \{(1,1);(1,3);(1,5);(2,2);(2,4);(3,1);(3,3);(3,5);(4,2);(4,4);(5,1);(5,3);(5,5)\}.$$

that R is Transitive, and finally, it is an equivalence relation on A.

■

■ **Example 7.7** Let R be a relation on $\mathbb{R}\times\mathbb{R}$ defined as $(a,b)\;R\;(a',b')$ if and only if $a+b' = b+a'$. Is R an equivalence relation?
The preceding definition of the relation is equivalent to:

$$(a,b)\;R\;(a',b') \iff a-a' = b-b'.$$

(i) It is obvious that R is reflexive.
(ii) Since the equality relation is symmetric, R is symmetric.
(iii) If $(a,b)\;R\;(a',b')$ and $(a',b')\;R\;(a",b")$, then we have:

$$\begin{cases} a-a' = b-b', \\ a'-a" = b'-b" \end{cases} \Longrightarrow (a-a')+(a'-a") = (b-b')+(b'-b") \Longrightarrow a-a" = b-b".$$

Then $(a,b)\;R\;(a",b")$ and R is transitive. Finally, R is an equivalence relation on $\mathbb{R}\times\mathbb{R}$.

■

■ **Example 7.8** Let R be a relation on $\mathbb{Z}$ defined as $x\,R\,y$ if and only if $x\,|\,y$. Is R an equivalence relation?
It is easy to see that R is not always symmetric. For instance, $2|8$, but 8 doesn't divide 2; thus R is not an equivalence relation. ■

■ **Example 7.9** Let R be a relation on $\mathbb{Z}$ defined as $a\,R\,b$ if and only if $7 \;\Big|\; (a-b)$. Is R an equivalence relation?
The preceding definition of the relation is equivalent to

$$a\,R\,b \iff \exists\, k \in \mathbb{Z}:\; a = b+7k.$$

(i) R is reflexive since $a-a = 0$ is a multiple of 7.
(ii) R is symmetric. Indeed, if $a\,R\,b$

$$\exists\, k \in \mathbb{Z}:\; a = b+7k \Longrightarrow b = a-7k = a+7(-k) = a+7k',\; k' \in \mathbb{Z} \Longrightarrow bR\,a.$$

(iii) R is transitive. Indeed, if $a\,R\,b$ and $b\,R\,c$, then we have:

$$\begin{cases} \exists\, k_1 \in \mathbb{Z}:\; a = b+7k_1, \\ \exists\, k_2 \in \mathbb{Z}:\; b = c+7k_2 \end{cases} \Longrightarrow a = (c+7k_2)+7k_1 = c+7(k_1+k_2) = c+7k_3 \Longrightarrow a\,R\,c.$$

Finally, R is an equivalence relation on $\mathbb{Z}$.

■

7.3 Equivalence Classes

An equivalence relation R on a set A divides A into distinguished subsets called **equivalence classes**. We now define this important definition associated with equivalence relation.

Definition 7.3.1 Suppose R is a relation on a set A. Given any element $a \in A$, the equivalence class containing a is the subset of A consisting of those elements that are related to a. Thus the equivalence class containing a is the set denoted as $[a]$ is,

$$[a] := \{x \in A : x\, R\, a\}.$$

$[a]$ also called an `equivalence class of` a `modulo` R.
Any element in $[a]$ is called a representative of $[a]$. In addition, A `modulo` R $= A/R = \{[a] :\ a \in A\}$ is the set of all equivalence classes of A.

■ **Example 7.10** Let $A = \{1,2,3,4,5\}$ and $R = \{(a,b) : a,b \in A \text{ and } a+b \text{ is even}\}$. So, $R = \{(1,1),(1,3),(1,5),(2,2),(2,4),(3,1),(3,3),(3,5),(4,2),(4,4),(5,1),(5,3),(5,5)\}$. The equivalence class containing 1 is the set $[1] = \{x \in A : x\, R\, 1\}$. Thus $[1] = \{1,3,5\}$. In addition, note that $[1] = [3] = [5]$. Similarly, $[2] = [4] = \{2,4\}$. In addition, it is clear that

$$A/R = \{[x],\ x \in A\} = \{[1],[2],[3],[4],[5]\} = \{[1],[2]\}.$$

■

■ **Example 7.11** Let R be a relation on $\mathbb{Z}$ defined in Example 7.9. Determine $[0]$, $[1]$.
In Example 7.9, we already proved that R defined by $a\, R\, b \iff 7 \mid a-b$ is an equivalence relation on $\mathbb{Z}$.

$$[0] = \{x \in \mathbb{Z} :\ x\, R\, 0\} = \{x \in \mathbb{Z} :\ 7 \mid x\} = \{7k,\ k \in \mathbb{Z}\} = 7\mathbb{Z},$$

and

$$[1] = \{x \in \mathbb{Z} :\ x\, R\, 1\} = \{x \in \mathbb{Z} :\ 7 \mid (x-1)\} = \{7k+1,\ k \in \mathbb{Z}\}.$$

■

■ **Example 7.12** Let R be a relation on $\mathbb{Z}$ defined as $x\, R\, y$ if and only if $3x - 5y$ is even. Then R is an equivalence relation.
Indeed,

- R is **reflexive**.
  If $x\, R\, x$, then $3x - 5x$ is even. Since, $3x - 5x = -2x = 2(-x) = 2k,\ k \in \mathbb{Z}$ is even, so R is reflexive.
- R is **symmetric**.
  If $x\, R\, y$, then $3x - 5y$ is even. Therefore, for some $l \in \mathbb{Z}$, we have $3x - 5y = 2l$, so, $3x - 3y = 3(x-y) = 2(l+y)$ is even.
  Now,
  $3y - 5x = 3(y-x) - 2x = -3(x-y) - 2x = -\underbrace{[3(x-y)+2x]}_{even+even}$
  is also even; therefore, $y\, R\, x$. Hence, R is symmetric.
- R is **transitive**
  Let $x\, R\, y$ and $y\, R\, z$. Then, $\exists\ m,n \in \mathbb{Z}$ such that $3x - 5y = 2m$ and $3y - 5z = 2n$. By adding these two equations, we get $3x - 2y - 5z = 2(m+n)$. Thus $3x - 5z = 2(y+m+n)$ is even as $(m+n+y) \in \mathbb{Z}$. So, $x\, R\, z$, and this makes R a transitive relation.

Hence, R is an equivalence relation.
Let us now determine the equivalence classes for the equivalence relation just discussed. We begin with the equivalence class containing 0, say. Then

$$[0] = \{x \in \mathbb{Z} : x\, R\, 0\} = \{x \in \mathbb{Z} : 3x \text{ is even}\},$$
$$= \{x \in \mathbb{Z} : x \text{ is even}\} = \{0, \pm 2, \pm 4, ...\}.$$

Since $[0]$ class doesn't consist of all integers, we may conclude that there must be other different equivalence classes. This is due to the fact that distinct equivalence classes partition the set involved, in this case $\mathbb{Z}$. Since $1 \notin [0]$, we move on to the next equivalence class $[1]$:

$$[1] = \{x \in \mathbb{Z} : x\, R\, 1\} = \{x \in \mathbb{Z} : 3x - 5 \text{ is even}\},$$
$$= \{x \in \mathbb{Z} : x \text{ is odd}\} = \{\pm 1, \pm 3, \pm 5, ...\}.$$

■

Since every integer belongs to exactly $[0]$ or $[1]$, partitioning $\mathbb{Z}$ into two classes. Hence $[0]$ and $[1]$ are the only equivalence classes. In other words,

$$\mathbb{Z}/A \;=\; \{[x],\; x \in \mathbb{Z}\} \;=\; \{[0],\; [1]\}.$$

■ **Example 7.13** Let T be a relation defined on $\mathbb{R}$ such that

$$a\,T\,b \iff a^2 - a = b^2 - b.$$

(i) Show that T is an equivalence relation.
(ii) Determine the equivalence classes $[0]$, $[2]$.
(iii) Discuss $n([x])$, $x \in \mathbb{R}$.

The given relation T can be re-expressed as follows:

$$a\,T\,b \iff a^2 - b^2 = a - b \iff (a-b)(a+b-1) = 0 \iff (a = b) \text{ or } (a+b = 1).$$

(i) 1. Since $a^2 - a = a^2 - a$, $a\ T\ a$ and T is reflexive.
2. Let us consider $a,\ b \in \mathbb{R}$ such that

$$a\ T\ b \Longrightarrow (a = b \text{ or } a+b = 1) \Longrightarrow (b = a \text{ or } b+a = 1) \Longrightarrow b\ T\ a.$$

Thus T is symmetric. We could proceed as follows:

$$a\ T\ b \Longrightarrow a^2 - b^2 = a - b \Longrightarrow b^2 - a^2 = b - a \Longrightarrow b\ T\ a.$$

3. Let us consider $a,\ b,\ c \in \mathbb{R}$ such that

$$\begin{cases} a\,T\,b & \iff & a^2 - b^2 = a - b, \\ b\,T\,c & \iff & b^2 - c^2 = b - c. \end{cases} \quad \text{Adding equations gives } a^2 - c^2 = a - c \Longrightarrow a\,T\,c.$$

Thus T is transitive and finally, T is an equivalence relation.

(ii) Using the definition of equivalence classes, we determine $[0]$, $[2]$:

$$[0] \;=\; \{x \in \mathbb{R} :\, x\,T\,0\} \;=\; \{x \in \mathbb{R} :\, x^2 - x = 0^2 - 0\} \;=\; \{x \in \mathbb{R} :\, x(x-1) = 0\} \;=\; \{0,\ 1\}$$

and

$$\begin{aligned} [2] \;=\; \{x \in \mathbb{R} :\; x\,T\,2\} &= \{x \in \mathbb{R} :\; x^2 - x = 2^2 - 2 = 2\} \\ &= \{x \in \mathbb{R} :\; x^2 - x - 2 = 0\} \\ &= \{x \in \mathbb{R} :\; (x-2)(x+1) = 0\} \\ &= \{-1,\ 2\}. \end{aligned}$$

(iii) Discuss $n([x])$, $x \in \mathbb{R}$.

$$\begin{aligned}[x] = \{t \in \mathbb{R} : xTt\} &= \{t \in \mathbb{R} : x^2 - x = t^2 - t\} \\ &= \{t \in \mathbb{R} : x^2 - t^2 = x - t\} \\ &= \{t \in \mathbb{R} : (x-t)(x+t-1) = 0\} \\ &= \{x,\ 1-x\}.\end{aligned}$$

Thus we can conclude first that

$$\mathbb{R}/T = \{[x],\ x \in \mathbb{R}\} = \Big\{\{x,\ 1-x\},\ x \in \mathbb{R}\Big\},$$

and that

$$n([x]) = \begin{cases} 2 & \text{if} \quad x \neq 1-x, \\ 1 & \text{if} \quad x = 1-x \end{cases} = \begin{cases} 2 & \text{if} \quad x \neq 1/2, \\ 1 & \text{if} \quad x = 1/2. \end{cases}$$

■

Ⓡ It is clear that for any equivalence relation on a set A, all equivalence classes are subsets of A.

Theorem 7.3.1 Let R be a relation on E. Then we have for all $x, y \in E$

$$[x] = [y] \iff xRy.$$

Proof. Let us prove both implications separately.

(i) $[x] = [y] \implies xRy$.

$$\text{Let } t \in [x] = [y] \implies \begin{cases} tRx, \\ tRy \end{cases} \underbrace{\implies}_{\text{by symmetry}} \begin{cases} xRt, \\ tRy \end{cases} \underbrace{\implies}_{\text{by transitivity}} xRy.$$

(ii) $xRy \implies [x] = [y]$.

$$\text{Let } t \in [x] \implies tRx \underbrace{\overset{xRy}{\implies}}_{\text{by transitivity}} tRy \implies t \in [y].$$

Thus $[x] \subseteq [y]$.
Using the symmetry of R, we get:

$$t \in [y] \implies tRy \underbrace{\overset{yRx}{\implies}}_{\text{by transitivity}} tRx \implies t \in [x].$$

Thus $[y] \subseteq [x]$. Finally, $[x] = [y] \iff xRy$.

■

■ **Example 7.14** Let R be a relation on $\mathbb{Z}$ defined in Example 7.9.

- Given $p \in \mathbb{Z}$, show the existence of r, $r \in \{0,1,2,3,4,5,6\}$ such that $[p] = [r]$.
- Determine $\mathbb{Z}/R$.

■

In Example 7.9, we already proved that R defined by $a\,R\,b \iff 7 \Big| a - b$ is an equivalence relation on $\mathbb{Z}$:

- Using the Euclidean division by 7 of any $p \in \mathbb{Z}$, we ensure the existence of $m \in \mathbb{Z}$ and $r \in \{0,1,2,3,4,5,6\}$ such that $p = 7m + r$; thus we obtain:

$$p = 7m + r \Longrightarrow p - r = 7m \Longrightarrow 7 \big| (p - r) \Longrightarrow p\,R\,r \Longrightarrow [p] = [r] \text{ (Theorem 7.3.1).}$$

- We proved above that for any $p \in \mathbb{Z}$, there exists $r \in \{0,1,2,3,4,5,6\}$ such that $[p] = [r]$. Thus we conclude that in $\mathbb{Z}$, we have only 7 distinct equivalence classes, and then

$$\mathbb{Z}/R = \{[0], [1], [2], [3], [4], [5], [6]\}.$$

Corollary 7.3.2 Consider a set A with an equivalence relation R; then for any $x, y \in A$, we have:

$$x \not R\, y \iff [x] \cap [y] = \emptyset.$$

Proof. Let us prove the two implications by contraposition.

(i) $x \not R\, y \Longrightarrow [x] \cap [y] = \emptyset$.

$$\begin{aligned} [x] \cap [y] \neq \emptyset &\Longrightarrow \exists t \in [x] \text{ and } t \in [y], \\ &\Longrightarrow \exists t : tRx \text{ and } tRy, && \text{(by symmetry)} \\ &\Longrightarrow \exists t : xRt \text{ and } tRy, && \text{(by transitivity)} \\ &\Longrightarrow xRy, \end{aligned}$$

which completes the proof.

(ii) $[x] \cap [y] = \emptyset \Longrightarrow x \not R\, y$.

$$\begin{aligned} xRy &\Longrightarrow x \in [x] \text{ and } x \in [y] \\ &\Longrightarrow x \in [x] \cap [y] \\ &\Longrightarrow [x] \cap [y] \neq \emptyset, \end{aligned}$$

which completes the proof.

■

Note that relations are sets (of ordered pairs), so the properties of sets also apply to relations. Let A be a set and R and S be relations on A:

- R and S are equal if for every $a,\ b \in A$, $a\,R\,b$ iff $a\,S\,b$.
- R is a subset of S if for every $a,\ b \in A$, $a\,R\,b$ implies $a\,S\,b$.

7.4 Relations Between Sets

Definition 7.4.1 A relation R from a set A to a set B is a subset of the Cartesian product $A \times B$ of A and B. We say the relation R is a set of ordered pairs (a,b), where the first coordinate of the pair belongs to A and the second belongs to B. That is, whenever $(a,b) \in R$, we say that a is related to b by R and write $a\,R\,b$. If the ordered pair $(a,b) \notin R$, then a is not related to b by R, and we write $a \not R\, b$.

In general, by a relation R from a set A to a set B, we simply mean some subset of the Cartesian product $A \times B$ of A and B.

■ **Example 7.15** Suppose $A = \{1,2,3\}$ and $B = \{a,b\}$ are two sets and $R = \{(1,b),(2,a),(3,b)\}$ is a subset of $A \times B$. ■

Theorem 7.4.1 The number of relations from set A to set B is given by

$$2^{n(A)\times n(B)}.$$

Proof. Since any relation R from A to B is considered a subset of $A\times B$, $R\in\mathscr{P}(A\times B)$, and therefore, the number of relations is the number of elements of the power set of $A\times B$. ■

Proposition 7.4.2 Let $n\in\mathbb{N}$. The relation $\equiv (mod\ n)$ on $\mathbb{Z}$ is an equivalence relation.

Proof. To show congruence modulo n, $\equiv (mod\ n)$ is an equivalence relation, we verify that $\equiv (mod\ n)$ is reflexive, symmetric and transitive on $\mathbb{Z}$:

1. **Reflexive**: Let $x\in\mathbb{Z}$. We observe that $n|0$, so $n|(x-x)$. By the definition of congruence modulo n, we have $x\equiv x\ (mod\ n)$. Thus $x\equiv x\ (mod\ n), \forall x\in\mathbb{Z}$. This shows that the relation $\equiv (mod\ n)$ is reflexive.
2. **Symmetric**: For $x,y\in\mathbb{Z}$, let $x\equiv y\ (mod\ n)$. Then $n|(x-y)$ by the definition of congruence modulo n that is, for some $a\in\mathbb{Z}$, such that $x-y=na$. Multiplying by -1, we have $y-x=n(-a)$. This shows that $n|(y-x)$, so $y\equiv x\ (mod\ n)$. Thus the relation $\equiv (mod\ n)$ is symmetric.
3. **Transitive**: Let $x,y,z\in\mathbb{Z}$. Suppose $x\equiv y\ (mod\ n)$ and $y\equiv z\ (mod\ n)$. Then by definition of congruence modulo n, $n|(x-y)$ and $n|(y-z)$. Therefore, there exist some $a,b\in\mathbb{Z}$ such that $x-y=na$ and $y-z=nb$. By adding these equations, we obtain $x-z=n(a+b)$. Consequently, $n|(x-z)$, and we have, $x\equiv z\ (mod\ n)$. Hence, we have shown that $x\equiv y\ (mod\ n)$ and $y\equiv z\ (mod\ n)$ implies that $x\equiv z\ (mod\ n)$. Therefore, $\equiv (mod\ n)$ is transitive.

This completes the proof. ■

■ **Example 7.16** Consider the equivalence relation $\equiv (mod\ 3)$ and let us determine its equivalence classes set $\mathbb{Z}$ modulo $\equiv$.

$$[0]=\{x\in\mathbb{Z}:\ x\equiv 0(mod\ 3)\} = \{x=3k,\ k\in\mathbb{Z}\} = 3\mathbb{Z},$$
$$[1]=\{x\in\mathbb{Z}:\ x\equiv 1(mod\ 3)\} = \{x=3k+1,\ k\in\mathbb{Z}\},$$
$$[2]=\{x\in\mathbb{Z}:\ x\equiv 2(mod\ 3)\} = \{x=3k+2,\ k\in\mathbb{Z}\},$$
$$[3]=\{x\in\mathbb{Z}:\ x\equiv 3(mod\ 3)\} = \{x=3k',\ k'\in\mathbb{Z}\} = [0],$$
$$[4]=\{x\in\mathbb{Z}:\ x\equiv 4(mod\ 3)\ \} = \{x=3k+4,\ k\in\mathbb{Z}\}=\{x=3k'+1,\ k'\in\mathbb{Z}\}=[1],$$
$$[5]=\{x\in\mathbb{Z}:\ x\equiv 5(mod\ 3)\ \} = \{x=3k+5,\ k\in\mathbb{Z}\}=\{x=3k'+2,\ k'\in\mathbb{Z}\}=[2].$$

Thus it is clear that

$$\mathbb{Z}/\equiv\ =\ \mathbb{Z}_3\ =\ \Big\{[0],\ [1],\ [2]\Big\}.$$

■

The next theorem gives a general result concerning the $\mathbb{Z}$ modulo $\equiv (mod\ m)$.

Theorem 7.4.3 Let $m\in\mathbb{N}$. The equivalence classes of the relation $\equiv (mod\ m)$ on $\mathbb{Z}$ are given by

$$\mathbb{Z}/R\ \equiv\ \mathbb{Z}_m\ =\ \Big\{[0],\ [1],\ [2],\ldots,[m-1]\Big\}.$$

Proof. See Question 7.14. ■

7.5 Chapter 7 Practice Problems

Question 1 Let $A = \{-1,2,3\}$ and $B = \{a,b\}$. Define all the relations between sets A and B.

Question 2 Find the domain and range of the relations defined in Question 1.

Question 3 Let $E = \{a, b\}$. Determine all the relations on E.

Question 4 Let $A = \{1,2,3,4,5,6,10,14,20,26\}$ and $R = \{(a,b) : a, b \in A \text{ and } a+b \text{ is divisible by } 3\}$ be a relation on A. Is R an equivalence relation?

Question 5 Let T be a relation defined on $\mathbb{R}$ such that

$$aTb \iff \cos^2(a) + \sin^2(b) = 1.$$

(i) Show that T is an equivalence relation.
(ii) Determine $[0]$, $[\pi/2]$.

Question 6 Given a set E, let A be a subset of E. We define a relation R on $\mathscr{P}(E)$ as follows

$$XRY \iff X \cup A = Y \cup A.$$

(a) Show that R is an equivalence relation.
(b) Determine $[\emptyset]$, $[E]$, $[A]$, and $[\overline{A}]$.

Question 7 Given a set E, let A and B be two subsets of E. We define a relation R on $\mathscr{P}(E)$ as follows:

$$ARB \iff A = \overline{B}.$$

Is R an equivalence relation on $\mathscr{P}(E)$?

Question 8 Let R be the relation defined on the set of real numbers by:

$$xRy \iff 2x + 3y \in \mathbb{Q}.$$

(i) Show that R is an equivalence relation.
(ii) Determine $[2]$, $[-1/3]$, $\sqrt{2}$.

Question 9 Let R be the relation defined on the $\mathbb{R} \times \mathbb{R}$ by

$$xRy \iff |x+2| = |y+2|.$$

(i) Show that R is an equivalence relation.
(ii) Discuss $n([x])$ when $x \in \mathbb{R}$.

Question 10 Let R be the relation defined on the $\mathbb{R} \times \mathbb{R}$ by:

$$(x,y)R(z,t) \iff x+z \geq y+t.$$

(i) Is R reflexive?
(ii) Is R symmetric?
(iii) Is R transitive?

Question 11 Let R be the relation defined on the $\mathbb{R} \times \mathbb{R}$ by

$$xRy \iff x^3 - y^3 = 3(x-y).$$

(i) Show that R is an equivalence relation.
(ii) Discuss $n([x])$ when $x \in \mathbb{R}$.

Question 12 Let R be the relation defined on the $\mathbb{R} \times \mathbb{R}$ by:

$$xRy \Longleftrightarrow x\,e^{y} = y\,e^{x}.$$

(i) Show that R is an equivalence relation.
(ii) Discuss $n([x])$ when $x \in \mathbb{R}$.

Question 13 Prove or disprove the following statement:
If R and S are two equivalence relations on a set A, then $R \cup S$ is also an equivalence relation on A.

Question 14 Prove or disprove. If $[a],\ [b]\ \in \mathbb{Z}_n$ then, $[a]+[b]=[a+b]$ and $[a][b]=[ab]$

7.6 Chapter 7 Additional Practice Problems with Solutions

Question 1 Let $A=\{1,2,3,4\}$ and $B=\{a,b,c\}$. Define the number of relations between sets A and B.

Solution. The number of all relations between A and B represents the number of subsets of the Cartesian product $A \times B$. Here, this number is equal to $2^{n(A\times B)} = 2^{n(A)\times n(B)} = 2^{4\times 3} = 2^{12}$.

Question 2 Let $A=\{1,2\}$. Determine all the relations on A.

Solution. There are $2^4=16$. It suffices to determine all subsets of $A\times A$. In other words,
$\{(1,1)\};\ \{(1,2)\};\ \{(2,1)\};\ \{(2,2)\};$
$\{(1,1);(1,2)\};\ \{(1,1);(2,1)\};\{(1,1);(2,2)\};\ \{(1,2);(2,1)\};\ \{(1,2);(2,2)\};\ \{(2,1);(2,2)\}$
$\{(1,1);(1,2);(2,1)\};\ \{(1,1);(1,2);(2,2)\};\ \{(1,1);(2,1);(2,2)\};\{(1,2);(2,1);(2,2)\};$
$\{(1,1);(1,2);(2,1);(2,2)\},\ \emptyset.$
All subsets of $A\times A$ are well-defined relations.

Question 3 Let $A=\{1,2,3,4,5\}$ and $R=\{(a,b) : a,b\in A \text{ and } a+b \text{ is odd}\}$ be a relation on A. Is R an equivalence relation?

Solution. Since $(1,1)\notin R=\{(1,2);(1,4);(2,1);(2,3);(2,5);(3,2);(3,4);(4,1);(4,3);(4,5)\}$, R is not reflexive and not an equivalence relation.

Question 4 Given a set E, let A be a subset of E. We define a relation R on $\mathscr{P}(E)$ as follows:

$$XRY \Longleftrightarrow X\cap A = Y\cap A.$$

(a) Show that R is an equivalence relation.
(b) Determine $[\emptyset]$, $[E]$, $[A]$, and $[\overline{A}]$

Solution. Let X,Y, and Z be subsets of E.

(a) R is an equivalence relation.

1. reflexive.
   Since $X\cap A = X\cap A$ is true, then $X\ R\ X$. Thus R is reflexive.

2. symmetric.
   Let $X\ R\ Y$. Then $X\cap A = Y\cap A$ which is equivalent to $Y\cap A = X\cap A$, then $Y\ R\ X$. Thus R is symmetric.

3. transitive.
   Let $\begin{cases} X\,R\,Y, \\ Y\,R\,Z \end{cases} \Longrightarrow \begin{cases} X\cap A = Y\cap A, \\ Y\cap A = Z\cap A, \end{cases} \Longrightarrow X\cap A = Z\cap A, \Longrightarrow X\ R\ Z.$

Thus R is transitive.

Finally, R is an equivalence relation on $\mathscr{P}(E)$.

(b) Determine $[\emptyset]$, $[E]$, $[A]$, and $[\overline{A}]$.

1. $[\emptyset] = \{X \subseteq E : X\,R\,\emptyset\} = \{X \subseteq E : X \cap A = \emptyset \cap A = \emptyset\}$.
   $X \cap A = \emptyset$ implies that $X \in \mathscr{P}(\overline{A})$. Thus $[\emptyset] = \mathscr{P}(\overline{A})$.
2. $[E] = \{X \subseteq E : X\,R\,E\} = \{X \subseteq E : X \cap A = E \cap A = A\} = \{X \subseteq E : X \supseteq A\}$.
3. $[A] = \{X \subseteq E : X\,R\,A\} = \{X \subseteq E : X \cap A = A \cap A = A\} = [E]$.
4. $[\overline{A}] = \{X \subseteq E : X\,R\,\overline{A}\} = \{X \subseteq E : X \cap A = \overline{A} \cap A = \emptyset\} = [\emptyset] = \mathscr{P}(\overline{A})$.

Question 5 Given a set E and let A and B be two subsets of E. We define a relation R on $\mathscr{P}(E)$ as follows:

$$A R B \iff A = B \text{ or } A = \overline{B}.$$

(a) Show that R is an equivalence relation.
(b) Determine $[\emptyset]$, $[E]$, $[A]$, and $[\overline{A}]$.

Solution. Let A, B, and C be subsets of E.

(a) R is an equivalence relation.

1. reflexive.
   Since $A = A$, $A\,R\,A$. Thus R is reflexive.

2. symmetric.
   Let $A\,R\,B$. Then $A = B$ or $A = \overline{B}$. If $A = B$, then $B = A$ and if $A = \overline{B}$, then $B = \overline{\overline{B}} = \overline{A}$ which is equivalent to $B\,R\,A$. Thus R is symmetric.

3. transitive.
   Let
   $$\begin{cases} A\,R\,B, \\ B\,R\,C, \end{cases} \implies \begin{cases} A = B \text{ or } A = \overline{B}, \\ B = C \text{ or } B = \overline{C}. \end{cases}.$$
   We will consider the four possible combinations:
   (i) $A = B$ and $B = C$, so $A = C$.
   (ii) $A = B$ and $B = \overline{C}$, so $A = \overline{C}$.
   (iii) $A = \overline{B}$ and $B = C$, so $A = \overline{B} = \overline{C}$
   (iv) $A = \overline{B}$ and $B = \overline{C}$, so $A = \overline{\overline{C}} = C$.
   Thus in all cases we obtain either $A = C$ or $A = \overline{C}$, which imply $A\,R\,C$. Thus R is transitive.

Finally, R is an equivalence relation on $\mathscr{P}(E)$.

(b)
1. $[\emptyset] = \{X \subseteq E : X\,R\,\emptyset\} = \{X \subseteq E : X = \emptyset \text{ or } X = E\} = \{\emptyset; E\}$.
2. $[E] = \{X \subseteq E : X\,R\,E\} = \{X \subseteq E : X = E \text{ or } X = \emptyset\} = \{\emptyset; E\} = [\emptyset]$.
3. $[A] = \{X \subseteq E : X\,R\,A\} = \{X \subseteq E : X = A \text{ or } X = \overline{A}\} = \{A; \overline{A}\}$.
4. $[\overline{A}] = \{X \subseteq E : X\,R\,\overline{A}\} = \{X \subseteq E : X = \overline{A} \text{ or } X = A\} = [A]$.

Question 6 Let R be the relation defined on the set of real numbers by:

$$xRy \iff x - y \in \mathbb{Q}.$$

(i) Show that R is an equivalence relation.
(ii) Determine $[1 - 2\sqrt{2}]$ and $[-1]$.

Solution. Let x, y, and z be three real numbers.

(i) R is an equivalence relation.

1. reflexive.
   Since $x - x = 0 \in \mathbb{Q}$, $x\,R\,x$. Thus R is reflexive.

2. symmetric.
   Let $x\,R\,y$. Then $x - y \in \mathbb{Q} \Longrightarrow y - x = -(x-y) \in \mathbb{Q} \Longrightarrow y\,R\,x$. Thus R is symmetric.

3. transitive.
   Let $\begin{cases} x\,R\,y, \\ y\,R\,z \end{cases} \Longrightarrow \begin{cases} x - y \in \mathbb{Q}, \\ y - z \in \mathbb{Q} \end{cases} \Longrightarrow (x-y)+(y-z) = x - z \in \mathbb{Q} \Longrightarrow$ $x\,R\,z$.

Thus R is transitive, and finally, R is an equivalence relation on $\mathbb{R}$.

(ii) Let us determine the following equivalence classes:

$$\begin{aligned} [1-2\sqrt{2}] &= \{x \in \mathbb{R} : x\,R\,(1-2\sqrt{2})\} \\ &= \{x \in \mathbb{R} : x - 1 + 2\sqrt{2} \in \mathbb{Q}\} \\ &= \{-2\sqrt{2} + r,\ r \in \mathbb{Q}\}, \end{aligned}$$

and $[-1] = \{x \in \mathbb{R} : x\,R\,-1\} = \{x \in \mathbb{R} : x + 1 \in \mathbb{Q}\} = \mathbb{Q}$.

Question 7 Let R be the relation defined on the set of natural numbers by:

$$xRy \Longleftrightarrow \exists\, p, q \in \mathbb{N} : y = px^q.$$

Determine whether R is an equivalence relation.

Solution. R is not symmetric; indeed, take $1\,R\,2$ but 2 is not in relation with 1.

$$1\,R\,2 \Longleftrightarrow \exists\, p = 2, q = 1 : 2 = 2\,(1^1) \text{ but } 1 = p\,2^q \geq 2p \geq 2 > 1$$

Thus R cannot be an equivalence relation.

Question 8 Let S be a relation defined on $\mathbb{R} \times \mathbb{R}$ such that

$$(a,b)\,S\,(a',b') \Longleftrightarrow \exists \alpha,\ \beta > 0 : a' = \alpha\, a, \text{ and } b' = \beta\, b.$$

(i) Show that S is an equivalence relation.
(ii) Determine $[(1,1)]$, $[(1,0)]$, and$[(0,-1)]$.
(iii) Discuss $n(\mathbb{R} \times \mathbb{R} \mid S)$; in other words determine all the equivalence classes of S.

Solution.

(i) Let x, y, z, t be real numbers.

1. S is reflexive it suffices to take $\alpha = \beta = 1 > 0 : x = 1x,\ \ y = 1y$, thus $(x,y)\,S\,(x,y)$.

2. is S symmetric?

$$\begin{aligned} (x,y)\ S\ (z,t) &\Longrightarrow \exists \alpha,\ \beta > 0 : z = \alpha x,\ \ t = \beta y \\ &\Longrightarrow x = \frac{1}{\alpha} z,\ \ y = \frac{1}{\beta} t \\ &\Longrightarrow \exists \alpha' = \frac{1}{\alpha},\ \beta' = \frac{1}{\beta} > 0 : x = \alpha' z,\ \ y = \beta' t, \\ &\Longrightarrow (z,t)\ S\ (x,y). \end{aligned}$$

Thus S is Symmetric.

3. is S transitive?

$$\begin{cases} (x,y)\; S\; (z,t), \\ (z,t)\; S\; (u,v) \end{cases} \Longrightarrow \begin{cases} \exists \alpha,\ \beta > 0:\ z = \alpha x,\ \ t = \beta y, \\ \exists \alpha',\ \beta' > 0:\ u = \alpha' z,\ \ v = \beta' t, \end{cases}$$

Then, $\exists \alpha",\ \beta" > 0:\ u = \alpha' \alpha x = \alpha" x,\ \ v = \beta' \beta y = \beta" y \Longrightarrow (x,y)\; S\; (u,v)$. Thus S is transitive, and finally, S is an equivalence relation.

(ii)

$$[(1,1)] = \{(x,y) \in \mathbb{R}\times\mathbb{R} :\ \exists \alpha,\ \beta > 0:\ x = \alpha \times 1,\ \ y = \beta \times 1\} = \{(x,y) \in \mathbb{R}\times\mathbb{R} :\ x = \alpha,\ y = \beta\}$$

Thus

$$[(1,1)] = (0,\ \infty) \times (0,\ \infty).$$

$$[(1,0)] = \{(x,y) \in \mathbb{R}\times\mathbb{R} :\ \exists \alpha,\ \beta > 0:\ x = \alpha \times 1,\ \ y = \beta \times 0\} = \{(x,0) \in \mathbb{R}\times\mathbb{R} :\ x > 0\}.$$

Thus

$$[(1,0)] = (0,\ \infty) \times \{0\}.$$

$$[(0,-1)] = \{(x,y) \in \mathbb{R}\times\mathbb{R} :\ \exists \alpha,\ \beta > 0:\ x = \alpha \times 0,\ \ y = \beta \times -1\} = \{(0,y) \in \mathbb{R}\times\mathbb{R} :\ y < 0\}.$$

Thus

$$[(0,-1)] = \{0\} \times (-\infty,\ 0).$$

(iii) We suggest in addition to (ii) to determine $[(-1,1)]$, $[(1,-1)]$, $[(-1,-1)]$ and then $[(0,1)]$, $[(-1,0)]$, $[(0,0)]$.
Directly, we observe[1] that the following family provides a partition to $\mathbb{R}\times\mathbb{R}$:

$$(-\infty,\ 0) \times (-\infty,\ 0);\ (-\infty,\ 0) \times \{0\};\ (0,\ \infty) \times (0,\ \infty);\ \{0\} \times (0,\ \infty);\ \{0\} \times (-\infty,\ 0);$$

$$(-\infty,\ 0) \times (0,\ \infty);\ (0,\ \infty) \times (-\infty,\ 0);\ (0,\ \infty) \times \{0\};\ \{(0,0)\}.$$

Obviously, these nine subsets form a partition of the plane $\mathbb{R}\times\mathbb{R}$:

$$n(\mathbb{R}\times\mathbb{R} \mid S) = 9.$$

Question 9 Let R be the relation defined on the $\mathbb{R}\times\mathbb{R}$ by:

$$(x,y)\, R\, (z,t) \Longleftrightarrow x + z \leq y + t.$$

(i) Is R reflexive?
(ii) Is R symmetric?
(iii) Is R transitive?

Solution. Let x, y, z, t be real numbers.
(i) Is R reflexive?

$$(x,y)\; R\; (x,y) \Longleftrightarrow x + x \leq y + y \Longleftrightarrow 2x \leq 2y \Longleftrightarrow x \leq y,$$

which is not always true. For example, $(4,3)\; R\; (4,3) \Longleftrightarrow 8 \leq 6$.

[1] $[(0,0)] = \{(0,0)\}$.

(ii) Is R symmetric?

$$(x,y)\ R\ (z,t), \Longleftrightarrow x+z \le y+t, \Longleftrightarrow z+x \le t+y \Longleftrightarrow (z,t)\ R\ (x,y).$$

Thus R is symmetric.

(iii) Is R transitive?

$$\begin{cases} (x,y)\ R\ (z,t), \\ (z,t)\ R\ (u,v) \end{cases} \Longleftrightarrow \begin{cases} x+z \le y+t, \\ z+u \le t+v \end{cases} \Longleftrightarrow x+2z+u \le y+2t+v,$$

which is not always true. Indeed, take

$$\begin{cases} (2,3)\ R\ (1,3), \\ (1,3)\ R\ (4.5,3), \end{cases} \text{ but } (2,3)\ R\ (4.5,3) \text{ is wrong. Thus } R \text{ is not transitive.}$$

Question 10 Let R be a relation on $\mathbb{Q}$ defined as $\frac{a}{b}\ R\ \frac{x}{y}$ if and only if $ay = bx$. Show that R is an equivalence relation and then determine $[1/1]$, $[1/2]$, $[2/3]$, $[4/7]$, and $[123/19]$.

Solution. It is direct to prove that R is reflexive and symmetric. Let us prove that R is transitive. Indeed, let x/y, z/t, and $u/v \in \mathbb{Q}$ such that:

$$\begin{cases} \frac{x}{y}\ R\ \frac{z}{t}, \\ \\ \frac{z}{t}\ R\ \frac{u}{v} \end{cases} \Longrightarrow \begin{cases} xt = yz, \\ \\ zv = ut \end{cases} \Longrightarrow \begin{cases} (xt)\,v = (yz)\ v\ = y\ (zv)\ =\ y(ut), \\ \\ zv = ut \end{cases} \Longrightarrow xv\ =\ yu \Longrightarrow \frac{x}{y}\ R\ \frac{u}{v}.$$

The equivalence class of a/b, where $ab \neq 0$ is given by

$$\begin{aligned} [a/b] &= \{x/y,\ y \neq 0, x, y \in \mathbb{Z} : a/b\ R\ x/y\}, \\ &= \{x/y,\ y \neq 0,\ x, y \in \mathbb{Z} : ay = bx\}, \\ &= \{x/y,\ y \neq 0,\ x, y \in \mathbb{Z} : a/b = x/y\}, \\ &= \{x/y,\ y \neq 0,\ x, y \in \mathbb{Z} : y/b = x/a\ = q,\ q \in \mathbb{Q}\}, \\ &= \left\{\frac{x}{y} :\ x = aq,\ y = bq,\ \ q \in \mathbb{Q}\right\}, \end{aligned}$$

and $[0] = \{0\}$. Thus our requested equivalence classes are:

$$\begin{aligned} [1/1] &= \left\{\frac{x}{y} :\ x = q,\ y = q,\ \ q \in \mathbb{Q}\right\} = \{1\}, \\ \left[\frac{1}{2}\right] &= \left\{\frac{x}{y} :\ x = q,\ y = 2q,\ \ q \in \mathbb{Q}\right\}, \\ \left[\frac{2}{3}\right] &= \left\{\frac{x}{y} :\ x = 2q,\ y = 3q,\ \ q \in \mathbb{Q}\right\}, \\ \left[\frac{4}{7}\right] &= \left\{\frac{x}{y} :\ x = 4q,\ y = 7q,\ \ q \in \mathbb{Q}\right\}, \\ \left[\frac{123}{19}\right] &= \left\{\frac{x}{y} :\ x = 123q,\ y = 19q,\ \ q \in \mathbb{Q}\right\}. \end{aligned}$$

R In the above equivalence classes, for instance in

$$\begin{bmatrix}2\\3\end{bmatrix} = \left\{\frac{x}{y} : x = 2q,\ y = 3q,\ q \in \mathbb{Q}\right\},$$

one may observe that we can assume $q \in \mathbb{Z}$ is sufficient. Indeed, if we consider $q = 5/7$,

$$\frac{x}{y} \in \begin{bmatrix}2\\3\end{bmatrix} \Longrightarrow \frac{x}{y} = \frac{2q}{3q} = \frac{2(5/7)}{3(5/7)} = \frac{10}{15} = \frac{2\times 5}{3\times 5}.$$

Thus we could replace $q \in \mathbb{Q}$ in these answers by $q \in \mathbb{Z}$.

Question 11 Prove or disprove the following statement: If a relation is symmetric and transitive, then it is also reflexive.

Solution. Let us consider the following counter-example: Let $E = \{1;2;3\}$ and R be a relation defined on E by $R = \{(1,1);(1,3);(3,1);(3,3)\}$.
It is clear that R is symmetric and transitive, but $(2,2) \notin R$. Thus R is not reflexive on E.

Question 12 Let $A = \{1,2,3,4\}$ and $B = \{a,b,c\}$. Define all the relations between sets A and B.

Solution. The number of all relations between sets A and B is given by $2^{4\times 3} = 2^{12}$.
The relations between sets A and B are all elements of $\mathscr{P}(A\times B)$.

Question 13 Describe the partition of $\mathbb{Z}$ resulting from the equivalence relation $(mod\ 5)$.

Solution. $\mathbb{Z} = [0]\cup[1]\cup[2]\cup[3]\cup[4]$, where $[r] = \{5k + r,\ k \in \mathbb{Z}\}$.

Question 14 Describe the addition and multiplication table for $\mathbb{Z}_6$.

Solution.

+	0	1	2	3	4	5
0	0	1	2	3	4	5
1	1	2	3	4	5	0
2	2	3	4	5	0	1
3	3	4	5	0	1	2
4	4	5	0	1	2	3
5	5	0	1	2	3	4

x	0	1	2	3	4	5
0	0	0	0	0	0	0
1	0	1	2	3	4	5
2	0	2	4	0	2	4
3	0	3	0	3	0	3
4	0	4	2	0	4	2
5	0	5	4	3	2	1

8. Functions

Chapter Key Terms

(a) *Functions*
(b) *Injective, surjective, and bijective Functions*
(c) *Composition of functions*
(d) *Inverse function*
(e) *Images and pre-images*
(f) *Set of Functions*

8.1 Functions

Let A and B be nonempty sets. A **function** f from A to B, written as $f : A \longrightarrow B$, is a relation from A to B, with the property that every element a of set A is associated with an (unique) element of set B. That is, a is the first coordinate of exactly one ordered pair in subset f of $A \times B$. In this case, the set A is the **domain** of f, denoted by $dom(f)$, and the set B is called the **co-domain** of f, denoted by $codom(f)$.
Let f be a function from A to B and $(a,b) \in f$. Then the ordered pair (a,b) contains a unique value b for each first coordinate a. For instance, if (a,b) and $(a,c) \in f$, then $b = c$. For an ordered pair $(a,b) \in f$, we call b as the **image** of a under the function f, written as $b = f(a)$, where a is called the **pre-image or antecedent** of b. The function f is sometimes called a mapping from A to B. The set of all those elements that appear as second coordinate of an ordered pair contained in f is called the **range** or **image**, denoted by $range(f)$ or $Im(f)$. That is, $range(f) = \{b \in B : (a,b) \in f\} = Im(f)$, or $range(f) = \{f(a) \in B : a \text{ in } A\} = Im(f)$.

■ **Example 8.1** A well-known function is $f(x) = x^2$ from $\mathbb{R}$ to $\mathbb{R}$. We note that its graph is the set of points $f = \{(x,x^2) : x \in \mathbb{R}\} \subseteq \mathbb{R} \times \mathbb{R}$. ■

■ **Example 8.2** Consider the function $f = \{(n, f(n)) : f(n) = |n| + 1, \text{ for all } n \in \mathbb{Z}\}$. We regard f as the set of points in its graph, that is $f = \{(n, |n| + 1) : n \in \mathbb{Z}\}$ is a subset of a $\mathbb{Z} \times \mathbb{N}$. The function

f indicates that for any given integer $n \in \mathbb{Z}$, the ordered pair $(n, |n|+1) \in f$ contains exactly one natural number: $|n|+1$ as second coordinate associated with n. For $2 \in \mathbb{Z}$, we have $(2,3) \in f$ and for $-2 \in \mathbb{Z}$, we have $(-2,3) \in f$. Note that the domain of f is $\mathbb{Z}$, and its codomain is $\mathbb{N}$. The range of f $=$ $range(f) = \{f(n) = |n|+1 : n \in \mathbb{Z}\} = \{1,2,3,4,...\}$. As noted earlier, the range is a subset of the codomain, but it does not equal the codomain set in general. In this example, we have the codomain of f equals its range. ∎

(R) A function is a specialized form of a relation. Indeed, every function is a relation but the converse is not true.

Definition 8.1.1 — Equal Functions. Let $f : A \longrightarrow B$ and $g : C \longrightarrow D$ be two functions. The functions f and g are equal if $A = C$, $B = D$, and $f(a) = g(a)$, $\forall a \in A$.

∎ **Example 8.3** Let $A = \{a,b,c\}$ and $B = \{1,2,3\}$. Let $f, g : A \longrightarrow B$ be two functions defined as $f = \{(a,1),(b,2),(c,1)\}$ and $g = \{(c,1),(b,2),(a,1)\}$, are equal functions. ∎

8.2 Injective, Surjective, and Bijective Functions

Definition 8.2.1 — injective function. The function $f : E \to F$ is said to be **injective** if different elements of E are mapped into different elements of F. The injective function is also called **one-to-one** .

Another way to characterize injectivity is through the use of mathematical expressions.

The function $f : E \to F$ is an injective function if

$$\forall\, x,\, x' \in E :\ f(x) = f(x') \implies x = x'.$$

Equivalently, by contraposition,

$$\forall\, x,\, x' \in E :\ x \neq x' \implies f(x) \neq f(x').$$

∎ **Example 8.4** Let $f : \mathbb{R} \to \mathbb{R}$ be a function defined as $f(x) = x^2$.

Thus f is not an injective function on $\mathbb{R}$. Indeed, it is sufficient to consider $f(-2) = f(2)$ but $-2 \neq 2$. ∎

∎ **Example 8.5** The function $g(x) = x^2$ is injective on $(-\infty,\, 0]$.

Indeed, $x,\, x' \leq 0 :\ g(x) = g(x') \implies x^2 = x'^2 \implies x = \pm x' \implies x = x'$. ∎

(R) If f is injective (one-to-one), then any horizontal line meets its graph in exactly one point.

■ **Example 8.6** The function $h(x) = x^2 - x + 2$ is not injective on $(0, \infty)$. ■

Consider $x, x' > 0$ such that $h(x) = h(x')$. Then

$$\begin{aligned} x^2 - x + 2 = x'^2 - x' + 2 &\implies x^2 - x = x'^2 - x', \\ &\implies (x^2 - x'^2) - (x - x') = 0, \\ &\implies (x - x')(x + x' - 1) = 0, \\ &\implies x = x' \text{ or } x = 1 - x'. \end{aligned}$$

Thus we cannot ensure that $x = x'$ whenever $h(x) = h(x')$. Therefore, h is not an injective function. Alternatively, you may choose to proceed in a different way based on the following results about polynomial injectivity.

Proposition 8.2.1 — **Injectivity of Quadratic Polynomials.** The quadratic polynomial function $P(x) = ax^2 + bx + c$, $a \neq 0$ is not injective on $\mathbb{R}$. Furthermore, the polynomial $P(x)$ is injective either on the left or on the right of the symmetry axis
$x = \alpha = \dfrac{-b}{2a}$ of the parabola.

Proof. Consider the following two cases:

$\boxed{Case(1) : a < 0}$

x	$-\infty$		α		$+\infty$
$P(x)$	$-\infty$	↗	β	↘	$-\infty$

$\boxed{Case(2) : a > 0}$

x	$+\infty$		α		$-\infty$
$P(x)$	$+\infty$	↘	β	↗	$+\infty$

Variations of the parabola, as shown in the tables, show clearly that P is not injective on $\mathbb{R}$, since there exist x_1 and x_2 such that $x_1 < \alpha < x_2$ and $P(x_1) = P(x_2)$. Whereas on $(-\infty, \alpha]$ or on $[\alpha, \infty)$, P is injective.

Theorem 8.2.2 A monotone function $f : [a, b] \to \mathbb{R}$ is injective.

Proof. Let us assume f is increasing on $[a, b]$ and for $x, x' \in [a, b]$, we have:

$$x < x' \implies f(x) < f(x').$$

Similarly,

$$x > x' \implies f(x) > f(x').$$

Therefore, we have

$$x \neq x' \implies f(x) \neq f(x') \underbrace{\iff}_{contraposition} f(x) = f(x') \implies x = x'.$$

Thus f is injective. Similarly, we prove the case where f is decreasing.

Theorem 8.2.3 If f is a continuous and injective function, then f is strictly monotone.

Proof. See Question 21.

8.2.1 Surjective Functions

Definition 8.2.2 — surjective function. The function $f: E \to F$ is said to be a **surjective** (**Onto**) function if every element of F has a pre-image in E. That is,

$$\forall\, y \in F, \;\; \exists\, x \in E: \; y = f(x).$$

■ **Example 8.7** The function $f(x) = x^2$ from $\mathbb{R} \to \mathbb{R}$ is not surjective on $\mathbb{R}$.
It is sufficient to consider $y = -4 < 0$; it can never be a square of any real number x. ■

■ **Example 8.8** The function $g(x) = x^2$ from $\mathbb{R} \to [0, \infty)$ is surjective.
That is,

$$\forall\, y \geq 0, \; \exists\, x = \pm\sqrt{y} \in \mathbb{R}: \; y = (\sqrt{y})^2 = f(x).$$

■

(R) Notice that x is not unique in the preceding example: $x = \pm\sqrt{y} \in \mathbb{R}$.

■ **Example 8.9** Let $h: E = \{a,b,c,d\} \to F = \{1,2,3\}$ be a function defined as

$$h = \{(a,1),\; (b,2),\; (c,1),\; (d,3)\}.$$

We observe that h is a surjective function, and $Im(h) = \{1,2,3\} = F$, and $n(F) = 3 < 4 = n(E)$. ■

■ **Example 8.10** The function $f: \mathbb{Z} \to \mathbb{Z}$ is surjective:

$$f(x) = \begin{cases} \dfrac{x}{2} & \text{if } \; x \text{ is even,} \\[2ex] \dfrac{1-x}{2} & \text{if } \; x \text{ is odd.} \end{cases}$$

■

If x is even or odd, we observe that for

$$y \in \mathbb{Z}, \;\; \exists\, x = 2y \text{ is even}: \;\; f(x) = f(2y) = \frac{2y}{2} = y,$$

and,

$$y \in \mathbb{Z}, \;\; \exists x = 1 - 2y \text{ is odd}: \;\; f(x) = f(1-2y) = \frac{1-(1-2y)}{2} = y.$$

Thus we have proven that for any $y \in \mathbb{Z}$, there always exists at least an x such that $y = f(x)$; therefore, f is surjective.

■ **Example 8.11** The following are some examples of functions that can be either injective, surjective, bijective, or neither.

- surjective and not injective.

$$f : \mathbb{R} \to [0, \infty); \; f(x) = x^2.$$

- not surjective and injective.

$$g : \mathbb{R} \to \mathbb{R}; \; g(x) = e^x.$$

- neither surjective nor injective.

$$h : \mathbb{R} \to \mathbb{R}; \; h(x) = |x|.$$

- surjective and injective.

$$k : [0, \pi] \to [-1, 1]; \; k(x) = \cos(x).$$

■

Theorem 8.2.4 A function $f : A \longrightarrow B$ is surjective if and only if

$$Range(f) = f(A) = B.$$

Proof. Since $f(A) \subseteq B$, it is sufficient to prove that $B \subseteq f(A)$.
If f is surjective, then

$$\forall y \in B, \; \exists \, x \in A : \; y = f(x) \implies y \in f(A) \implies B \subseteq f(A).$$

Conversely, f is surjective if $Range(f) = f(A) = B$. ■

8.2.2 Bijective Functions

Definition 8.2.3 — Bijective Ffunction. The function $f : E \longrightarrow F$ is said to be **bijective** (one-to-one correspondence) if and only if f is both injective and surjective. In other words, if f is bijective, then every element of F has a unique pre-image in E. That is,[a]

$$\forall \, y \in F, \; \exists ! \, x \in E : \; y = f(x).$$

[a]The symbol $\exists !$ stands for there exists a unique value.

■ **Example 8.12** The function $f : \mathbb{R} \to \mathbb{R}$ defined as $f(x) = x^2$ is not a bijective function.
Indeed, we saw previously that it is not surjective (since $y = -4 < 0$, and it can never be a square of any real number x). ■

■ **Example 8.13** The function $g(x) = x^2$ is a bijection from $[0, \infty) \to [0, \infty)$.
That is,

$$\forall \, y \geq 0, \; \exists ! \, x = \sqrt{y} \geq 0 : \; y = (\sqrt{y})^2 = g(x).$$

■

■ **Example 8.14** Let $h : E = \{a,b,c\} \to F = \{1,2,3\}$ be a function defined as

$$h = \{(a,1),\ (b,2),\ (c,3)\}.$$

Clearly, h is a bijective function. Moreover, $Im(h) = \{1,2,3\} = F$ and $n(F) = 3 = n(E)$. ■

■ **Example 8.15** The function $f : \mathbb{N} \to \mathbb{Z}$ is bijective:

$$f(x) = \begin{cases} \dfrac{x}{2} & \text{if } x \text{ is even,} \\ \\ \dfrac{1-x}{2} & \text{if } x \text{ is odd.} \end{cases}$$

This is the same function as in Example 8.10, where it was shown that f is surjective, so, it is sufficient here, to show that f is injective as well. To this end, let us consider two positive integers x_1 and x_2 such that $f(x_1) = f(x_2)$. Then we have the following three cases:

1. Case- (1): If x_1 and x_2 are even, then

$$\frac{x_1}{2} = \frac{x_2}{2} \implies x_1 = x_2.$$

2. Case- (2): If $x_1,\ x_2$ are odd, then we have

$$\frac{1-x_1}{2} = \frac{1-x_2}{2} \implies x_1 = x_2.$$

3. Case- (3): Without Loss Of Generality, let x_1 be even and x_2 be odd, then $x_1 \neq x_2$:

$$f(x_1) = \frac{x_1}{2} > 0, \text{ and } f(x_2) = \frac{1-x_2}{2} \leq 0 \implies f(x_1) \neq f(x_2).$$

Thus f is injective. Therefore, f is a bijection. ■

■ **Example 8.16** Consider

$$g : \mathbb{R} \times \mathbb{R} \longrightarrow \mathbb{R} \times \mathbb{R};\ g(x,y) = (xy,\ x+y).$$

Let us verify whether g is a bijection. We note that $g(2,0) = (0,2) = g(0,2)$ and $(2,0) \neq (0,2)$. Thus g is not injective and therefore cannot be bijective. ■

■ **Example 8.17** Consider

$$f(x) = \begin{cases} 3-x & \text{if } x \leq 1, \\ 1+\dfrac{1}{x} & \text{if } x > 1. \end{cases}$$

We begin by determining whether f is surjective. To accomplish this, we must determine $Range(f)$. If $x \leq 1 \implies -x \geq -1 \implies f(x) = 3-x \geq 2 \implies f(x) \in [2,\ \infty)$. If $x > 1 \implies 0 < \dfrac{1}{x} < 1 \implies$ $0+1 < 1+\dfrac{1}{x} < 1+1 = 2 \implies f(x) \in (1,\ 2)$.

In both cases, $Range(f) = (1,\ \infty) \neq \mathbb{R}$. Thus f is not surjective, which also implies f is not bijective.

■

Theorem 8.2.5 If $f : [a, b] \to \mathbb{R}$ is continuous and strictly monotone, then f is bijective.

Proof. See Question 22. ■

8.3 Composition

8.3.1 Definition and Properties

Definition 8.3.1 — Composition Function. Let $f : A \to B$ and $g : B \to C$, be two functions. We define the **composition** of f and g, denoted by $\mathbf{g} \circ \mathbf{f}$ to be the function defined by

$$g \circ f : \begin{cases} A \longrightarrow C, \\ x \longmapsto (g \circ f)(x) = g[f(x)]. \end{cases}$$

■ **Example 8.18** If $f(x) = x^2$ and $g(x) = 3x - 5$, then

$$(g \circ f)(x) = g[f(x)] = g(x^2) = 3(x^2) - 5 = 3x^2 - 5,$$

and

$$(f \circ g)(x) = f[g(x)] = f(3x-5) = (3x-5)^2 = 9x^2 - 30x + 25.$$

Notice that

$$3x^2 - 5 \neq 9x^2 - 30x + 25.$$

■

(R) In general, the composition of functions is not commutative. That is,

$$g \circ f \neq f \circ g.$$

Proposition 8.3.1 The composition of functions is associative.

$$g \circ (f \circ h) = (g \circ f) \circ h.$$

Proof.

$$[g \circ (f \circ h)](x) = g[f \circ h](x) = g[f(h(x))] = g\Big(f(h(x))\Big)$$

and

$$[(g \circ f) \circ h](x) = (g \circ f)[h(x)] = g[f(h(x))] = g\Big(f(h(x))\Big).$$

■

■ **Example 8.19** Let $f(x) = \sqrt{x+2}$, $g(x) = e^x$, and $h(x) = \sin(x)$,

$$\begin{aligned} [(g \circ f) o h](x) &= (g \circ f)[h(x)], \\ &= (g \circ f)(\sin(x)), \\ &= g(f(\sin(x))), \\ &= g\left(\sqrt{\sin(x)+2}\right), \\ &= e^{\sqrt{\sin(x)+2}}. \end{aligned}$$

Similarly,

$$[g o (f o h)](x) = e^{\sqrt{\sin(x)+2}}.$$

■

Proposition 8.3.2 The domain of the definition of $g \circ f$ is given by:

$$D_{g\circ f} = \{x \in D_f : f(x) \in D_g\}.$$

Proof. Left as an exercise. ■

■ **Example 8.20** Let $f(x) = 2\sqrt{x-3}$ and $g(x) = \dfrac{1}{x^2-4}$.

Clearly, $D_f = [3, \infty)$ and $D_g = \mathbb{R} \setminus \{-2, +2\}$.

Let us determine the following domains:

1. $D_{f\circ f}$.

$$\begin{aligned} x \in D_{f\circ f} &\iff x \in D_f \text{ and } f(x) = 2\sqrt{x-3} \in D_f, \\ &\iff x \geq 3 \text{ and } 2\sqrt{x-3} \geq 3, \\ &\iff x \geq 3 \text{ and } \sqrt{x-3} \geq 3/2, \\ &\iff x \geq 3 \text{ and } x-3 \geq 9/4, \\ &\iff x \geq 3 \text{ and } x \geq 3+9/4 = \frac{21}{4}. \end{aligned}$$

Thus $D_{f\circ f} = \left[\dfrac{21}{4}, \infty\right)$.

2. $D_{g\circ f}$.

$$\begin{aligned} x \in D_{g\circ f} &\iff x \in D_f \text{ and } f(x) = 2\sqrt{x-3} \in D_g, \\ &\iff x \geq 3 \text{ and } 2\sqrt{x-3} \neq \pm 2, \\ &\iff x \geq 3 \text{ and } 2\sqrt{x-3} \neq +2, \\ &\iff x \geq 3 \text{ and } x-3 \neq 1, \\ &\iff x \geq 3 \text{ and } x \neq 4. \end{aligned}$$

Thus $D_{g\circ f} = [3, \infty) \setminus \{4\}$.

3. $D_{g\circ g}$.

$$\begin{aligned} x \in D_{g\circ g} &\iff x \in D_g \text{ and } g(x) = \frac{1}{x^2-4} \in D_g, \\ &\iff x \neq \pm 2 \text{ and } \frac{1}{x^2-4} \neq \pm 2, \\ &\iff x \neq \pm 2 \text{ and } x^2 - 4 \neq \frac{\pm 1}{2}, \\ &\iff x \neq \pm 2 \text{ and } x^2 \neq 4 \pm \frac{1}{2}, \\ &\iff x \neq \pm 2 \text{ and } x \neq \pm\sqrt{\frac{8 \pm 1}{2}}. \end{aligned}$$

Thus $D_{g\circ g} = \mathbb{R} \setminus \left\{\pm 2, \pm\dfrac{3\sqrt{2}}{2}, \pm\sqrt{\dfrac{7}{2}}\right\}$.

4. $D_{f\circ g}$.

$$\begin{aligned} x \in D_{f\circ g} &\iff x \in D_g \text{ and } g(x) = \frac{1}{x^2-4} \in D_f, \\ &\iff x \neq \pm 2 \text{ and } \frac{1}{x^2-4} \geq 3, \\ &\iff x \neq \pm 2 \text{ and } \frac{1}{x^2-4} - 3 \geq 0, \\ &\iff x \neq \pm 2 \text{ and } \frac{13-3x^2}{x^2-4} \geq 0, \\ &\iff x \in [-t, -2) \cup (2, t], \ t = \sqrt{\frac{13}{3}}. \end{aligned}$$

Thus $D_{f\circ g} = [-t,\ -2) \cup (2,\ t]$ and $t = \sqrt{\dfrac{13}{3}}$.

The solution of the inequality $\dfrac{13-3x^2}{x^2-4} \geq 0$ is left for readers (see Chapter 1 for quadratic inequalities).

■

■ **Example 8.21** We consider the functions f and g from Example 8.20 and determine the composition function $h(x) = (g\circ f)(x)$ to find its domain:

$$h(x) = (g\circ f)(x) \;=\; g[f(x)] = g\left[2\sqrt{x-3}\right] = \frac{1}{4(x-3)-4} = \frac{1}{4x-16} \Longrightarrow D_h = \mathbb{R}\setminus\{4\}.$$

Unfortunately, $f(1) = 2\sqrt{1-3}$ is not defined and then $h(1) = (g\circ f)(1) = g\,[f(1)]$ is also not defined, which implies $x = 1 \notin D_h$. When comparing the domain of h with the one obtained for $(g\circ f)$ in Example 8.20, we note that

$$\boxed{D_{g\circ f} \neq D_h}$$

Hence, we need to be aware of this common mistake. ■

■ **Example 8.22** Let $g(x) = \begin{cases} 3-x & \text{if} \quad x \leq 1, \\ 1+\dfrac{1}{x} & \text{if} \quad x > 1. \end{cases}$

We find $(g\circ g)(x)$ by considering the following two cases:

1. $\boxed{x \leq 1}$

$$-x \geq -1 \Longrightarrow g(x) = 3-x \geq 3-1 = 2 > 1 \Longrightarrow g\,[g(x)] = 1+\frac{1}{g(x)} = 1+\frac{1}{3-x} = \frac{4-x}{3-x}.$$

2. $\boxed{x > 1}$

$$\frac{1}{x} > 0 \Longrightarrow g(x) = 1+\frac{1}{x} > 1 \Longrightarrow g\,[g(x)] = 1+\frac{1}{g(x)} = 1+\frac{1}{1+\dfrac{1}{x}} = \frac{2x+1}{x+1}.$$

Therefore,

$$(g\circ g)(x) = \begin{cases} \dfrac{4-x}{3-x} & \text{if} \quad x \leq 1, \\[2ex] \dfrac{2x+1}{x+1} & \text{if} \quad x > 1. \end{cases}$$

■

8.3.2 Composition and Injectivity

Theorem 8.3.3 Let $f : A \to B$ and $g : B \to C$ be two functions:
(a) If f and g are both injective, then $g\circ f$ is also injective.
(b) If $g\circ f$ is injective, then f is injective.

Proof.

(a) Assume that f and g are both injective functions. For $x_1, x_2 \in A$, we have

$$\begin{aligned}(g\circ f)(x_1) = (g\circ f)(x_2) &\implies g[f(x_1)] = g[f(x_2)], \quad (g \text{ is injective})\\ &\implies f(x_1) = f(x_2), \quad (f \text{ is injective})\\ &\implies x_1 = x_2.\end{aligned}$$

Thus $g\circ f$ is also injective.

(b) Let $g\circ f$ be injective function. For $x_1, x_2 \in A$, we have

$$\begin{aligned}f(x_1) = f(x_2) &\implies g[f(x_1)] = g[f(x_2)], \quad (g \text{ is a function})\\ &\implies (g\circ f)(x_1) = (g\circ f)(x_2), \quad (g\circ f \text{ is injective})\\ &\implies x_1 = x_2,\end{aligned}$$

so f is injective.

■

8.3.3 Composition and Surjectivity

Theorem 8.3.4 Let $f : A \to B$ and $g : B \to C$ be two functions:

(a) $g\circ f$ is surjective whenever f and g are surjective.

(b) g is surjective whenever $g\circ f$ is surjective.

Proof.

(a) Assume that f and g are surjective. For any $z \in C$, there exists $y \in B$ such that $z = g(y)$ because g is surjective. Now, since f is surjective, there exists $x \in A$ such that $y = f(x)$. That is,

$$\forall z \in C,\ \exists x \in A:\ z = g[f(x)] = (g\circ f)(x) \implies g\circ f \text{ is surjective.}$$

(b) Assume that $g\circ f$ is surjective. Then for any $z \in C$, there exists at least one $x \in A$ such that $z = (g\circ f)(x) = g[f(x)]$. Now, let $y = f(x) \in B$; then

$$\forall z \in C,\ \exists y \in B:\ z = g(y).$$

Thus g is surjective.

■

■ **Example 8.23** Consider $f(x) = e^x - 2$ from $\mathbb{R} \to \mathbb{R}$ and $g(x) = x^2 + 1$ from $\mathbb{R} \to [1, \infty)$. We note that $g\circ f$ is surjective, but it doesn't require f to be surjective. ■

8.4 Inverse Functions

8.4.1 Definition and Properties

Definition 8.4.1 — identity function. The **identity** function on a set E is defined as follows:

$$\begin{aligned}Id : E &\longrightarrow E,\\ x &\longmapsto Id(x) = x.\end{aligned}$$

Definition 8.4.2 — inverse function. Let $f : E \longrightarrow F$ be a **bijective** function. The **inverse** of f is a function $f^{-1} : F \longrightarrow E$ such that:

- $(f\circ f^{-1})(y) = Id_F(y) = y$ and
- $(f^{-1}\circ h)(x) = Id_E(x) = x.$

Proposition 8.4.1 If $f : E \longrightarrow F$ is a **bijective** function, then:
(i) f^{-1} is unique and
(ii) $y = f(x) \iff x = f^{-1}(y)$.

Proof. Left as an exercise (see Question 23). ■

R It is evident that
$$Id_E\, o\, Id_E = Id_E \iff Id_E^{-1} = Id_E.$$

Example 8.24 Let $f : [1, \infty) \longrightarrow [0, \infty)$ be a function defined as $f(x) = \sqrt{x-1}$. Note that f is a bijective function, so, it is always the case that

$$\forall\, y \geq 0,\ \exists!\, x = y^2 + 1 \geq 1 :\ f(x) = f(y^2+1) = \sqrt{y^2} = |y| = y.$$

We determine the inverse function f^{-1} as follows:
1. Swap x and y values in the given function: $y = \sqrt{x-1}$.
2. We obtain $x = \sqrt{y-1}$.
3. Solve the preceding equation for y, that is, $x^2 = y - 1 \implies y = x^2 + 1$.
4. if the solution is unique, then
$$f^{-1}(x) = y = x^2 + 1.$$

Thus we obtain the inverse of f, $f^{-1}(x) = x^2 + 1$. ■

Example 8.25 Find the inverse of f if $f(x) = \sin(x), \forall\, x \in [0, \pi)$.
We determine the inverse function f^{-1} as shown in the following way:

1. Swap x and y values in $f(x) = \sin(x)$.
2. Obtain $x = \sin(y)$.
3. Solve above equation for y, that is $y = \sin^{-1}(x)$.
4. For $x_1 = \sin(y_1) = 1/2 = \sin(y_2)$, we have $y_1 = \pi/3 \neq y_2 = 2\pi/3$.

Thus the solution is not unique. We conclude that the inverse function, f^{-1}, does not exit. Readers may check again on $[-\pi/2, \pi/2]$ if f^{-1} does exist. ■

Theorem 8.4.2 Let $f : E \to \mathbb{R}$ be injective. Then there exists a bijection $g : E \to f(E)$, which implies the existence of $f^{-1} : f(E) \to E$.

Proof. It is sufficient to take $g = f$ from E to $f(E)$, thus surjectivity is obtained by construction, and injectivity is assumed. Therefore, the inverse of f exists from $f(E)$ to E. ■

Theorem 8.4.3 Let $f : A \to B$ and $g : B \to C$ be two functions.
1. If f is bijective, then f^{-1} is bijective, and $\left(f^{-1}\right)^{-1} = f$.
2. If f and g are bijective then $g \circ f$ is a bijective functions, and $(g \circ f)^{-1} = f^{-1} \circ g^{-1}$.

Proof.

1. Suppose f is bijective, then $f \circ f^{-1} = Id_B$ and $f^{-1} \circ f = Id_A$ are also bijective functions. Thus $\left(f^{-1}\right)^{-1} = f$.

2. Assume that f and g are both bijective functions, then f and g are both injective and surjective. Therefore, $g \circ f$ is also both injective and surjective. Therefore, $g \circ f$ is bijective. That is,

$$(f^{-1} \circ g^{-1}) \circ (g \circ f) = f^{-1} \circ (g^{-1} \circ g) \circ f = f^{-1} \circ (Id_B) \circ f = f^{-1} \circ f = Id_A.$$

Hence, $(g \circ f)^{-1} = f^{-1} \circ g^{-1}$.

■

8.5 Image and Pre-image of a Set

The idea of defining an image of a single point is generally extended to the concept of subsets for a given function. We can associate some special sets with a function $f : A \longrightarrow B$. If $X \subseteq A$, then $f(X)$ has a special meaning to it; it stands for the collection of all those points, as a subset of B that is the image of all points in X. Likewise, the pre-image set $f^{-1}(Y)$ for some $Y \subseteq B$ has a meaning as a subset of A, even if f is not invertible. Now, we write their precise definitions.

Definition 8.5.1 — Image of a Fuction. Suppose $f : A \longrightarrow B$ is a function, and let $X \subseteq A$. Then the **image** of set X is the set $f(X) \subseteq B$ that contains $f(x)$ for all $x \in X$. We write the image as $f(X) = \{f(x) : x \in X\}$. Similarly, if $Y \subseteq B$, the **pre-image** of Y is the set $f^{-1}(Y) = \{x \in A : f(x) \in Y\} \subseteq A$.

Here, we should not think of f^{-1} as the inverse function. The inverse function exists only when f is bijective. In other words, the pre-image $f^{-1}(Y)$ of Y is the set of all elements in A that f^{-1} sends for the elements of Y to A. We might think of $f^{-1}(X)$ as a kind of a sealed copy of Y in A. The image $f(X)$ of X is the set containing a wrapped collection of all elements in Y that are sent under f into the set X. Let us present some examples of the relationships between subsets and their images and pre-images.

■ **Example 8.26** Let $f : \mathbb{R} \to \mathbb{R}$ be a function defined as $f(x) = x^2$. Let $A = \{1,2,3\}$, and $B = \{1,4,9\}$. We observe that $f(A) = B$, but $f^{-1}(B) = \{-2,-1,1,2,3,-3\} \neq A$. This shows $f(f^{-1}(B)) = B$. Moreover, $f([-1,3]) = [0,9]$ but $f^{-1}([0,9]) = [-3,3]$. ■

The next example shows another case where $f(f^{-1}(B)) \neq B$.

■ **Example 8.27** Let $f : \mathbb{R} \to \mathbb{R}$ be a function defined as $f(x) = x^2$. Let $A = \{1,2,3\}$, and $B = \{-1,1,4,9\}$. We observe that $f^{-1}(B) = \{-2,-1,1,2,3,-3\} \neq A$ and $f(f^{-1}(B)) = \{4,1,9\} \neq B$.
■

Theorem 8.5.1 For $A \subseteq X$ and $C \subseteq Y$, let $f : X \longrightarrow Y$ be a function. Then
(a) $f(f^{-1}(C)) \subseteq C$.
(b) $A \subseteq f^{-1}(f(A))$.

Proof.

(a) $f(f^{-1}(C)) \subseteq C$.
Let us consider $y \in f(f^{-1}(C))$, then by definition of image, there exists some $x \in f^{-1}(C)$ such that $y = f(x)$. This shows $y = f(x) \in C$. Therefore it is always true that $f(f^{-1}(C)) \subseteq C$.

(b) $A \subseteq f^{-1}(f(A))$.
By contraposition, if $x \notin f^{-1}(f(A))$ then, $f(x) \notin f(A)$, which implies that $x \notin A$. Thus the proof is complete.

■

Theorem 8.5.2 For $A, B \subseteq X$ and $C, D \subseteq Y$, let $f : X \longrightarrow Y$ be a function. Then
(a) $f(A \cup B) = f(A) \cup f(B)$,
(b) $f(A \cap B) \subseteq f(A) \cap f(B)$,
(c) $f^{-1}(A \cap B) = f^{-1}(A) \cap f^{-1}(B)$,
(d) $f^{-1}(A \cup B) = f^{-1}(A) \cup f^{-1}(B)$.

Proof. Proof of parts (b) and (d) are given in Question 14.
(a) (i) $f(A \cup B) \subseteq f(A) \cup f(B)$.

$$\begin{aligned} \text{Let } y \in f(A \cup B) &\implies \exists x \in A \cup B : \ y = f(x) \\ &\implies \exists x \in A \text{ or } x \in B : \ y = f(x), \\ &\implies \exists \ (x \in A, \ y = f(x)) \text{ or } (x \in B : \ y = f(x)), \\ &\implies y \in f(A) \text{ or } y \in f(B), \\ &\implies y \in f(A) \cup f(B). \end{aligned}$$

Thus $f(A \cup B) \subseteq f(A) \cup f(B)$.

(ii) $f(A) \cup f(B) \subseteq f(A \cup B)$.
Let $y \in f(A) \cup f(B) \Longrightarrow y \in f(A)$ or $y \in f(B)$. If $y \in f(A) \Longrightarrow \exists x \in A : \ y = f(x)$, and since $x \in A \Longrightarrow x \in A \cup B$, then we have $y = f(x) \in f(A \cup B)$. Similarly, if $y \in f(B) \Longrightarrow \exists x \in B : \ y = f(x)$, and since $x \in B \Longrightarrow x \in A \cup B$, then we have: $y = f(x) \in f(A \cup B)$. Thus $f(A) \cup f(B) \subseteq f(A \cup B)$, and finally $f(A) \cup f(B) = f(A \cup B)$.
(c) $f^{-1}(A \cap B) = f^{-1}(A) \cap f^{-1}(B)$.

(i) $f^{-1}(A \cap B) \subseteq f^{-1}(A) \cap f^{-1}(B)$.

$$\begin{aligned} \text{Let } x \in f^{-1}(A \cap B) &\implies f(x) \in A \cap B, \\ &\implies f(x) \in A \text{ and } f(x) \in B, \\ &\implies x \in f^{-1}(A) \text{ and } x \in f^{-1}(B), \\ &\implies x \in f^{-1}(A) \cap f^{-1}(B). \end{aligned}$$

Thus $f^{-1}(A \cap B) \subseteq f^{-1}(A) \cap f^{-1}(B)$.

(ii) $f^{-1}(A) \cap f^{-1}(B) \subseteq f^{-1}(A \cap B)$.

$$\begin{aligned} \text{Let } x \in f^{-1}(A) \cap f^{-1}(B) &\implies x \in f^{-1}(A) \text{ and } x \in f^{-1}(B), \\ &\implies f(x) \in A \text{ and } f(x) \in B, \\ &\implies f(x) \in A \cap B, \\ &\implies x \in f^{-1}(A \cap B). \end{aligned}$$

Thus $f^{-1}(A) \cap f^{-1}(B) \subseteq f^{-1}(A \cap B)$, and finally $f^{-1}(A) \cap f^{-1}(B) = f^{-1}(A \cap B)$.

■

Theorem 8.5.3 Let $A \subseteq X$ and $C \subseteq Y$. If $f : X \longrightarrow Y$ is a given function, then the following is true.
(a) If f is injective, then $f^{-1}(f(A)) = A$.

(b) If f is surjective, then $f(f^{-1}(C)) = C$.

Proof.

(a) For inclusion $A \subseteq f^{-1}(f(A))$, see Theorem 8.5.1(b).
For the converse, we will use the injectivity of the function f. Let $x \in f^{-1}(f(A))$, then $f(x) = y \in f(A)$. For $y \in f(A)$, there exists $t \in A$ such that $f(t) = y$. Since, $f(x) = y = f(t)$, and f is injective, we have $t = x$. Thus $x \in A$. That is, $f^{-1}(f(A)) \subseteq A$. Hence, if f is injective, we have $f^{-1}(f(A)) = A$.

(b) As $C \subseteq Y$, for inclusion $f(f^{-1}(C)) \subseteq C$, see Theorem 8.5.1(a).
For the converse, assume f is a surjective function. Let $y \in C$, then there exists $x \in X$ such that $f(x) = y \in C$. So $x \in f^{-1}(C)$. Thus $y = f(x) \in f(f^{-1}(C))$, and consequently, $C \subseteq f(f^{-1}(C))$. Hence, if f is surjective, then $f(f^{-1}(C)) = C$.

■

Theorem 8.5.4 Let A, B be subsets of X. If $f : X \longrightarrow Y$ is a given function, then

$$\text{If } f \text{ is injective} \implies f(A \cap B) = f(A) \cap f(B).$$

Proof. We will only show that $f(A) \cap f(B) \subseteq f(A \cap B)$, since the converse inclusion is already proved in Theorem 8.5.2(b). We begin by assuming $y \in f(A) \cap f(B)$. Then, $y \in f(A)$, and $y \in f(B)$. It follows that there exists a point $a \in A$ such that $f(a) = y$. Similarly, there exists a point $b \in B$ such that $f(b) = y$. Since f is an injective function, and $f(a) = y = f(b)$, which implies $a = b$. Therefore, $a \in A \cap B$. Then, $f(a) = y \in f(A \cap B)$. ■

Corollary 8.5.5 Let $f : X \longrightarrow Y$. Then for any $A \subseteq X$ and $C \subseteq Y$, we have the following

(a) f is injective if and only if $f^{-1}(f(A)) = A$.

(b) f is surjective if and only if $f(f^{-1}(C)) = C$.

Proof.

(a) f is injective; then $f^{-1}(f(A)) = A$ was already proven in Theorem 8.5.3. To prove the converse, let us consider $x,\ x' \in X$ such that $f(x) = f(x')$; if we set $A = \{x\} \subset X$, then we have:

$$\{x\} = f^{-1}(f(\{x\})) = f^{-1}(f(x)),$$

and since $f(x) = f(x')$ implies $x' \in f^{-1}(f(x))$, we conclude $x' \in \{x\}$, which finally gives $x = x'$ and f is injective.

(b) For (b), see Question 9.

■

8.6 Cardinality of Sets

8.6.1 Equivalent and Finite Sets

Definition 8.6.1 — Equivalent Set. Two sets, A and B, are equivalent, if they have the same number of elements, and we write $A \approx B$.

Example 8.28 The sets
- $A = \{a^2-1,\ a^2,\ a^2+1\} \ \approx\ B = \{2021, 2022, 2023\}$
- $E = \{21, 11, 68, 13, 10, 29\} \ \not\approx\ F = \{a, b, c, d, e\}$.

■

Additionally, we also define equivalent sets as following:

Definition 8.6.2 — Equivalent Sets. Two sets A and B are said to be **equivalent** ($A \approx B$) if there exists a bijection from A to B.

Example 8.29 For $A = [0,\ 1]$ and $B = [a,\ b]$, let $f : A \longrightarrow B$ be a function defined as

$$f(x) = a + (b-a)x, \quad x \in A.$$

To determine $A = [0,\ 1] \ \approx\ B = [a,\ b]$, we need to show that f is a bijective function:
- f is injective
  Let $x_1,\ x_2 \in [0,\ 1]$, such that
  $$f(x_1) = f(x_2) \implies a + (b-a)x_1 = a + (b-a)x_2 \implies (b-a)x_1 = (b-a)x_2 \implies x_1 = x_2.$$
- f is surjective
  Let $y \in [a,\ b]$; then
  $$a \le y \le b \implies 0 \le y - a \le b - a \implies 0 \le \frac{y-a}{b-a} \le 1.$$
  That is,
  $$\forall\, y \in [a,\ b],\ \exists\, \tilde{x} = \frac{y-a}{b-a} \in [0,\ 1] :\ f(\tilde{x}) = a + (b-a)\tilde{x} = a + (y-a) = y.$$
  Therefore, f is a bijective function.

■

Example 8.30 $\mathbb{N} \cup \{0\} \ \approx\ \mathbb{N}$.
To determine this equivalence, let us define a function $g : \mathbb{N} \cup \{0\} \to \mathbb{N}$ such that $g(t) = t+1$, and verify that g is a bijective function as follows:
- g is injective:

  Let $n_1,\ n_2 \in \mathbb{N} \cup \{0\}$ such that
  $$g(n_1) = g(n_2) \implies 1 + n_1 = 1 + n_2 \implies n_1 = n_2.$$
- g is surjective: Let $n \in \mathbb{N} \implies n \ge 1 \implies n - 1 \ge 0$. That is, $\forall\, n \in \mathbb{N},\ \exists\ t = n-1 \in \mathbb{N} \cup \{0\}$ such that $g(t) = t + 1 = (n-1) + 1 = n$. Thus g is bijective.

■

Example 8.31 The set of odd integers is equivalent to the set of even integers.
Let $h :\ ODD = \{2k-1,\ k \in \mathbb{Z}\} \to EVEN = \{2k,\ k \in \mathbb{Z}\}$ such that $h(x) = x+1$. It is obvious that h is injective. We will prove that it is surjective as well. Let y be an even number; then,

$$y = 2p \in EVEN \implies \exists \tilde{x} = y - 1 = 2p - 1 \in ODD :\ h(\tilde{x}) = \tilde{x} + 1 = (2p-1) + 1 = 2p = y.$$

Hence, h is surjective, and therefore, h is a bijection. ■

Theorem 8.6.1 The relation $\approx$ defined on sets is an equivalence relation.

Proof.

- $\approx$ is reflexive. obvious since the identity function from A to A such that $I(x) = x$ is bijective. Thus we have $A \approx A$ for any set A.
- $\approx$ is symmetric. let A and B be equivalent sets ($A \approx B$), then there exists a bijection $f: A \longrightarrow B$. Since, it is well-known that the inverse function $g = f^{-1}: B \longrightarrow A$ is also a bijection, we conclude $B \approx A$.
- $\approx$ is transitive. Let A, B, and C be three sets such that $A \approx B$ and $B \approx C$; then there exist two bijective functions $f: A \longrightarrow B$, and $k: B \longrightarrow C$. It is sufficient to prove that the composition $k \circ f: A \longrightarrow C$ is also a bijection:
  - Let $x, x' \in A$

$$\begin{aligned}(k \circ f)(x) = (k \circ f)(x') &\Longrightarrow k[f(x)] = k[f(x')], \\ &\Longrightarrow f(x) = f(x'), \qquad (k \text{ is one-to-one}) \\ &\Longrightarrow x = x', \qquad (f \text{ is one-to-one}).\end{aligned}$$

    Thus $k \circ f$ is one-to-one.
  - Let

$$\begin{aligned}z \in C &\Longrightarrow \exists\, y \in B: z = k(y) \qquad (k \text{ is onto}) \\ &\Longrightarrow \exists\, y \in B, \exists\, x \in A: z = k(y),\ y = f(x) \qquad (f \text{ is onto}) \\ &\Longrightarrow \exists\, y \in B, \exists\, x \in A: z = k(y) = k[f(x)] = (k \circ f)(x), \\ &\Longrightarrow \exists\, x \in A: z = (k \circ f)(x),\end{aligned}$$

    Thus $k \circ f$ is onto and finally $\approx$ is an equivalence relation.

■

Theorem 8.6.2 Given four sets A, B, C, and D such that $A \cap B = \emptyset$, $C \cap D = \emptyset$, $A \approx C$, and $B \approx D$. Then, $A \cup B \approx C \cup D$.

Proof. See Question 31. ■

Theorem 8.6.3 Let A, B, C and D be non-empty sets such that $A \approx C$ and $B \approx D$. Then

$$A \times B \approx C \times D.$$

Proof. Since $A \approx C$ and $B \approx D$, there exist bijections $\psi: A \longrightarrow C$ and $\varphi: B \longrightarrow D$. Let us define the function $\theta: A \times B \longrightarrow C \times D$, where $f(x, y) = (\psi(x), \varphi(y))$. We need to prove that θ is bijective.

- Let $(x, y), (x', y') \in A \times B$ such that $\theta(x, y) = \theta(x', y')$ which implies that,

$$(\psi(x), \varphi(y)) = (\psi(x'), \varphi(y')) \Longrightarrow \begin{cases} \psi(x) = \psi(x'), \\ \varphi(y) = \varphi(y'), \end{cases} \underbrace{\Longrightarrow}_{\psi,\ \varphi \text{ one-to-one}} \begin{cases} x = x', \\ y = y'. \end{cases}$$

  Then θ is one-to-one.
- Let $(z, z') \in C \times D$, then using the fact that ψ is onto, we have

$$z \in C \Longrightarrow \exists\, x \in A: z = \psi(x),$$

  and since φ is onto,

$$z' \in D \Longrightarrow \exists\, x' \in B: z' = \varphi(x').$$

Thus

$$\forall\,(z,z') \in C \times D,\ \exists\,(x,x') \in A \times B:\ (z,z') = \big(\psi(x), \varphi(x')\big) = \theta(x,x')$$

which implies that θ is onto, and thus bijective. Finally, $A \times B \approx C \times D$.

■

Definition 8.6.3 — Finite Sets. For each natural number k, let $N_k = \{1,2,3,...,k\}$:
A set E is **finite** if $E = \emptyset$ or $E \approx N_k$ for some $k \in \mathbb{N}$.
A set E is **infinite** if E is not a finite set.

■ **Example 8.32** It is easy to see that the sets $\{r,s,t,u,w\}$ and $\{\pi, \sqrt{2}, e\}$ are finite sets because we can easily prove that $\{r,s,t,u,w\} \approx N_5$ and $\{\pi, \sqrt{2}, e\} \approx N_3$. ■

Definition 8.6.4 — Cardinal Number. Given a finite set A

- If $A \neq \emptyset$, then $A \approx N_k$ for some natural number k, and A has **cardinal number**[a]. **k** denoted $\mathbf{n(A) = k}$.
- If $A = \emptyset$, A has **cardinal number** 0, i.e., $\mathbf{n(A) = 0}$.

[a]We may come cross different notations for the cardinal number: $n(A) = card(A) = [A] = \bar{\bar{A}}$

Theorem 8.6.4 If S is finite and $S \approx T$, then T is finite.

Proof. If $S = \emptyset$, then $n(S) = 0$, and since $T \approx S$, $n(T) = n(S) = 0$, and then $T = \emptyset$.
If S is finite, then there exists $k \in \mathbb{N}$ such that $S \approx N_k$, and since $T \approx S$, then $T \approx N_k$ by transitivity, and finally, T is finite. ■

Proposition 8.6.5 Let S be a set with $n(S) = k$ and $x \notin S$; then $T = S \cup \{x\}$ is finite and $n(T) = k+1$.

Proof.

- $S \neq \emptyset$, and there exists $k \in \mathbb{N}:\ S \approx N_k$, and since $\{x\}$ is equivalent to any singleton, we may write that $\{x\} \approx \{k+1\}$, which implies that $S \cup \{x\} \approx N_k \cup \{k+1\} = N_{k+1}$ and $n(S \cup \{x\}) = n(N_{k+1}) = k+1 = n(S)+1$.
- If $S = \emptyset$, then $n(S) = 0$ and $n(S \cup \{x\}) = n(\{x\}) = 1 = 0+1$.

■

Proposition 8.6.6 For every $k \in \mathbb{N}$, every subset of N_k is finite.

Proof. Let us use a mathematical induction on k and let $P(k)$: every subset of N_k is finite.
`Base Step`: For $k = 1$, let $X \subseteq N_1$, then $X = \emptyset$ and $n(X) = 0$ or $X = \{a\}$ and $n(X) = 1$, so $P(1)$ is true.
`Inductive Step`: We assume $P(k)$ is true, and let us prove that $P(k+1)$ is also true.
It is obvious that any subset Y of $N_{k+1} = \{1,2,3,...,k,(k+1)\} = N_k \cup \{k+1\}$ is either a subset of N_k and Y is finite by $P(k)$ or is the union of a subset of N_k and $\{k+1\}$, which is also a finite set according to Proposition 8.6.5. Thus $P(k+1)$ is true, and finally, $P(n)$ is true for any $n \in \mathbb{N}$. ■

Theorem 8.6.7 Every subset of a finite set is finite.

Proof. See Question 32. ■

■ **Example 8.33** Show that the set $A = \{x \in \mathbb{N} : x^2 - 7x < 0\}$ is finite.

It is clear that $x^2 - 7x = x(x-7) = 0$ for $x = 0$ or $x = 7$ and the sign of the quadratic polynomial is negative between the roots. In other words, $x^2 - 7x < 0$ if and only if $x \in (0,\ 7)$ and if we restrict to only integers, we get $A = \{1, 2, 3, 4, 5, 6\}$ which is clearly finite ($A \approx N_6$). ■

■ **Example 8.34** Show that if A is a finite set then $A \cap B$ is also finite for any set B.

A is finite then $\exists\ \varphi : A \to \mathbb{N}_k$ bijective for some integer k and since $A \cap B \subseteq A$, then the restriction $\varphi\,|_{A\cap B} : A \cap B \to \varphi(A \cap B) \subseteq \mathbb{N}_k$ remains bijective and then there exists a positive integer $m \leq k$ such that $A \cap B \approx N_m$ and then $A \cap B$ is finite. ■

Theorem 8.6.8 — Cardinality of Finite Sets.

(a) If A and B are disjoint finite sets, then $A \cup B$ is finite, and

$$n(A \cup B) = n(A) + n(B).$$

(b) If A and B are finite sets, then $A \cup B$ is finite, and

$$n(A \cup B) = n(A) + n(B) - n(A \cap B).$$

(c) If $A_1,\ A_2, ..., A_n$ are finite sets, then

$$\bigcup_{j=1}^{n} A_j \text{ is finite.}$$

Proof.

(a) Since A and B are finite sets, $\exists p,\, q \in \mathbb{N}$ such that $A \approx N_p$ and $B \approx N_q$. Now, we will build a finite set $F = \{p+1, p+2, ..., p+q\}$, which is equivalent to N_q. Indeed, it suffices to consider the bijection $f : F \to N_q$, where $f(x) = x - p$. Furthermore, Theorem 8.6.2, ensures $A \cup B \approx N_p \cup F = N_{p+q}$ and then $n(A \cup B) = n(A) + n(B)$.

(b) To be able to use the part (a), we will express $A \cup B$ as a disjoint union of A and $B - A$. In other words,

$$A \cup B = A \cup (B - A) \text{ and } A \cap (B - A) = \emptyset.$$

Since B is finite, $B - A$ is also finite, and using the result in (a), we obtain $n(A \cup B) = n(A) + n(B - A)$. Thus to complete the proof, we need to determine $n(B - A)$. To this end, let us observe that $B = (B - A) \cup (A \cap B)$, and since $(B - A) \cap (A \cap B) = \emptyset$, $(B - A)$ and $(A \cap B)$ are finite sets, we can apply again (a) to get

$$n(B) = n(B - A) + n(A \cap B) \implies n(B - A) = n(B) - n(A \cap B).$$

Finally, $n(A \cup B) = n(A) + n(B - A) = n(A) + n(B) - n(A \cap B)$.

(c) We will use an induction on k, where $P(k) : \bigcup_{j=1}^{k} A_j$ is a finite set.

`Base Step:` $n = 1$, A_1 is finite, and then $P(1)$ is true.

`Inductive Step:` Assume $P(k)$ is true for $k \leq n$, and show that $P(k+1)$ is also true:

$$\bigcup_{j=1}^{k+1} A_j = \underbrace{\bigcup_{j=1}^{k} A_j}_{P(k)} \cup A_{k+1}$$

By applying (b), we conclude that $\bigcup_{j=1}^{k+1} A_j$ is also finite, and then $P(k+1)$ is true. Finally,

$$\bigcup_{j=1}^{n} A_j$$

is finite for $n \in \mathbb{N}$.

■

Corollary 8.6.9 If A is finite, then A is not equivalent to any of its proper subsets.

Proof. Since any proper subset B of A is strictly contained in A, $n(A) > n(B)$. (see Question 34). We conclude that it is not possible to build a one-to-one correspondence between B and A, which completes the proof. ■

8.6.2 Infinite Sets

In the previous section, we defined infinite sets as those that are never equivalent to any N_k, but there is another characterization given by the contrapositive of Corollary 8.6.9.

Corollary 8.6.10 If A is equivalent to one of its proper subsets, then A is infinite.

Theorem 8.6.11 $\mathbb{N}$ is infinite.

Proof. We will build a bijection between $\mathbb{N}$ and E a proper subset of $\mathbb{N}$ that is the set of even natural numbers. The bijection is given by $f(x) = 2x, \;\; x \in \mathbb{N}$ which implies that $\mathbb{N} \approx E$, and then $\mathbb{N}$, is infinite according to Corollary 8.6.9. ■

We let the curious readers prove this theorem by contradiction.[1]

Theorem 8.6.12 [Cantor Theorem] For any given set A, there exists no surjective function from A onto the power set of A (denoted as $\mathscr{P}(A)$).

Proof. The proof below is based on the "magic idea of Cantor[2]". We need to consider two possible cases:

- If A is a finite set.
  Obvious since $n(A) < n(\mathscr{P}(A)) = 2^{n(A)}$. Thus no surjection is possible.
- If A is an infinite set.
  We will prove it by contradiction. To this goal, let us assume the existence of a surjection $f : A \to \mathscr{P}(A)$ and let us define a subset E of A as follows:

$$E = \{x \in A \; : \; x \notin f(x)\}.$$

[1] The proof is elaborated in Question 36.
[2] Goerg Cantor (1845–1918) a German Mathematician.

Since E is a subset of A, so we conclude directly that $E \in \mathscr{P}(A)$ and then,

$$\exists\, z \in A \;:\; f(z) = E.$$

Here we consider the two possibilities
(i) $z \in E \implies z \notin f(z) = E$ which is a contradiction.
(ii) $z \notin E \implies z \in f(z) = E$ which is a contradiction.
Thus, we proved that $E \notin \mathscr{P}(A)$ which is a contradiction and then what we assumed is wrong and the function f cannot be surjective.

■

Definition 8.6.5 — Denumerable Set. A set S is said to be **denumerable** if $S \approx \mathbb{N}$.
A denumerable set has a **cardinality** of $\aleph_0 = n(S)$, which is also known as a **cardinal number**.

■ **Example 8.35** The following sets are denumerable:

1. set of even numbers,
2. set of odd numbers,
3. set of multiples of 3,
4. set $\mathbb{N} \setminus \{1,\ 100\}$.

■

■ **Example 8.36** The set $S = \{5k,\ k \in \mathbb{N}\}$ is denumerable.

Indeed, let us consider the function $\psi : S \to \mathbb{N}$ defined as $\psi(x) = x/5$. We will examine whether ψ is surjective. To demonstrate ψ is surjective, take an arbitrary natural number $m \in \mathbb{N}$. We can find an element $x = 5m$ in the set S such that $\psi(x) = \psi(5m) = (5m)/5 = m$. This shows that for any natural number m, there exists an element x in S such that $\psi(x) = m$. Therefore, ψ is surjective.
Now, let us consider two elements x and y in S such that $\psi(x) = \psi(y)$. This implies $x/5 = y/5$, which simplifies to $x = y$. Thus, ψ is injective.
In conclusion, since ψ is both surjective and injective, it is bijective. Consequently, set S is denumerable. ■

Theorem 8.6.13 $\mathbb{Z}$ is denumerable.

Proof. It is sufficient to build a bijection between $\mathbb{N}$ and $\mathbb{Z}$. To this end, let us consider the function

$$f(x) = \begin{cases} \dfrac{x}{2} & \text{if } x \text{ is even} \\[2ex] \dfrac{1-x}{2} & \text{if } x \text{ is odd.} \end{cases}$$

We have to show the following
(i) f is injective (one-to-one). Let $x, x' \in \mathbb{N}$. If

$$x,\ x' \text{ both are even}: \; f(x) = f(x') \implies \frac{x}{2} = \frac{x'}{2} \implies x = x',$$

$$x,\ x' \text{ both are odd}: \; f(x) = f(x') \implies \frac{1-x}{2} = \frac{1-x'}{2} \implies x = x',$$

$$x, \text{ is even and } x' \text{ is odd}: \; f(x) = f(x') \implies \frac{x}{2} = \frac{1-x'}{2} \implies x = 1 - x' \leq 0.$$

Thus we have $f(x') \neq f(x)$ because $f(x) = \dfrac{x}{2} \leq 0$ and $f(x') \geq 0$. Finally f is injective.

(ii) f is surjective (onto). Let us consider $y \in \mathbb{Z}$. Thus if

$$y \leq 0 \implies x = 1 - 2y \text{ is an odd positive integer in } \mathbb{N}: \; f(x) = f(1-2y) = \frac{1-(1-2y)}{2} = y;$$

and if

$$y > 0 \implies x = 2y \text{ is an even positive integer in } \mathbb{N}: \; f(x) = f(2y) = \frac{2y}{2} = y.$$

Finally, f is surjective, which implies f is bijective, and then $\mathbb{Z} \approx \mathbb{N}$.

■

Theorem 8.6.14 The set $\mathbb{N} \times \mathbb{N}$ is denumerable.

Proof. We will discuss a classical proof given in many textbooks. To this end, let us consider the function

$$F : \mathbb{N} \times \mathbb{N} \to \mathbb{N} \text{ defined by } F(p,q) = 2^{p-1}(2q-1).$$

We need to show that is injective and surjective.

- Let $(p,q), (s,t) \in \mathbb{N} \times \mathbb{N}: \; F(p,q) = F(s,t)$, so

$$2^{p-1}(2q-1) = 2^{s-1}(2t-1)$$

  1. If $p > s \geq 1$, so

$$2^{(p-1)-(s-1)}(2q-1) = 2t-1 \implies \underbrace{2^{p-s}(2q-1)}_{even} = \underbrace{2t-1}_{odd},$$

     which is impossible.
  2. If $s > p \geq 1$, then

$$2^{s-1}(2t-1) = 2^{p-1}(2q-1) \implies 2^{(s-1)-(p-1)}(2t-1) = 2q-1 \implies \underbrace{2^{s-p}(2t-1)}_{even} = \underbrace{2q-1}_{odd},$$

     which is impossible.

  This implies that $p = s \geq 1$, and in this case, we have:

$$2^{p-1}(2q-1) = 2^{p-1}(2t-1) \implies 2q-1 = 2t-1 \implies q = t.$$

  Finally, we obtain:

$$F(p,q) = F(s,t) \implies (p,q) = (s,t) \implies F \text{ is injective.}$$

- Let $m \in \mathbb{N}$, then m is either odd or even
  1. m is odd, so $\exists\, k \in \mathbb{N}: \; m = 2k-1 = 2^0(2k-1) = 2^{1-1}(2k-1) = F(1,k)$. Thus we obtain the following:

$$\text{For all odd } m, \; \exists\, (1,k) \in \mathbb{N} \times \mathbb{N}: \; F(1,k) = m.$$

2. m is even, so we can express its prime decomposition as follows:

$$m = 2^{n_1} \underbrace{p_1^{n_2} p_2^{n_3} \dots p_{l-1}^{n_l}}_{odd} = 2^{n_1}(2k-1), \quad k \geq 1,$$

i.e.,

$$\text{for all } m \text{ (even)}, \ \exists\, (n_1+1,k) \in \mathbb{N}\times\mathbb{N} : \ F(n_1+1,k) = m.$$

Finally, F is surjective, and then F is bijective, which completes the proof. ■

Theorem 8.6.15 Let A and B be two denumerable sets; then their Cartesian product $A \times B$ is also denumerable.

Proof. A and B are denumerable, then $A \approx \mathbb{N}$ and $B \approx \mathbb{N}$. Thus using the previous result, we obtain:

$$(A \times B) \approx \mathbb{N}\times\mathbb{N} \approx \mathbb{N} \implies A \times B \ \text{ is denumerable.}$$

■

8.6.3 Countable Sets

Definition 8.6.6 — countable sets.

- A set S is said to be **countable** if S is **finite or denumerable**.
- We say S is **uncountable** if S is not countable.

Lemma 1 If set $S \subset T$ and S is uncountable, then T is also uncountable.

Proof. Direct by contradiction. ■

■ **Example 8.37** $\mathbb{Z}$ is countable.
We already proved that $\mathbb{Z}$ is denumerable in Theorem 8.6.13. Here, we propose to use another bijection

$$f : \mathbb{Z} \longrightarrow \mathbb{N}, \quad f(x) = \begin{cases} 2x & \text{if } \ x > 0, \\ 1-2x & \text{if } \ x \leq 0. \end{cases}$$

■

■ **Example 8.38** $\mathbb{Q}$ is countable.
It is sufficient to prove that the function f below is a surjection.

$$f : \mathbb{Z}\times\mathbb{N} \longrightarrow \mathbb{Q}, \quad f(p,q) = \begin{cases} p/q & \text{if } \ q \neq 0, \\ 0 & \text{if } \ q = 0. \end{cases}$$

Let $\frac{m}{n} \in \mathbb{Q}$, $n \neq 0$, so taking $p = m$ and $q = n$ such that $f(p,q) = \frac{p}{q}$, which implies f is surjective. Since f is surjective and $\mathbb{Z}\times\mathbb{N}$ is countable (by Example 8.37 and Theorem 8.6.15), then we conclude that $\mathbb{Q}$ is also countable. ■

Theorem 8.6.16 The open interval $(0,\ 1)$ and $\mathbb{R}$ are uncountable sets.

Proof. See [5] and [8]. ■

We conclude this chapter with a discussion of cardinality concerning infinite sets. The notion of infinity introduces intricacy into the concept of cardinality, especially when dealing with infinite sets. The primary focus is to elucidate the notion that there exist various forms of infinity, and not all infinities are of equal size. It is possible for two sets, A and B, to both possess infinite cardinalities, yet the cardinality of set A may be less than the cardinality of set B.

Definition 8.6.7 — Cardinal Number. A set S has cardinal number **c** (or cardinality c) if S is equivalent to $(0,\ 1)$. We write $n(S) = c$, which stands for continuum.

Definition 8.6.8 Two sets A and B share identical cardinality, denoted as $n(A) = n(B)$, when there exists a bijective function $f : A \to B$.
In cases where no such bijective function exists, the sets possess distinct cardinalities, meaning $n(A) \neq n(B)$.

To demonstrate that $n(A) = n(B)$, there can exist numerous bijective functions from set A to set B. We only require the identification of a single bijective function to draw our conclusion.

Theorem 8.6.17 There exists a bijection $f : \mathbb{N} \to \mathbb{Z}$. Therefore $n(\mathbb{N}) = n(\mathbb{Z})$.

Proof. Left to readers with a hint.: prove that f given by

$$f(n) = \begin{cases} 0 & \text{if} \quad n = 1 \\ p & \text{if} \quad n = 2p \\ -p & \text{if} \quad n = 2p+1. \end{cases}$$

is a bijection from $\mathbb{N}$ to $\mathbb{Z}$. ■

The cardinality of the natural numbers is denoted as $\aleph_0$. That is, $n(\mathbb{N}) = \aleph_0$. Thus any countable infinite set has cardinality $\aleph_0$. Thus, we have:

$$n(\mathbb{N}) = \aleph_0 = n(\mathbb{Z}).$$

It's crucial to acknowledge that Georg Cantor was the pioneering figure who rigorously demonstrated that

$$n(\mathbb{Z}) \neq n(\mathbb{R})$$

by providing the argument that it is impossible to construct a surjective function between the set of natural numbers and the set of real numbers.

■ **Example 8.39**

$$n((0,\ \infty)) = n((0,\ 1)) = c.$$

One may try to consider the function $f : (0,\ \infty) \to (0,\ 1)$ given by:

$$f(x) = \frac{x}{x+1}.$$

- When we examine two positive values, x and x', such that $f(x) = f(x')$, it is evident that this implies $x = x'$, thereby establishing the injective property of the function f.
- Now, let us consider any $y \in (0,\ 1)$, then we have:

$$y = \frac{x}{x+1} \Longleftrightarrow x = \frac{y}{1-y} > 0.$$

  Thus, we proved that:

$$\forall\, y \in (0,\ 1),\ \exists\, x = \frac{y}{1-y} < 0\ :\ y = \frac{x}{x+1} = f(x)$$

  which implies that f is surjective and finally f is a bijection from $(0,\ \infty) \to (0,\ 1)$.

■

■ **Example 8.40**

$$n(\mathbb{R}) = n((0, \infty)) = c.$$

It is sufficient to consider the function $f : \mathbb{R} \to (0, \infty)$ given by

$$f(x) = e^x.$$

■

Thus, we may conclude that

$$n(\mathbb{R}) = n((0, \infty)) = n((0, 1)) = c.$$

■ **Example 8.41** Determine the cardinality of the infinite set $S = \{3^m, m \in \mathbb{N}\}$. Since m is running in $\mathbb{N}$, so the set $S = \{3^1, 3^2, 3^3, 3^4, cdots, 3^n, cdots\}$ is clearly infinite. The simple function

$$f : S \to \mathbb{N} \text{ defined by } f(x) = f(3^m) = m$$

is clearly a bijection and then S is denumerable and then $n(S) = \aleph_0$. ■

Theorem 8.6.18

$$n(\mathbb{Q}) = \aleph_0, \quad n(\mathbb{R} \setminus \mathbb{Q}) = c.$$

Definition 8.6.9 Let A and B be two sets.

(i) $n(A) = n(B)$ is equivalent to the existence of a bijection from A to B.

(ii) $n(A) < n(B)$ is equivalent to the existence of a injection from A to B but no surjection from A to B.

■ **Example 8.42**

$$n(\mathbb{N}) < n(\mathbb{R}).$$

It is enough to identify an injective function that is not surjective between the sets $\mathbb{N}$ and $\mathbb{R}$. For example, we can take the function $f(n) = n$ for all integers n. It is clear that f is not surjective, but its injective nature is evident. ■

Thus, we may write:

$$\aleph_0 < c.$$

The previous theorem of Cantor is also cast equivalently as follows:

Theorem 8.6.19 The cardinality of a given set A is always strictly less than the cardinality of its power set. In other terms

$$n(A) < n(\mathscr{P}(A)).$$

Proof. It is adequate to establish an injection from set A to its power set $\mathscr{P}(A)$. For instance, we can define the function $f(x) = \{x\}$, which is clearly injective. Additionally, applying Cantor's Theorem 8.6.12, we can deduce the absence of a surjective function between set A and its power set $\mathscr{P}(A)$. This conclusion serves as the final step in our proof. ■

Thanks to Cantor for this important result which induce the following inequalities between certain infinite sets:

$$n(A) < n(\mathscr{P}(A)) < n\Big(\mathscr{P}(\mathscr{P}(A))\Big) \cdots$$

thus, we have:

$$\aleph_0 = n(\mathbb{N}) < n(\mathscr{P}(\mathbb{N})) < n\Big(\mathscr{P}(\mathscr{P}(\mathbb{N}))\Big) \cdots$$

The next result (is given without proof) provides a nice result linking cardinality of $\mathbb{R}$ and of $\mathscr{P}(\mathbb{N})$.

Theorem 8.6.20 The sets $\mathbb{R}$ and $\mathscr{P}(\mathbb{N})$ have the same cardinality.

$$c = n(\mathbb{R}) = n(\mathscr{P}(\mathbb{N})) > n(\mathbb{N}) = \aleph_0.$$

(R) According to $n(\mathscr{P}(A)) = 2^{n(A)}$, we may conclude:

$$c = 2^{\aleph_0}.$$

■ **Example 8.43** Arrange in increasing order (if possible) the following

$$n(0,\ 2),\ n(\mathbb{R}-\mathbb{Q}),\ n(\{0,2\}),\ n(\mathbb{R}),\ n(\phi),\ n(\mathbb{Q}),\ n(\mathscr{P}(\mathbb{R})).$$

The order:

$$n(\phi) < n(\{0,2\}) < n(\mathbb{Q}) < n(0,\ 2) = n(\mathbb{R}-\mathbb{Q}) = n(\mathbb{R}) < n(\mathscr{P}(\mathbb{R})).$$

■

We close this section by introducing some operations within the continuum c of infinite uncountable sets. In set theory, the operations involving the cardinality of the continuum can be a bit counterintuitive when compared to standard arithmetic. Here's how the operations $c+c$ and $c.c$ work:

- $c+c=c$: This means that if you take the union of two sets, each with cardinality c, the resulting set still has cardinality c. In other words, adding a set with the cardinality of the continuum to itself does not increase its cardinality.
- $c.c$: Similarly, if you take the Cartesian product of two sets, each with cardinality c, the resulting set still has cardinality c. This means that the product of the continuum with itself has the same cardinality as the continuum.

These results can be surprising, as they differ from typical arithmetic operations. In set theory, the size of infinity "c" has some unique properties that can lead to these seemingly counterintuitive outcomes. These properties are established within the framework of set theory and are essential in understanding the cardinal arithmetic of infinite sets.

Theorem 8.6.21 Let c be the cardinality of the continuum then

$$c+c=c, \quad c.c=c, \quad \aleph_0.c = c.$$

8.7 Chapter 8 Practice Problems

Question 1 Let $f : E \longrightarrow E$ be a function defined on $E = \{-3,-2,0,1,2,3,6,8,10\}$ by

$$f = \{(10,-3),(1,0),(-2,6),\ (2,0),\ (3,1),\ (-3,10),\ (6,8),(8,6),(0,-2)\}.$$

Find

$$f(\{-2,0,3,10\}),\ f(\{-3,8,6\}),\ \text{and } f(\{1\}).$$

Question 2 Let $f : A \longrightarrow A$ be a function defined on $A = \{-3,-2,0,1,2,3,6,8,10\}$ by

$$f = \{(10,-3),(1,0),(-2,6),\ (2,0),\ (3,1),\ (-3,10),\ (6,8),(8,6),(0,-2)\}.$$

Find

$$f^{-1}(\{1,6,0\}),\ f^{-1}(\{6,10,-3\}).$$

Question 3 Find the range of the function $f : \mathbb{N}\times\mathbb{N} \longrightarrow \mathbb{Z}$ defined as $f(m,n) = 7m-21n$.

Question 4 Find the Domain of

$$f(x) = \sqrt{x^3+1},\ g(x) = \frac{1-\sin(x)}{2-\cos(x)},\ \text{and } h(x) = \sqrt{15-7x-2x^2}.$$

Question 5 Find the Domain of $l(x) = \sqrt{6x-1-9x^2}$ and $m(x) = \sqrt{2\sin(4x)-3}$.

Question 6 Find the Range of $l(x) = \sqrt{6x-1-9x^2}$.

Question 7 Find the Domain and Range of $h(x) = \sqrt{15-7x-2x^2}$.

Question 8 Let $f : \mathbb{R} \longrightarrow \mathbb{R}$ be a function such that $f(x) = |x-1|$. Find

$$f^{-1}(\{7\}),\ f^{-1}(\{-5\}),\ f^{-1}(\{x :\ x \le 0\}),\ \text{and } f^{-1}([4,\ 25]).$$

Question 9 Let $f : \mathbb{R} \longrightarrow \mathbb{R}$ be a function such that $f(x) = x^2$. Find

$$f^{-1}(\{25\}),\ f^{-1}(\{-9\}),\ f^{-1}(\{x :\ x \le 0\}),\ \text{and } f^{-1}([2,\ 3]).$$

Question 10 Let $f : \mathbb{R} \longrightarrow \mathbb{R}$ be a function such that $f(x) = x^3$. Find

$$f^{-1}(\{25\}),\ f^{-1}(\{-9\}),\ f^{-1}(\{x :\ x \le 0\}),\ \text{and } f^{-1}([2,\ 3]).$$

Question 11 Let $f : X \longrightarrow Y$ be a function, and let $A,\ B \subseteq X$. Prove that

$$A \subseteq B \implies f(A) \subseteq f(B).$$

Question 12 Let $U = \{a,b,c,d,e,f,g\}$, and let $A = \{a,b,c,f\}$, $B = \{b,c,d,g\}$, and $C = \{a,d,e,f\}$. Find

$$\chi_{A\cap B},\ \chi_{A\cup C},\ \chi_{(A\cup C)-B}.$$

Question 13 Let $U = \mathbb{R}\setminus[-1,\ 1]$ and $X = [-2,\ 5)$. Determine

$$\chi_X(3),\ \chi_X(0),\ \chi_X(5),\ \chi_X(4),\ \text{and } \chi_X(10),$$

where χ_X denotes the characteristic function of the given set X defined by

$$\chi_X(x) = \begin{cases} 1 & \text{if } x \in X, \\ 0 & \text{if } x \notin X. \end{cases}$$

Question 14 For $A,B \subseteq X, A',B' \subseteq Y$, let $f : X \longrightarrow Y$ be a function,

a. $f(A \cap B \cap C) \subseteq f(A) \cap f(B) \cap f(C)$.

b. $f^{-1}(A' \cup B' \cup C') = f^{-1}(A') \cup f^{-1}(B') \cup f^{-1}(C')$.

Question 15 Verify whether the following are injective, surjective, or bijective functions.

(a) Let $f : \mathbb{Z} \longrightarrow \mathbb{Z}$ be a function defined as $f(x) = 2 - 5x, \forall x \in \mathbb{Z}$.

(b) Let $f : \mathbb{N} \longrightarrow \mathbb{Q}$ be a function defined as $f(n) = \dfrac{n}{n+1}, \forall n \in \mathbb{N}$.

Question 16 Find a bijection between $\mathbb{N} \longrightarrow \mathbb{N} \setminus \{1,\ 2\}$.

Question 17 In each part, find a function $f : \mathbb{N} \longrightarrow \mathbb{N}$ that has the desired properties:

1. surjective, but not injective.

2. injective, but not surjective.

3. neither surjective nor injective.

4. bijective.

Question 18 Is $f(x) = 3x^3 - 6x^2 + x$ injective if the domain and the range of f are equal to $\mathbb{R}$?

Question 19 Suppose that $f : A \longrightarrow B$ is a function. Prove that if f is injective, then there is a function $g : B \longrightarrow A$ such that $g \circ f = Id_A$.

Question 20 Suppose that $f : A \longrightarrow B$ is a function. Prove that if f is surjective, then there is a function $h : B \longrightarrow A$ such that $f \circ h = Id_B$.

Question 21 Is the function $h : \mathbb{R} \to \mathbb{R}$ given by $h(x) = x^2(x^3 + 27)$ bijective?

Question 22 Is the function $f : \mathbb{R} \to \mathbb{R}$ given by $f(x) = \dfrac{e^{|x|} - e^{-x}}{e^x + e^{-x}}$ injective? Is-it surjective?

Question 23 Study if g is a bijection on $\mathbb{R}$, if

$$g(x) = \begin{cases} \dfrac{1}{2x-1} & \text{if } \ x < 1/2, \\ \sqrt{2x-1} & \text{if } \ x \geq 1/2. \end{cases}$$

Question 24 If g is defined by

$$g(x) = \begin{cases} 1 - \dfrac{1}{1-x} & \text{if } \ x > 2, \\ 2x - 2 & \text{if } \ x \leq 2. \end{cases}$$

(a) Is the function g surjective (onto)?

(b) By sketching the graph of $y = g(x)$, prove f is not injective.

Question 25 Let A and B be two given sets with $n(A) = 2$ and $n(B) = 3$. How many injective functions are there from $A \to B$?

Question 26 Determine the expression of $(g \circ g)(x)$, where g is given in Question 23.

Question 27 Determine $(f \circ f)(x)$ if g is given in Question 24.

Question 28 Given two sets A and B, prove that

$$n(A) \le n(B) \iff \exists C \subseteq B \;:\; n(A) = n(C).$$

Question 29 Prove that $(0,1)$ and $(0,\infty)$ have the same cardinality.

Question 30 Prove that $n(E) = \aleph_0$ if $E = \{2k \;:\; k \in \mathbb{Z}\}$.

Question 31 Prove or disprove: Every infinite set is a subset of a countably infinite set.

Question 32 Prove or disprove: If $A \subseteq B$ and A is an infinite countable set and B is uncountable, then $B - A$ is also uncountable.

8.8 Chapter 8 Additional Practice Problems with Solutions

Question 1 Let $f :\; E \longrightarrow E$ be a function defined on $E = \{0,1,2,3,4,5\}$ by

$$f = \{(0,0),(1,5),\; (2,2),\; (3,1),\; (4,4),\; (5,2)\}.$$

Find

$$f(\{1,0,3\}),\;\; f(\{3,4,5\}),\; \text{and} f(\{0,5\}).$$

Solution.

- $f(\{1,0,3\}) \;=\; \{f(1), f(0), f(3)\} = \{5,\, 0,\, 1\}$.
- $f(\{3,4,5\}) \;=\; \{f(3), f(4), f(5)\} = \{1,\, 4,\, 2\}$.
- $f(\{0,\, 5\}) \;=\; \{f(0), f(5)\} = \{0,\, 2\}$.

Question 2 Let $f :\; A \longrightarrow A$ be a function defined on $A = \{1,2,3,4,5\}$ by

$$f = \{(1,3),\; (2,2),\; (3,5),\; (4,3),\; (5,2)\}.$$

Find

$$f^{-1}(\{1,2,3\}),\;\; f^{-1}(\{4,5\}).$$

Solution.

- $f^{-1}(\{1,2,3\}) \;=\; \{2,\, 5,\, 4,\, 1\}$, but $y = 1$ has no antecedent $\tilde{x}$ such that $f(\tilde{x}) = 1$.
- $f^{-1}(\{4,5\}) \;=\; \{3\}$, but $y = 4$ has no antecedent $\tilde{x}$ such that $f(\tilde{x}) = 4$.

Question 3 Find the range of the function $f : \mathbb{Z} \times \mathbb{Z} \longrightarrow \mathbb{Z}$ defined as $f(m,n) = 5m - 15n$.

Solution.

$$R_f \;=\; \{p \in \mathbb{Z} :\; \exists (m,n) \in \mathbb{Z} \times \mathbb{Z} :\; f(m,n) = p\} \;=\; \{p \in \mathbb{Z} :\; \exists (m,n) \in \mathbb{Z} \times \mathbb{Z} :\; 5m - 15n = p\}.$$

In other words, $p \in R_f \Longrightarrow p = 5(m-3n) \in 5\mathbb{Z}$. Now, let us try to prove that $5\mathbb{Z} \subseteq R_f$.
Let us consider

$$p \in 5\mathbb{Z} \Longrightarrow \exists q \in \mathbb{Z} :\; p = 5q \;=\; 5(4q \;-\; 3q) = 5(m-3n) \in R_f, \text{ with } m = 4q,\; n = q \in \mathbb{Z}.$$

Thus $5\mathbb{Z} \subseteq R_f$, and then finally, $R_f \;=\; 5\mathbb{Z}$.

Question 4 Find the domain and range of $h(x) = \sqrt{15 - 7x - 2x^2}$.

Solution.

- $D_h = \{x \in \mathbb{R} :\; 15 - 7x - 2x^2 \ge 0\}$. Here, $a = -2,\; b = -7,$ and $c = 15$.
  $\Delta = b^2 - 4ac = 49 - 4(-2)(15) = 49 + 120 = 169 = 13^2$, and then $x_{1,2} = \dfrac{7 \pm 13}{2(-2)} \in \{-5;\, \frac{3}{2}\}$.
  Since $a = -2 < 0$, the quadratic expression is non-negative if $x \in [-5,\; 3/2]$, and thus $D_h \;=\; [-5,\; 3/2]$.

- $R_h = \{h(x) : x \in D_h\}$. Since $a = -2 < 0$, the quadratic expression has a maximum value $\beta = \frac{-\Delta}{4a} = \frac{-169}{-8} = \frac{169}{8}$, reached at $\alpha = \frac{-b}{2a} = \frac{-7}{4}$. And finally

$$R_h = \left[0, \sqrt{\frac{169}{8}}\right].$$

Question 5 Let $f : \mathbb{R} \longrightarrow \mathbb{R}$ be a function such that $f(x) = |x|$. Find

$$f^{-1}(\{7\}),\ f^{-1}(\{-5\}),\ f^{-1}(\{x : x \leq 0\}), \text{ and } f^{-1}([4, 25]).$$

Solution.

- $f^{-1}(\{7\}) = \{-7; +7\}$, because $|x| = 7$ has two solutions ± 7.
- $f^{-1}(\{-5\}) = \emptyset$, because $|x| = -5$ has no solution.
- $f^{-1}(\{x : x \leq 0\}) = \{0\}$, because $|x| \leq 0 \Longleftrightarrow |x| = 0$, which has only $x = 0$ as a solution.
- $f^{-1}([4, 25]) = [-25, -4] \cup [4, 25]$.
  Students may tackle such questions through basic algebra:

$$x \in f^{-1}([4, 25]) \Longleftrightarrow f(x) = |x| \in [4, 25].$$

  Here, we need to consider two cases:

  - If $x \geq 0$, then $4 < x < 25$.
  - If $x < 0$, then $4 < -x < 25$.

  Thus $-25 < x < -4$.

Question 6 Let $f : X \longrightarrow Y$ be a function and let A and B be two subsets of X. Prove that

$$f(A - B) \supseteq f(A) - f(B) \text{ and } f(A \cup B) = f(A) \cup f(B).$$

Solution.

- $f(A - B) \supseteq f(A) - f(B)$. Let us consider

$$\begin{aligned} y \in f(A) - f(B)) &\Longrightarrow y \in f(A) \text{ and } y \notin f(B), \\ &\Longrightarrow (\exists\, x \in A : y = f(x)) \text{ and } (\forall x' \in B : y \neq f(x')), \\ &\Longrightarrow (\exists\, x)\ (x \in A,\ y = f(x)) \text{ and } x \notin B, \\ &\Longrightarrow (\exists\, x)\ (x \in A \text{ and } x \notin B : y = f(x)), \\ &\Longrightarrow (\exists\, x)\ (x \in A - B \text{ and } y = f(x)), \\ &\Longrightarrow y \in f(A - B). \end{aligned}$$

- $f(A \cup B) = f(A) \cup f(B)$. We will prove both inclusions:
  1. $f(A \cup B) \subseteq f(A) \cup f(B)$.

$$\begin{aligned} y \in f(A \cup B) &\Longrightarrow \exists\, x \in (A \cup B) : y = f(x), \\ &\Longrightarrow (\exists x)\ (x \in A) \text{ or } (x \in B) : y = f(x), \\ &\Longrightarrow (\exists x)\ (x \in A : y = f(x)) \text{ or } (x \in B : y = f(x)), \\ &\Longrightarrow y \in f(A) \text{ or } y \in f(B), \\ &\Longrightarrow y \in f(A) \cup f(B). \end{aligned}$$

  2. $f(A) \cup f(B) \subseteq f(A \cup B)$.

$$\begin{aligned} y \in f(A) \cup f(B) &\Longrightarrow y \in f(A) \text{ or } y \in f(B), \\ &\Longrightarrow (\exists x)\ (x \in A : y = f(x)) \text{ or } (\exists x')\ (x' \in B : y = f(x')), \\ &\Longrightarrow (\exists x)\ (x \in A \cup B : y = f(x)) \text{ or } (\exists x')\ (x' \in B \cup A : y = f(x')), \\ &\Longrightarrow y \in f(A \cup B) \text{ or } y \in f(B \cup A), \\ &\Longrightarrow y \in f(A \cup B). \end{aligned}$$

Question 7 Let $f: X \longrightarrow Y$ be a function, and let A be a subset of Y. Prove that

$$f^{-1}(\overline{A}) = \overline{f^{-1}(A)}.$$

Solution.

(i) $f^{-1}(\overline{A}) \subseteq \overline{f^{-1}(A)}$.

$$\begin{aligned} x \in f^{-1}(\overline{A}) &\Longrightarrow f(x) \in \overline{A}, \\ &\Longrightarrow f(x) \notin A, \\ &\Longrightarrow x \notin f^{-1}(A), \\ &\Longrightarrow x \in \overline{f^{-1}(A)}. \end{aligned}$$

(ii) $\overline{f^{-1}(A)} \subseteq f^{-1}(\overline{A})$.

$$\begin{aligned} x \in \overline{f^{-1}(A)} &\Longrightarrow x \notin f^{-1}(A), \\ &\Longrightarrow \forall\, y \in A,\ x \neq f^{-1}(y), \\ &\Longrightarrow \forall\, y \in A,\ f(x) \neq y, \\ &\Longrightarrow f(x) \notin A, \\ &\Longrightarrow f(x) \in \overline{A}, \\ &\Longrightarrow x \in f^{-1}(\overline{A}). \end{aligned}$$

Question 8 Let $f: X \longrightarrow Y$ be a function, and let A and B be two subsets of Y. Prove that

(i) $f^{-1}(A \cap B) = f^{-1}(A) \cap f^{-1}(B)$.

(ii) $f^{-1}(A - B) = f^{-1}(A) - f^{-1}(B)$.

Solution.

(i) $f^{-1}(A \cap B) = f^{-1}(A) \cap f^{-1}(B)$.

1. $f^{-1}(A \cap B) \subseteq f^{-1}(A) \cap f^{-1}(B)$.

$$\begin{aligned} x \in f^{-1}(A \cap B) &\Longrightarrow f(x) \in A \cap B, \\ &\Longrightarrow f(x) \in A \text{ and } f(x) \in B, \\ &\Longrightarrow x \in f^{-1}(A) \text{ and } x \in f^{-1}(B), \\ &\Longrightarrow x \in f^{-1}(A) \cap f^{-1}(B). \end{aligned}$$

2. $f^{-1}(A) \cap f^{-1}(B) \subseteq f^{-1}(A \cap B)$.

$$\begin{aligned} x \in f^{-1}(A) \cap f^{-1}(B) &\Longrightarrow x \in f^{-1}(A) \text{ and } x \in f^{-1}(B), \\ &\Longrightarrow f(x) \in A \text{ and } f(x) \in B, \\ &\Longrightarrow f(x) \in A \cap B, \\ &\Longrightarrow x \in f^{-1}(A \cap B). \end{aligned}$$

(ii) $f^{-1}(A - B) = f^{-1}(A) - f^{-1}(B)$.

1. $f^{-1}(A - B) \subseteq f^{-1}(A) - f^{-1}(B)$. We will use $f^{-1}(\overline{B}) = \overline{f^{-1}(B)}$ in the following fifth line.

$$\begin{aligned} x \in f^{-1}(A - B) &\Longrightarrow f(x) \in A - B, \\ &\Longrightarrow f(x) \in A \text{ and } f(x) \notin B, \\ &\Longrightarrow f(x) \in A \text{ and } f(x) \in \overline{B}, \\ &\Longrightarrow x \in f^{-1}(A) \text{ and } x \in f^{-1}(\overline{B}), \\ &\Longrightarrow x \in f^{-1}(A) \text{ and } x \in \overline{f^{-1}(B)}, \\ &\Longrightarrow x \in f^{-1}(A) \text{ and } x \notin f^{-1}(B), \\ &\Longrightarrow x \in f^{-1}(A) - f^{-1}(B). \end{aligned}$$

2. $f^{-1}(A) - f^{-1}(B) \subseteq f^{-1}(A-B)$.

$$\begin{aligned} x \in f^{-1}(A) - f^{-1}(B) &\Longrightarrow x \in f^{-1}(A) \text{ and } x \notin f^{-1}(B), \\ &\Longrightarrow f(x) \in A \text{ and } f(x) \notin B, \\ &\Longrightarrow f(x) \in A - B, \\ &\Longrightarrow x \in f^{-1}(A-B). \end{aligned}$$

Question 9 Show that a function $f: X \longrightarrow Y$ is surjective (onto) if and only if

$$f\left(f^{-1}(C)\right) = C, \ \forall\, C \subseteq Y.$$

Solution. f is surjective $f(f^{-1}(C)) = C$, which was already proven in Theorem 8.5.3.
To prove the converse, we need to prove that $\forall\, y \in Y, \exists\, x \in X: \ y = f(x)$. To this, let us consider $y \in Y$ and define the singleton $C = \{y\}$. Using $f\left(f^{-1}(C)\right) = C$, we get $f\left(f^{-1}(\{y\})\right) = \{y\}$. Thus $f^{-1}(\{y\})$ is not empty, which implies the existence of $x \in f^{-1}(\{y\}) \subset X: \ f(x) = y$. Finally, f is surjective.

Question 10 Let $U = \{a,b,c,d,e\}$, and let $A = \{a,b,c\}, \ B = \{c,d\}$, and $C = \{a,d,e\}$. Find

$$\chi_A, \ \chi_B, \text{ and } \chi_C.$$

Solution. The characteristic function of a given set is defined as follows:

$$\chi_M(x) = \begin{cases} 1 & \text{if } x \in M, \\ 0 & \text{if } x \notin M. \end{cases}$$

(i) $\chi_A(a) = \chi_A(b) = \chi_A(c) = 1$ and $\chi_A(d) = \chi_A(e) = 0$.
(ii) $\chi_B(c) = \chi_B(d) = 1$ and $\chi_B(a) = \chi_B(b) = \chi_B(e) = 0$.
(iii) $\chi_C(a) = \chi_C(d) = \chi_C(e) = 1$ and $\chi_C(b) = \chi_C(c) = 0$.

Question 11 Prove that

(i) $\chi_{A\cap B} = \chi_A \chi_B$.
(ii) $\chi_{A\cup B} = \chi_A + \chi_B - \chi_{A\cap B}$.
(iii) $\chi_{A-B} = \chi_A - \chi_{A\cap B}$.

Solution.

(i) $\chi_{A\cap B} = \chi_A \chi_B$. Let

$$\begin{aligned} x \in A \cap B &\Longrightarrow x \in A \text{ and } x \in B, \\ &\Longrightarrow \chi_A(x) = 1 \text{ and } \ \chi_B(x) = 1. \end{aligned}$$

And since $x \in A \cap B$, $\chi_{A\cap B}(x) = 1 = \chi_A(x) \times \chi_B(x)$. Now let us consider the case where

$$\begin{aligned} x \notin A \cap B &\Longrightarrow x \in \overline{A \cap B}, \\ &\Longrightarrow x \in \overline{A} \text{ or } x \in \overline{B}, \\ &\Longrightarrow x \notin A \text{ or } x \notin B, \\ &\Longrightarrow \chi_A(x) = 0 \text{ or } \ \chi_B(x) = 0, \end{aligned}$$

and then $\chi_{A\cap B}(x) = 0 = \chi_A(x) \times \chi_B(x)$.

(ii) Let $x \in A \cup B$. We have to consider three cases:
1. $x \in A$ and $x \in B$. Then $x \in A \cap B$, and we have $\chi_A(x) = \chi_B(x) = \chi_{A\cap B}(x) = 1$, which implies $\chi_{A\cup B}(x) = 1 = 1 + 1 - 1 = \chi_A(x) + \chi_B(x) - \chi_{A\cap B}(x)$.

2. $x \in A$ and $x \notin B$. Then $x \notin A \cap B$, and we have $\chi_A(x) = 1$, $\chi_B(x) = \chi_{A\cap B}(x) = 0$, which imply $\chi_{A\cup B}(x) = 1 = 1+0-0 = \chi_A(x) + \chi_B(x) - \chi_{A\cap B}(x)$.

3. $x \notin A$ and $x \in B$, then $x \notin A \cap B$, and we have $\chi_A(x) = \chi_{A\cap B}(x) = 0$ and $\chi_B(x) = 1$, which imply $\chi_{A\cup B}(x) = 1 = 0+1-0 = \chi_A(x) + \chi_B(x) - \chi_{A\cap B}(x)$.

To complete the proof, we need to consider the case $x \notin (A \cup B)$. Thus $x \notin A$ and $x \notin B$, which imply we have $\chi_A(x) = 0$ and $\chi_B = 0$. Then $\chi_{A\cap B}(x) = 0$ and $\chi_{A\cup B}(x) = 0 = 0+0-0 = \chi_A(x) + \chi_B(x) - \chi_{A\cap B}(x)$.

(iii) $\chi_{A-B} = \chi_A - \chi_{A\cap B}$.

1. If $x \in A - B \Longrightarrow x \in A$ and $x \notin B$, then $x \notin A \cap B$. Thus $\chi_A(x) = 1$ and $\chi_{A\cap B}(x) = 0$, which imply that

$$\chi_{A-B}(x) = 1 = 1-0 = \chi_A(x) - \chi_{A\cap B}.$$

2. If

$$\begin{aligned} x \notin (A-B) &\Longrightarrow x \notin A \cap \overline{B}, \\ &\Longrightarrow x \in \overline{A \cap \overline{B}}, \\ &\Longrightarrow x \in \overline{A} \cup B, \\ &\Longrightarrow x \notin A \text{ or } x \in B, \\ &\Longrightarrow \chi_A(x) = 0, \text{ and } \chi_B(x) = 1, \\ &\Longrightarrow \chi_{A\cap B}(x) = 0, \\ &\Longrightarrow \chi_{A-B}(x) = 0 = \chi_A(x) - \chi_{A\cap B}(x). \end{aligned}$$

Question 12 Prove that

$$f\left(\bigcup_{i=1}^{n} A_i\right) = \bigcup_{i=1}^{n} f(A_i).$$

Solution.

(i)

$$\begin{aligned} y \in f\left(\bigcup_{i=1}^{n} A_i\right) &\Longrightarrow \exists\, x \in \bigcup_{i=1}^{n} A_i : y = f(x), \\ &\Longrightarrow (\exists x)\ (\exists i)\ i \in \{1,2,\dots,n\} : x \in A_i, \text{ and } y = f(x), \\ &\Longrightarrow (\exists x)\ (\exists i)\ i \in \{1,2,\dots,n\} : y = f(x) \in f(A_i), \\ &\Longrightarrow y \in \bigcup_{i=1}^{n} f(A_i). \end{aligned}$$

(ii)

$$\begin{aligned} y \in \bigcup_{i=1}^{n} f(A_i) &\Longrightarrow (\exists j)\ j \in \{1,2,\dots,n\} : y \in f(A_j), \\ &\Longrightarrow (\exists j)\ j \in \{1,2,\dots,n\}, (\exists x) : x \in A_j, \text{ and } y = f(x), \\ &\Longrightarrow (\exists x)\ x \in \bigcup_{i=1}^{n} A_i,\ y = f(x), \\ &\Longrightarrow y \in f\left(\bigcup_{i=1}^{n} (A_i)\right). \end{aligned}$$

Question 13 Prove that

$$f\left(\bigcap_{i=1}^{n} A_i\right) \subseteq \bigcap_{i=1}^{n} f(A_i).$$

Solution. Let us consider

$$\begin{aligned} y \in f\left(\bigcap_{i=1}^{n} A_i\right) &\Longrightarrow \exists x \in \bigcap_{i=1}^{n} A_i : y = f(x), \\ &\Longrightarrow (\exists x)\ (\forall i)\ i \in \{1, 2, ..., n\} : x \in A_i,\ y = f(x) \in f(A_i), \\ &\Longrightarrow y = f(x) \in \bigcap_{i=1}^{n} f(A_i). \end{aligned}$$

Question 14 Prove that

$$f^{-1}\left(\bigcup_{i=1}^{n} B_i\right) = \bigcup_{i=1}^{n} f^{-1}(B_i).$$

Solution.

(i) Let us consider

$$\begin{aligned} x \in f^{-1}\left(\bigcup_{i=1}^{n} B_i\right) &\Longrightarrow f(x) \in \bigcup_{i=1}^{n} B_i, \\ &\Longrightarrow (\exists i)\ i \in \{1, 2, ..., n\},\ f(x) \in B_i, \\ &\Longrightarrow (\exists i)\ i \in \{1, 2, ..., n\},\ x \in f^{-1}(B_i), \\ &\Longrightarrow x \in \bigcup_{i=1}^{n} f^{-1}(B_i). \end{aligned}$$

(ii) Let us consider

$$\begin{aligned} x \in \bigcup_{i=1}^{n} f^{-1}(B_i) &\Longrightarrow (\exists i)\ i \in \{1, 2, ..., n\},\ x \in f^{-1}(B_i), \\ &\Longrightarrow (\exists i)\ i \in \{1, 2, ..., n\},\ f(x) \in B_i, \\ &\Longrightarrow f(x) \in \bigcup_{i=1}^{n} B_i, \\ &\Longrightarrow x \in f^{-1}\left(\bigcup_{i=1}^{n} B_i\right). \end{aligned}$$

Question 15 Prove that

$$f^{-1}\left(\bigcap_{i=1}^{n} B_i\right) = \bigcap_{i=1}^{n} f^{-1}(B_i).$$

Solution.

(i) Let us consider

$$\begin{aligned} x \in f^{-1}\left(\bigcap_{i=1}^{n} B_i\right) &\Longrightarrow f(x) \in \bigcap_{i=1}^{n} B_i, \\ &\Longrightarrow (\forall i)\ i \in \{1, 2, ..., n\},\ f(x) \in B_i, \\ &\Longrightarrow (\forall i)\ i \in \{1, 2, ..., n\},\ x \in f^{-1}(B_i), \\ &\Longrightarrow x \in \bigcap_{i=1}^{n} f^{-1}(B_i). \end{aligned}$$

(ii) Let us consider

$$
\begin{aligned}
x \in \bigcap_{i=1}^{n} f^{-1}(B_i) &\Longrightarrow (\forall i)\ i \in \{1,2,...,n\},\ x \in f^{-1}(B_i),\\
&\Longrightarrow (\forall i)\ i \in \{1,2,...,n\},\ f(x) \in B_i,\\
&\Longrightarrow f(x) \in \bigcap_{i=1}^{n} B_i,\\
&\Longrightarrow x \in f^{-1}\left(\bigcap_{i=1}^{n} B_i\right).
\end{aligned}
$$

Question 16 Let $f:\ X \longrightarrow Y$ be a function, and let $A \subseteq X$ and $B \subseteq Y$. Prove:

(i) $A \subseteq \left(f^{-1} o f\right)(A)$.

(ii) $B \supseteq \left(f \circ f^{-1}\right)(B)$.

Solution.

(i) $A \subseteq \left(f^{-1} o f\right)(A)$.

$$x \in A \Longrightarrow f(x) \in f(A) \Longrightarrow x \in f^{-1}(f(A)).$$

(ii) $B \supseteq \left(f \circ f^{-1}\right)(B)$.

$$y \in \left(f \circ f^{-1}\right)(B) = f\left(f^{-1}(B)\right) \Longrightarrow \exists\, x \in f^{-1}(B):\ y = f(x)\ \in\ B.$$

Question 17 Give an example of a function $f:\ X \longrightarrow Y$ and two subsets A and B of X such that

$$f(A \cap B)\ \neq\ f(A)\ \cap\ f(B).$$

Solution. Take $f:\ \{0,1\} \to \{0,1\}$ with $f(0) = f(1) = 0$ and consider $A = \{0\}$ and $B = \{1\}$. Here we have: $A \cap B = \emptyset$ and $f(A) \cap f(B)\ =\ \{0\} \neq \emptyset = f(A \cap B)$.

Question 18 Verify whether the following are injective or surjective functions.

(a) Let $f : \mathbb{Z} \longrightarrow \mathbb{Z}$ be a function defined as $f(x) = 4x - 1, \forall x \in \mathbb{Z}$.

(b) Let $f : \mathbb{N} \longrightarrow \mathbb{Q}$ be a function defined as $f(n) = \dfrac{1}{n}, \forall n \in \mathbb{N}$.

Solution.

(a) It is clear that f is injective. Indeed, let us consider $x,\ x' \in \mathbb{Z}:\ 4x - 1 = 4x' - 1$ then easily, $x = x'$.

Since for any integer x, we have $f(x) = 4x - 1$ is an odd integer, we conclude that any even integer cannot be an image for any integer. Thus f is not surjective. In other words, if y is an even integer, then

$$\exists\, k \in \mathbb{Z}:\ y = 2k = 4x - 1 \Longrightarrow 1 + 2k = 4x,\ \text{ which is impossible (odd=even).}$$

(b) Directly, we can prove that f is injective, but since $f(n) = \dfrac{1}{n}$ on $\mathbb{N}$, it is easy to observe that $y = 0 \in \mathbb{Q}$ cannot be an image of any natural number, and then f is not surjective.

Question 19 In each part, find a function $f : \mathbb{N} \longrightarrow \mathbb{N}$ that has the desired properties:

1. surjective, but not injective.

$$f(n) = \left\lfloor \frac{n}{3} \right\rfloor.$$

It is easy to see that f is not injective because $f(2) = f(1) = 0$ but $1 \neq 2$. f is surjective because $\forall m \in \mathbb{N},\ \exists n = 3m \in \mathbb{N} :\ f(n) = m$.

2. injective, but not surjective.

$$f(n) = 1 + \frac{1}{n}.$$

Injectivity is trivial, and for showing that f is not surjective, it is sufficient to observe that $m = 1$ cannot be the image for any integer n, since $\frac{1}{n}$ can never be zero.

3. neither surjective nor injective.

$$f(n) = 1 + (n-2)(n-1).$$

It is easy to show that f is not injective ($f(1) = f(2) = 1$) and it is not surjective, since $f(n)$ is always odd. Thus any even natural number cannot be an image of any natural number by f.

4. bijective.

$$f(n) = n.$$

Question 20 Let $f : X \to Y$ be a function. Show that

$$(a) \Longrightarrow (b) \Longrightarrow (c) \Longrightarrow (a).$$

(a) f is injective.
(b) $f(A \cap B) = f(A) \cap f(B)$.
(c) $\forall A, B \subseteq X$ and $A \cap B = \emptyset$ we have $f(A) \cap f(B) = \emptyset$.

Solution.

- $(a) \Longrightarrow (b)$.
  f is injective $\Longrightarrow f(A \cap B) = f(A) \cap f(B)$. Since it is always true that $f(A \cap B) \subseteq f(A) \cap f(B)$, we will prove the other inclusion. To this, let us consider

$$\begin{aligned} y \in f(A) \cap f(B) &\Longrightarrow y \in f(A) \text{ and } y \in f(B), \\ &\Longrightarrow (\exists\, x \in A, y = f(x)) \text{ and } (\exists\, x' \in B, y = f(x')), \\ &\Longrightarrow (\exists\, x \in A)(\exists\, x' \in B) :\ y = f(x) = f(x'), \\ &\Longrightarrow (\exists\, x \in A)(\exists\, x' \in B) :\ x = x' \text{ and } y = f(x), \qquad (f \text{ is injective}) \\ &\Longrightarrow x \in A \cap B,\ y = f(x), \\ &\Longrightarrow y \in f(A \cap B). \end{aligned}$$

- $(b) \Longrightarrow (c)$.
  $f(A \cap B) = f(A) \cap f(B) \Longrightarrow \forall A, B \subseteq X,\ A \cap B = \emptyset$. We have $f(A) \cap f(B) = f(A \cap B) = \emptyset$. Directly the given hypotheses in (c), we have $A \cap B = \emptyset \Longrightarrow f(A \cap B) = \emptyset = f(A) \cap f(B)$.

- $(c) \Longrightarrow (a)$.
  Let $x,\ x' \in X,\ x \neq x'$. Then, we can take $A = \{x\}$ and $B = \{x'\}$ which satisfy $A \cap B = \emptyset$, and by the hypothesis, we conclude that $f(A) \cap f(B) = \emptyset = \{f(x)\} \cap \{f(x')\}$; therefore $f(x) \neq f(x')$, and f is injective.

Question 21 Show that if f is continuous and injective from $[a,\ b] \rightarrow \mathbb{R}$, then f is strictly monotone.

Solution. By contradiction, let us assume that f is not strictly monotone. This means that f is neither strictly decreasing nor strictly increasing, which implies the following

(i) $\exists\, x_1,\ x_2 \in [a,\ b],\ x_1 \neq x_2 :\ \ f(x_1) = f(x_2)$, which contradicts the fact that f is injective.

(ii) $\exists\, x_1, x_2,\ x_3 \in [a,\ b],\ \ x_1 < x_2 < x_3$ such that $f(x_1) < f(x_2)$ and $f(x_2) > f(x_3)$. Let us consider the following two cases:

1. $f(x_1) < f(x_3) < f(x_2)$. According to the continuity of f on $[x_1,\ x_2] \subset [a,\ b]$, the (IVT) guarantees the existence of $t \in (x_1,\ x_2)$ such that $f(t) = f(x_3)$ and $t \neq x_3$, which contradicts the fact that f is injective.

2. $f(x_3) < f(x_1) < f(x_2)$. Similarly, the IVT guarantees the existence of $t \in (x_2,\ x_3)$ such that $f(t) = f(x_1)$ and $t \neq x_1$, which contradicts that f is injective.

(iii) $\exists\, x_1, x_2,\ x_3 \in [a,\ b],\ \ x_1 < x_2 < x_3$ such that $f(x_1) > f(x_2)$ and $f(x_2) < f(x_3)$. Similarly, we may consider the following two cases:

1. $f(x_1) > f(x_3) > f(x_2)$. According to the continuity of f on $[x_1,\ x_2] \subset [a,\ b]$, the IVT guarantees the existence of $t_1 \in (x_1,\ x_2)$ such that $f(t) = f(x_3)$ and $t \neq x_3$, which contradicts the fact that f is injective.

2. $f(x_3) > f(x_1) > f(x_2)$. Similarly, the IVT guarantees the existence of $t_2 \in (x_2,\ x_3)$ such that $f(t_2) = f(x_1)$ and $t_2 \neq x_1$, which contradicts the fact that f is injective.

Thus what we assumed is wrong.

Question 22 Show that if f is continuous and strictly monotone on $[a,\ b] \rightarrow \mathbb{R}$, then f is bijective from $[a,\ b]$ on $[f(a),\ f(b)]$ or $[f(b),\ f(a)]$.

Solution. Let us assume f is strictly increasing.

1. Let $x, x' \in [a,\ b] : x \neq x' \Longrightarrow x < x'$ or $x > x'$, and since f is increasing, we conclude $f(x) < f(x')$ or $f(x) > f(x')$, which imply $f(x) \neq f(x')$, and then f is injective.

2. Since f is strictly increasing, $R_f \subseteq [f(a),\ f(b)]$. In addition, for any $y \in [f(a),\ f(b)]$, by applying the IVT we prove the existence of $c \in (a,\ b) :\ f(c) = y$. Therefore, $[f(a),\ f(b)] \subseteq R_f$, implying $R_f = [f(a),\ f(b)]$; thus f is surjective and f is bijective on $[a,\ b]$.

Question 23 Let $f : E \longrightarrow F$ be a bijective function. Show the following

(i) The inverse f^{-1} is unique.

(ii) $y = f(x) \iff x = f^{-1}(y)$.

Solution.

(i) The inverse f^{-1} is unique.

Assume $\exists f_1 \neq f_2 :\ F \rightarrow E :\ f_1 \circ f = Id_E = f_2 \circ f;\ \ f \circ f_1 = Id_F = f \circ f_2$, then

$$f_2 = Id_E \circ f_2 = (f_1 \circ f) \circ f_2 = f_1 \circ (f \circ f_2) = f_1 \circ Id_F = f_1,$$

which contradicts the fact that $f_1 \neq f_2$. Thus the inverse of f is unique.

(ii) $y = f(x) \iff x = f^{-1}(y)$. Using the bijectivity of f, we get:

$$y = f(x) \iff f^{-1}(y) = f^{-1}(f(x)) = (f^{-1} \circ f)(x) = Id_E(x) = x.$$

Question 24 Is the function $h : \mathbb{R} \to \mathbb{R}$ given by $h(x) = x(x^2 - 4)$ bijective?

Solution. The answer is no. Indeed, $h(0) = h(2)$ and $0 \neq 2$ imply that h is not one-to-one (injective), so h is not bijective.

Question 25 Study if g is a bijection on $\mathbb{R}$, if

$$g(x) = \begin{cases} \dfrac{1}{x-1} & \text{if } x < 1, \\ \sqrt{x-1} & \text{if } x \geq 1. \end{cases}$$

Solution.

1. g is injective:

   - Let $x,\ x' < 1$; then

   $$\begin{aligned} g(x) = g(x') &\implies \frac{1}{x-1} = \frac{1}{x'-1}, \\ &\implies x - 1 = x' - 1 \\ &\implies x = x'. \end{aligned}$$

   - Let $x,\ x' \geq 1$; then

   $$\begin{aligned} g(x) = g(x') &\implies \sqrt{x-1} = \sqrt{x'-1}, \\ &\implies x - 1 = x' - 1, \\ &\implies x = x'. \end{aligned}$$

   - Let $x < 1$ and $x' \geq 1$; then

   $$x - 1 < 0 \implies \frac{1}{x-1} < 0 \implies g(x) \in (-\infty,\ 0)$$

   and

   $$x' - 1 \geq 0 \implies \sqrt{x'-1} \geq 0 \implies g(x') \in [0,\ \infty).$$

   Thus we obtain:

   $$x < 1 \leq x' \implies g(x) < 0 \leq g(x') \implies g(x) \neq g(x').$$

   Thus g is injective.

2. g is surjective:
   In the last case, we proved that the range $R_g = (-\infty,\ 0) \cup [0,\ \infty) = \mathbb{R}$, g is surjective on $\mathbb{R}$.

Question 26 If g is defined by

$$g(x) = \begin{cases} 3-x & \text{if } x \leq 1, \\ 1+1/x & \text{if } x > 1. \end{cases}$$

(a) Is g surjective (onto)?
(b) Show that $x_1 \neq x_2 \Longrightarrow g(x_1) \neq g(x_2)$. What can you conclude?

Solution. Hint: determine R_g as we did in the previous question.

Question 27 Let A and B be two given sets with $n(A) = 2$ and $n(B) = 3$. How many functions are there from $A \to B$?

Solution. Any function from A to B is defined by forming the images of the elements of A with f. Thus each element has 3 different possibilities and then the requested number of functions is 9.

Question 28 Let A and B be two given sets with $n(A) = 2$ and $n(B) = 3$. How many surjective functions are there from $A \to B$?

Solution. No surjection can be defined, since at most $n(range) = 2$, where $n(B) = 3$.

Question 29 Determine the expression of $(g \circ g)(x)$ if

$$g(x) = \begin{cases} \dfrac{1}{x-1} & \text{if } x < 1, \\ \sqrt{x-1} & \text{if } x \geq 1. \end{cases}$$

Solution. Since $(g \circ g)(x) = g(g(x))$, we need to consider these cases.

1. If $x < 1 \Longrightarrow g(x) = \dfrac{1}{x-1}$ thus $g(g(x)) = g\left(\dfrac{1}{x-1}\right)$. Now, we need to check whether $\dfrac{1}{x-1} < 1$. Since

$$x-1 < 0 \Longrightarrow \frac{1}{x-1} < 0 < 1 \Longrightarrow (g \circ g)(x) = \frac{1}{\dfrac{1}{x-1} - 1} = \frac{x-1}{2-x}.$$

2. If $x \geq 1 \Longrightarrow g(x) = \sqrt{x-1} \Longrightarrow g(g(x)) = g\left(\sqrt{x-1}\right)$. Now, we need to check whether $\sqrt{x-1} \geq 1$. Let us consider the following:

$$\sqrt{x-1} < 1 \Longrightarrow x-1 < 1 \Longrightarrow x < 2.$$

Then for $1 \leq x < 2$, we have

$$(g \circ g)(x) = g\left(\sqrt{x-1}\right) = \frac{1}{\sqrt{x-1}-1} = \frac{\sqrt{x-1}+1}{(x-1)-1} = \frac{\sqrt{x-1}+1}{x-2}.$$

Thus if $x \geq 2$, we have $\sqrt{x-1} \geq 1$, and then

$$(g \circ g)(x) = g\left(\sqrt{x-1}\right) = \sqrt{\sqrt{x-1}-1},$$

and finally,

$$(g \circ g)(x) = \begin{cases} \dfrac{x-1}{2-x} & \text{if } x < 1, \\ \dfrac{\sqrt{x-1}+1}{x-2} & \text{if } 1 \leq x < 2, \\ \sqrt{\sqrt{x-1}-1} & \text{if } x \geq 2. \end{cases}$$

Question 30 Determine $(f \circ f)(x)$ if

$$f(x) = \begin{cases} 3-x & \text{if } x \leq 1, \\ 1+1/x & \text{if } x > 1. \end{cases}$$

Solution. Since $(f \circ f)(x) = f(f(x))$, we need to consider these cases.

1. If $x \leq 1 \implies f(x) = 3-x$; thus $f(f(x)) = f(3-x)$, and now, we need to check whether $3-x \leq 1$. Since

$$x \leq 1 \implies -x \geq -1 \implies 3-x \geq 2 > 1 \implies f(3-x) = 1 + \frac{1}{3-x}.$$

2. If $x > 1 \implies f(x) = 1+1/x \implies f(f(x)) = f(1+1/x)$, and now, since $x > 1 > 0$, we have $1+1/x > 1$ and then,

$$(f \circ f)(x) = 1 + \frac{1}{1+1/x} = \frac{2x+1}{x+1}.$$

Finally,

$$(f \circ f)(x) = \begin{cases} \dfrac{4-x}{3-x} & \text{if } x \leq 1, \\ \dfrac{2x+1}{x+1} & \text{if } x > 1. \end{cases}$$

Question 31 Given four sets A, B, C, and D such that $A \cap B = \emptyset$, $C \cap D = \emptyset$, $A \approx C$, and $B \approx D$. Then, $A \cup B \approx C \cup D$.

Solution. Hint: use the function $f : A \cup B \to C \cup D$ defined as follows: $f(x) = \begin{cases} g(x) & \text{if } x \in A, \\ h(x) & \text{if } x \in B, \end{cases}$, where g and h are two bijections on A and B respectively.

Question 32 Every subset of a finite set is finite.

Solution. Let B be a finite set. If $B = \emptyset$, then the unique subset is itself, and it is finite. Assume $B \neq \emptyset$, and since it is finite, then there exists a bijection $f : N_k \to B$ for some integer k. Let A be a non-empty subset of B, then the restriction function of f to A named $\tilde{f} : A \to R_{\tilde{f}}$ is a bijection, and then $A \approx R_{\tilde{f}}$. Now, since $R_{\tilde{f}}$ is a subset of N_k and according to Proposition 8.6.6, we conclude $R_{\tilde{f}}$ is finite, and then A is also finite.

Question 33 Prove Theorem 8.4.2.

Solution. Hint: use the fact that any function from $E \to f(E)$ is always surjective.

Question 34 Given two sets A and B, prove that

$$A \subseteq B \implies n(A) \leq n(B).$$

Solution. We can examine the function f from set A to set B defined as $f(x) = x$ for every $x \in A$. It is evident that f is an injective (one-to-one) function. Therefore, we can conclude that $n(A) \leq n(B)$, which serves to complete the proof.

Question 35 Prove that we have

$$n((0,\ 1)) = n([0,\ 1]).$$

Solution. By applying the result in 34:

$$(0,\ 1) \subseteq [0,\ 1] \Longrightarrow n((0,\ 1)) \leq n([0,\ 1]),$$

and

$$[0,\ 1] \subseteq (-2,\ 2) \Longrightarrow n([0,\ 1]) \leq n((-2,\ 2)).$$

Now, by using the equivalence between $(0,\ 1)$ and any interval $(-2,\ 2)$, we have $n((0,\ 1)) = n((-2,\ 2))$ and then we conclude that

$$n((-2,\ 2)) = n((0,\ 1)) \leq n([0,\ 1])) \leq n((-2,\ 2)) \Longrightarrow n((0,\ 1)) = n([0,\ 1]).$$

Question 36 Prove Theorem 8.6.11. (Hint: use contradiction.)

Solution. Assume that $\mathbb{N}$ is finite; then there exists a bijection f from $\mathbb{N}_k \to \mathbb{N}$ for some $k \in \mathbb{N}$. If we set $m = 1 + \max_{i \in \mathbb{N}_k} f(i)$,[3] then it is clear that $\forall\, i = 1, \ldots, k:\ m > f(i)$. Thus

$$\exists\, m \in \mathbb{N},\ \forall\, i \in N_k:\ f(i) \neq m$$

f is not surjective, which contradicts the assumption and so what we assumed is wrong, and $\mathbb{N}$ is infinite.

Question 37 If A and B are both countable infinite, then $A \cup B$ is countable.

Solution. Suppose A and B are both countable infinite, then we know we can express A and B in list form as

$$A = \{a_1, a_2, a_3, \cdots\}, \quad B = \{b_1, b_2, b_3, \cdots\}.$$

Now, by mixing the above elements, we build up an infinite sequence of elements and then $A \cup B$ is countable.

Question 38 Prove or disprove: If S is uncountable, then $n(S) = n(\mathbb{R})$.

Solution. It is not true. One may take $S = \mathscr{P}(\mathbb{R})$ which has a cardinality strictly greater than $c = n(\mathbb{R})$ by applying Theorem 8.6.19.

Question 39 Prove or disprove: If $A \subseteq B \subseteq C$ where A and C are two infinite countable sets, then set B is infinite and countable.

[3]This is a classical idea used in mathematical analysis and in number theory to elaborate proofs by contradiction.

Solution. This statement is true. To explain further, when we have an infinite set A satisfying $A \subseteq B$, it implies that B is an infinite set. Furthermore, as C is countable, there exists a bijective function between C and $\mathbb{N}$ (the set of natural numbers). Restricting this bijection to $B \subseteq C$ maintains its bijective nature, which guarantees that B is also countable.

Question 40 Prove that if A and B are finite sets with n(A)=n(B), then any injection $f : A \to B$ is also a surjection.

Solution. We want to prove that any injection $f : A \to B$ is also a surjection.
To prove this, let us assume for the sake of contradiction that there exists an injection $f : A \to B$ that is not a surjection. That is, there exists some element $y \in B$ such that there is no element $x \in A$ such that $y = f(x)$.
Since f is an injection, it means that for every pair of distinct elements x_1 and x_2 in A, we have $f(x_1) \neq f(x_2)$. This implies that f maps each element of A to a distinct element in B, and since $n(A) = n(B)$, there should be a one-to-one correspondence between the elements of A and B. However, if there exists an element $y \in B$ such that $f(x) \neq y$ for any $x \in A$, it would mean that we have an element in B that is not in the range of f, contradicting the one-to-one correspondence between sets A and B.
Hence, our assumption that there exists an injection f that is not a surjection is false. Therefore, any injection from A to B is also a surjection when A and B are finite sets with $n(A) = n(B)$.

Question 41 Show this is not necessarily true if A and B are not finite in the above question.

Solution. To provide a counterexample to the statement when both sets A and B are infinite, we need to consider different sets. For this purpose, we can take set $A = \mathbb{N}$, and set B as the set of prime natural numbers.
We can define the function $f : A \to B$ as follows: $f(n)$ is equal to the nth prime number.
It is evident that f is injective because it assigns a unique prime number to each natural number. However, f is not a surjection because there exists a prime number in set B that are not within the range of f. For instance, the first prime number, 2, cannot be the result of applying f to any natural number, which means there is no natural number n such that $f(n) = 2$.

Bibliography

[1] T. Andreescu and M. Tetiva, *Sums and products*. XYZ Press, (2018).

[2] R. Hammack, *Book of Proof*. Virginia Commonwealth University, Mathematics Textbook Series (2009).

[3] A. Hamdi, *Let us do math: Basic mathematics level one: More than 365 solved examples and exercises. from grade 8 to university*. Editions Universitaires Européennes, (2018).

[4] A. Hamdi, *Let us do math- basic mathematics - level two: first steps to excel in math –about* $\lfloor 100\pi \rfloor$ *solved problems including olympiad questions*. Editions Universitaires Europ eenne, (2021).

[5] A. N. Kolmogorov and S. V. Fomin, *Introductory real analysis*. Dover Books on Mathematics (1975).

[6] R. B. Manfrino, J. A. G. Ortega and R. V. Delgado, *Topics in algebra and analysis*. Birkhäuser, (2015).

[7] K. Plofker, *Mathematics in India*. Princeton University Press (2009).

[8] W. Rudin, *Principles of mathematical analysis*. International series in pure and applied mathematics, McGraw-Hill (1976).

[9] S. Parmanand, *The so-called Fibonacci numbers in ancient and medieval India.* Academic Press. 12 (3): 232 (1985).

[10] D. Smith, M. Eggen, and R. St. Andre, *A Transition to advanced mathematics*. Cengage-Learning, 8th Edition, (2014).

[11] J. Stewarts, *Calculus*, Eighth Edition, Cengage Learning, (2015).

Index

www.ingramcontent.com/pod-product-compliance
Ingram Content Group UK Ltd.
Pitfield, Milton Keynes, MK11 3LW, UK
UKHW050728071026
15590UKWH00012B/2259

9 781793 564283